Fuzzy Neural Intelligent Systems

Mathematical Foundation and the Applications in Engineering

Hongxing Li
C.L. Philip Chen
Han-Pang Huang

Fuzzy Neural Intelligent Systems

Mathematical Foundation and the Applications in Engineering

CRC Press
Boca Raton   London   New York   Washington, D.C.

Library of Congress Cataloging-in-Publication Data

Li, Hong-Xing, 1953-
  Fuzzy neural intelligent systems : mathematical foundation and the applications in engineering / Hongxing Li, C.L. Philip Chen, Han-Pang Huang.
     p. cm.
  Includes bibliographical references and index.
  ISBN 0-8493-2360-6 (alk. paper)
  1. Neural networks (Computer science) 2. Fuzzy systems. 3. Engineering--Data processing. I. Chen, C.L. Philip. II. Huang, Han-Pang. III. Title.

  QA76.87. L5 2000
  006.3′2—dc21   00-044485
                                                                                              CIP

© 2001 by CRC Press LLC

No claim to original U.S. Government works
International Standard Book Number 0-8493-2360-6
Library of Congress Card Number 00-044485
Printed in the United States of America 1 2 3 4 5 6 7 8 9 0
Printed on acid-free paper

Preface

Fuzzy systems and neural networks have been regarded as the main branches of soft computing. Most research works have been focused on the development of theories and design of systems and algorithms for specific applications. These works have shown that neuro-fuzzy systems indeed demonstrate their exceptional intelligent capability for computing and learning. However, we may be aware that there is little theoretical support for existing neuro-fuzzy systems, especially their mathematical foundaion. From the literature, a neuro-fuzzy system is defined as a combination of fuzzy systems and neural networks such that the parameters of fuzzy systems are determined by neural network learning algorithms. The intention is to take the advantage of neural network methods to improve or to create a fuzzy system. On the other hand, a fuzzy neural network is defined as the use of fuzzy methods to enhance or to improve the learning capabilities of a neural network. Unfortunately, little work has been done in the fuzzy neural network area. The main features of this book give a layout of mathematical foundation for fuzzy neural networks and a better way of combining neural networks with fuzzy logic systems.

This book was written to provide engineers, scientists, researchers, and students interested in fuzzy systems, neural networks, and fuzzy neural integrated systems a systematic and comprehensive structure of concepts and applications. The required mathematics for reading this book are not beyond linear algebra and engineering mathematics.

This book contains 19 chapters and consists of three major parts. Part I (Chapters 1-5, 10, 11) covers the fundamental concepts and theories for fuzzy systems and neural networks. Part II (Chapters 6-8, 12, 13) provides the foundation and important topics in fuzzy neural networks. Part III (Chapters 14-19) gives extensive case examples for neuro-fuzzy systems, fuzzy systems, neural network systems, and fuzzy-neural systems. In short, Chapter 1 briefly introduces fundamental knowledge of fuzzy systems. These include fuzzy sets, fuzzy relations, resolution theorem, representation theorem, extension principle, fuzzy clustering, fuzzy logic, fuzzy inference, fuzzy logic systems, etc. Chapter 2 discusses determination of membership functions for a fuzzy logic system. Chapter 3 reveals mathematical essence and structures of neural networks. Chapter 4 studies structures of functional-link neural networks and fuzzy functional-link neural networks. Chapter 5 describes flat neural networks, computational algorithms, and their applications. Chapter 6 describes the structure of fuzzy neural networks in detail, from the multifactorial functions point of view. Chapter 7 discloses mathematical essence and structures of feedback neural networks and fuzzy neural networks, where it is indicated that stable points of a feedback can be, in essence, regarded as fixed points of a function. Extending the idea from Chapters 6 and 7, Chapter 8 introduces generalized additive weighted multifactorial functions and the applications to fuzzy inference and neural networks. Chapter 9 discusses interpolation mechanisms of fuzzy control including some innovative methods and important results. Chapter 10 shows the relations between fuzzy logic controllers and PID controllers mathematically. Chapter 11 discusses adaptive fuzzy control by using variable universe. Chapters 12 and 13 introduce factor spaces theory and study of neuron models and neural networks formed by factor spaces. Chapter 14 gives the foundation of neuro-fuzzy systems. Chapter 15 explores the nature of data and discusses the importance of data preprocessing. Chapters 16 to 18 give engineering applications of both fuzzy neural and neuro-fuzzy systems. Chapter 18 shows the application of hybrid neural network and fuzzy systems. Chapter 19 gives the on-line learning and DSP implementation of fuzzy neural systems, followed by myoelectric applications.

The materials of this book can be used as different graduate courses (15-week semester courses):

- Introduction to Fuzzy and Neural Systems: Chapters 1-8, 12, 13.
- Introduction to Intelligent Control: Chapters 1-5 10,15.
- Advanced Intelligent Control: Chapters 9, 11, 14-19.

Of course, this book can also be used as a self-study textbook and reference book.

Acknowledgments

We are indebted to many people who directly or indirectly assisted in the preparation of the text. In particular, we would like to thank Professors L. A. Zadeh, C. S. George Lee, Yoh-Han Pao, S. S. Lu, and N. H. McClamroch. Special thanks goes to Dr. Pei-Ling H. Lee for her continuous encouragement during the last few years. The graduate students who worked on the project over the past few years also contributed to the text. They are: C. C. Liang, Y. C. Lee, C. Y. Juang, W. M. Lee, K. P. Wong, C. H. Lin, C. Y. Chiang, J. Y. Wang, Q. He, Z. H. Miao, and Q. F. Cai.

We would also like to extend our appreciation to Beijing Normal University, Wright State University, National Taiwan University, National Natural Science Foundation of China, Dr. Steven R. LeClair at Wright-Patterson Air Force Base, and National Science Council of Taiwan, for their sponsorship of our research activities in fuzzy neural systems, manufacturing automation and robotics, and related areas. We also thank Cindy Carelli, Steve Menke, and Helena Redshaw at CRC Press for their skillful coordination of the production of the book.

Finally, we would like to thank our wives and children, HX: Qian Xuan and child Ke-Yu; CLP: Cindy, and children, Oriana, Melinda, and Nelson; HP: Li-Chu, and children, Jong-Pyng, Quann-Ru, for their understanding and constant encouragement, without which this book would not have been possible.

H. X. Li
Beijing
C. L. Philip Chen
Dayton
H. P. Huang
Taipei

Table of Contents

Chapter 1

Foundation of Fuzzy Systems

Fuzzy concepts derive from fuzzy phenomena that commonly occur in the natural world. The concepts formed in human brains for perceiving, recognizing, and categorizing natural phenomena are often fuzzy concepts. Boundaries of these concepts are vague. We shall first introduce the basic concept of fuzzy systems in this chapter. We start with definitions of fuzzy sets and fuzzy operators and then we give some extension principles and theorems that will be used as the foundation throughout this book.

1.1 Definition of Fuzzy Sets

Fuzzy concepts derive from fuzzy phenomena that commonly occur in the natural world. For example, "rain" is a common natural phenomenon that is difficult to describe precisely since it can "rain" with varying intensity anywhere from a light sprinkle to a torrential downpour. Since the word "rain" does not adequately or precisely describe the wide variations in the amount and intensity of any rain event, rain is considered a "fuzzy" phenomenon.

The concepts formed in human brains for perceiving, recognizing, and categorizing natural phenomena are often fuzzy. Boundaries of these concepts are vague. The classifying (dividing), judging, and reasoning they produce are also fuzzy concepts. For instance, "rain" might be classified as "light rain", "moderate rain", and "heavy rain" in order to describe the degree of "raining". Unfortunately, it is difficult to say when rain is light, moderate, or heavy. The concepts of light, moderate, and heavy are prime examples of fuzzy concepts themselves and are examples of fuzzy classifying. If it is raining today, you can call it light rain, or moderate rain, or heavy rain based on the relative amount of rainfall: This is fuzzy judging. If you are predicting a good, fair, or poor harvest based on the results of fuzzy judging, you are using fuzzy reasoning.

The human brain has the incredible ability of processing fuzzy classification, fuzzy judgment, and fuzzy reasoning. The natural languages are ingeniously permeated with inherent fuzziness so that we can express rich information content in a few words.

Historically, as reflected in classical mathematics, we commonly seek "precise and crisp" descriptions of things or events. This precision is accomplished by expressing phenomena in numerical values. However, due to fuzziness, classical mathematics can encounter substantial difficulties. People in ancient Greece discussed such a problem: How many seeds in a pile constitute a heap? Because "heap" is a fuzzy concept, they could not find a unique number that could be judged as a heap.

In fact, we often come into contact with fuzziness. There exist many fuzzy concepts in everyday life, such as a "tall" man, a "fat" man, a "pretty" girl, "cloudy" skies, "dawn", and "dusk", etc. We may say that fuzziness is absolute, whereas crispness or preciseness is relative. The so-called crispness or preciseness is only separated from fuzziness by simplification and idealization. The separation is significant because people can conveniently describe, in some situations, by means of exact models with pure mathematical expressions. But the knowledge domain is getting increasingly complex and deep. The complication has two striking features: (1) There are many factors associated with problems of interest. In practice, only a subset of factors is considered, and the originally crisp things are transformed into fuzzy things. (2) The degree of difficulty in dealing with problems is increased and the fuzziness is accumulated incrementally, with the result that the fuzziness cannot be neglected. The polarity between fuzziness and preciseness (or crispness) is quite a striking contradiction for the development of today's science. One of the effective means of resolving the contradiction is fuzzy set theory, a bridge between high preciseness and high complexity.

People routinely use the word "concept". For example, the object "man" is a concept. A concept has its intension and extension; by intension we mean attributes of the object, and by extension, we mean all of the objects defined by the concepts. The extension of the concept "set" has been interpreted as the set formed by all of the objects defined by the concept. That is, sets can be used to express concepts. Since set operations and transformations can express judging and reasoning, modern mathematics becomes a formal language for describing and expressing certain areas of knowledge.

Set theory was founded in 1874 by G. Cantor, a German mathematician. One of the important methods used by Cantor in creating sets is the comprehension principle, which means that for any p, a property, all the objects with and only with p can be included together to form a set denoted by the symbol

$$A = \{a \mid p(a)\},$$

where A expresses the set and "a" an object in A. Generally, "a" is referred to as an element or a number of A. The expression $p(a)$ represents the element, a satisfies p, and $\{\ \}$ represents all the elements that satisfy p subsumed to form a set. In logic, the comprehension principle is stated as

$$(\forall\, a)(a \in A \iff p(a)).$$

Cantor's set theory has made great contributions to the foundations of mathematics. Unfortunately, it has also given rise to some restrictions on the use of mathematics. In fact, according to Cantor's claim, the objects that form the set are definite and distinct from each other. Thus, the property p used to form the set must be crisp: for any object, it must be precise whether the property p is satisfied or not. This is the application of the law of the excluded middle. From this law, the concept (e.g., property, proposition) expressed as a set is either true or false. Reasoning only with true or false forms a kind of bivalent logic. However, concepts in human brains hardly have crisp extensions. For example, the concept "tall man" (property p) cannot form a set based on Cantor's set theory because, for any man, we cannot unequivocally determine if the man satisfies the property p (i.e., tall man.)

A concept without a crisp extension is a fuzzy concept. We now ask if a fuzzy concept can be rigidly described by Cantor's notion of sets or the bivalent (true/false or two-valued) logic. We will show that the answer is negative via the "baldhead paradox". Since one single hair change does not distinguish a man from his baldheaded status, we have the following postulate:

Postulate. If a man with n (a natural number) hairs is baldheaded, then so is a man with $n + 1$ hairs.

Based on the postulate, we can prove the following paradox:

Baldhead Paradox. Every man is baldheaded.
Proof. By mathematical induction,
1) A man having only one hair is baldheaded.
2) Assume that a man with n hairs is baldheaded.
3) By the postulate, a man with $n + 1$ hairs is also baldheaded.
4) By induction, we have the result: every man is baldheaded. **Q.E.D.**

The cause of the paradox is due to the use of bivalent logic for inference, whereas in fact, bivalent logic does not apply in this case.

Qualitative and quantitative values are closely related to each other. As the quantity changes so does the quality. In the baldheaded example, one cannot define a man as baldheaded because one cannot establish an absolute boundary by means of the number of hairs. But the tiny increase or decrease in the number of hairs (changes in quantity) does influence the change in quality, which cannot be described in words like "true/yes" or "false/no".

"True" and "false", regarded as logical values, can be respectively denoted by 1 (=100%) and 0 (=1-100%). Logical values are a kind of measure for the degree of truth. The "baldheaded paradox" shows us that it is not enough to use only two values, 1 and 0, for fuzzy concepts; we have to use other logical values between 1 and 0 to express different degrees of truth. Therefore, in order to enable mathematics to describe fuzzy phenomena, it is of prime importance to reform Cantor's set concept, namely, to define a new kind of set called a fuzzy set. Starting with Zadeh's fuzzy sets theory [1], tremendous research, to name a few, has demonstrated success of the fuzzy sets and systems in both theoretical and practical areas [2-12].

Definition 1   A fuzzy set A on the given universe U is that, for any $u \in U$, there is a corresponding real number $\mu_A(u) \in [0,1]$ to u, where $\mu_A(u)$ is called the grade of membership of u belonging to A.

This means that there is a mapping,

$$\mu_A : U \longrightarrow [0,1], \quad u \mapsto \mu_A(u)$$

and this mapping is called the membership function of A.

Just as Cantor sets can be completely described by characteristic functions, fuzzy sets can also be described by membership functions. If the range of μ_A admits only two values 0 and 1, then μ_A degenerates into a usual set characteristic function.

$$A = \{u \in U \mid \mu_A(u) = 1\}.$$

Therefore, Cantor sets are special cases of fuzzy sets.

All of the fuzzy sets on U will be denoted by $\mathcal{F}(U)$, and the power set of U in the sense of Cantor is denoted by $\mathcal{P}(U)$. Obviously, $\mathcal{F}(U) \supset \mathcal{P}(U)$. When $A \in \mathcal{F}(U) \setminus \mathcal{P}(U)$, A is called a proper fuzzy set; there exists at least one element $u_0 \in U$ such that $\mu_A(u_0) \notin \{0,1\}$.

Example 1   Zadeh defines the fuzzy sets "young" and "old", denoted by Y and O, respectively, over the universe $U = [0, 100]$ as follows:

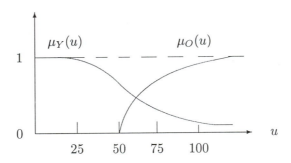

Figure 1   Membership function of "young" and "old"

Example 2   Let $U = \{1, 2, \cdots, 9\}$ and A be the set of "natural numbers close to 5". The membership function of A is defined as follows:

$\mu_A(u)$	0	0.2	0.6	0.9	1.0	0.9	0.6	0.2	0
u	1	2	3	4	5	6	7	8	9

Example 3   Let U be the set of real numbers and A be the set of "real numbers considerably larger than 10". Then a membership function of A is defined as $\mu_A(u) = 0$, $u < 10$, and

$$\mu_A(u) = [1 + (u - 10)^{-2}]^{-1}, \quad u \geq 10$$

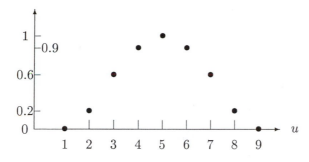

Figure 2   Natural numbers close to 5

There are at least three forms for representing a fuzzy set. We will use Example 2 to illustrate these forms.

1)   Zadeh's form. A is represented by

$$A = \frac{0}{1} + \frac{0.2}{2} + \frac{0.6}{3} + \frac{0.9}{4} + \frac{1.0}{5} + \frac{0.9}{6} + \frac{0.6}{7} + \frac{0.2}{8} + \frac{0}{9}$$

where the numerator of each fraction is the grade of the membership of the corresponding element in the denominator, and the plus sign $(+)$ is simply a notation without the usual meaning of addition. When the grade of membership for some element is zero, then that element may be omitted from the expression. For example, the terms $\frac{0}{1}$ and $\frac{0}{9}$ in the expression may be omitted.

2) A is represented by a set of ordered pairs. The first entry of the pair is an element of the universe, and the second entry of the pair is the grade of membership of the first entry. In this form, A is represented as

$$A = \{(1,0),(2,0.2),(3,0.6),(4,0.9),(5,1.0),(6,0.9),(7,0.6),(8,0.2),(9,0)\}.$$

3) A is represented by a vector called the fuzzy vector.

$$A = (0,0.2,0.6,0.9,1.0,0.9,0.6,0.2,0).$$

Note 1   Both Zedah's form and the ordered pairs representation may be extended to an infinite universe U by the following form:

$$A = \int_{u \in U} \mu_A(u)/u, \qquad A = \{(u,\mu_A(u)) \mid u \in U\},$$

where the sign $\int$ is not an integration in the usual sense; rather it is a notation that represents the grade of membership of u in a continuous universe U. For example,

A in Example 3 can be expressed by

$$A = \int_{u \in \Re} [1 + (u - 10)^{-2}]^{-1}/u$$

$$A = \left\{ \left(u, [1 + (u - 10)^{-2}]^{-1} \right) \,\middle|\, u \in \Re \right\},$$

where $\Re$ is the set of all real numbers.

1.2   Basic Operations of Fuzzy Sets

Let A and B be members of $\mathcal{F}(U)$. We now define the basic fuzzy set operations on A and B, such as inclusion, equality, union, and intersection, and the complement A^c of A as follows:

$$A \supset B \iff (\forall u \in U)(\mu_A(u) \geq \mu_B(u))$$
$$A = B \iff A \supset B \text{ and } A \subset B, \text{ or equivalently, } (\forall u \in U)((\mu_A(u) = \mu_B(u))$$
$$C = A \cup B \iff (\forall u \in U)(\mu_C(u) = \mu_A(u) \vee \mu_B(u))$$
$$C = A \cap B \iff (\forall u \in U)(\mu_C(u) = \mu_A(u) \wedge \mu_B(u))$$
$$C = A^c \iff (\forall u \in U)(\mu_C(u) = 1 - \mu_A(u))$$

where $\vee$ and $\wedge$ are max and min operators, respectively.

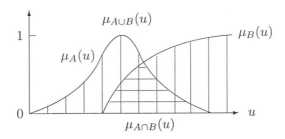

Figure 3   Membership functions and union and intersection of two fuzzy sets

The union and intersection operations can be extended to any index set T:

$$\mu_{\cup_{t \in T} A_t}(u) = \bigvee_{t \in T} \mu_{A_t}(u)$$

$$\mu_{\cap_{t \in T} A_t}(u) = \bigwedge_{t \in T} \mu_{A_t}(u),$$

where $\vee$ and $\wedge$ means sup and inf, respectively.

The operations of fuzzy sets can be illustrated by a graph of their membership functions as shown in Figure 3 and Figure 4.

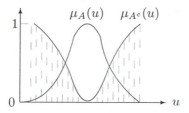

Figure 4   The membership function of a complement fuzzy set

Example 4   Let the fuzzy sets young (Y) and old (O) be defined as in Example 1. Then the fuzzy sets "young or old" $(Y \cup O)$, "young and old" $(Y \cap O)$, and "not young" (Y^c) are defined as follows:

$$\mu_{Y \cup O}(u) = \begin{cases} 1, & 0 \le u \le 25 \\ \left[1 + (\frac{u - 25}{5})^2\right]^{-1}, & 25 < u \le 51 \\ \left[1 + (\frac{u - 25}{5})^{-2}\right]^{-1}, & 51 < u \le 100 \end{cases}$$

$$\mu_{Y \cap O}(u) = \begin{cases} 1, & 0 \le u \le 25 \\ \left[1 + (\frac{u - 50}{5})^{-2}\right]^{-1}, & 25 < u \le 51 \\ \left[1 + (\frac{u - 25}{5})^2\right]^{-1}, & 51 < u \le 100 \end{cases}$$

$$\mu_{Y^c}(u) = \begin{cases} 0, & 0 \le u \le 25 \\ 1 - \left[1 + (\frac{u - 25}{5})^2\right]^{-1}, & 25 < u \le 100. \end{cases}$$

The union, intersection, and complement operations of fuzzy sets have the following properties:

1. Idempotency.

$$A \cup A = A, \quad A \cap A = A$$

2. Commutativity.

$$A \cup B = B \cup A, \quad A \cap B = B \cap A$$

3. Associativity.

$$(A \cup B) \cup C = A \cup (B \cup C), \quad (A \cap B) \cap C = A \cap (B \cap C)$$

4. Absorption.

$$(A \cap B) \cup A = A, \quad (A \cup B) \cap A = A$$

5. Distributivity.

$$A \cap (B \cup C) = (A \cap B) \cup (A \cap C)$$
$$A \cup (B \cap C) = (A \cup B) \cap (A \cup C)$$

6. Bipolarity.

$$A \cup U = U, \quad A \cap U = A, \quad A \cup \Phi = A, \quad A \cap \Phi = \Phi$$

7. Reflexivity.

$$(A^c)^c = A$$

8. De Morgan's Law.

$$(A \cup B)^c = A^c \cap B^c, \quad (A \cap B)^c = A^c \cup B^c$$

Note that the complementary law of Cantor sets does not apply to the fuzzy sets, i.e.,

$$A \cup A^c \neq U, \quad A \cap A^c \neq \Phi.$$

In fact, for any $u_0 \in U$ and $0 < \mu_A(u_0) < 1$, we have

$$\mu_A(u_0) \vee (1 - \mu_A(u_0)) < 1 \quad \text{and} \quad \mu_A(u_0) \wedge (1 - \mu_A(u_0)) > 0$$

Hence,

$$\mu_{A \cup A^c}(u_0) \neq 1 \quad \text{and} \quad \mu_{A \cap A^c}(u_0) \neq 0.$$

Let U and V be two universes and $A \in \mathcal{F}(U)$ and $B \in \mathcal{F}(V)$ be two fuzzy sets in U and V. Then we can form a new fuzzy set $A \times B \in \mathcal{F}(U \times V)$ whose membership function is defined by: $\forall (u, v) \in U \times V$,

$$\mu_{A \times B}(u, v) = \mu_A(u) \wedge \mu_B(v).$$

This is called the direct product (or Cartesian product) of A and B.

1.3   The Resolution Theorem

In an actual application, people often like to make a crisp decision, even though the decision is based on fuzzy concepts. Therefore, we need to establish a "bridge" between fuzzy sets and Cantor sets. For a Cantor set $A \in \mathcal{P}(U)$, an element $u \in U$ is an element of A only when $\chi_A(u) = 1$, where χ_A is the characteristic function of A. However, for a fuzzy set, such a "threshold" is too high; it must be lowered to a real number $\lambda \in [0, 1]$. For a given $\lambda \in [0, 1]$, u is said to be an element of A if and only if $\mu_A(u) \geq \lambda$. In other words, for each $\lambda \in [0, 1]$, a crisp Cantor set is formed with the property that each element in that set has a grade (confidence level) of λ or higher.

Let $A \in \mathcal{F}(U)$, $\lambda \in [0, 1]$. Define crisp sets

$$A_\lambda = \{u \in U \mid \mu_A(u) \geq \lambda\}$$

and

$$A_{\overline{\lambda}} = \{u \in U \mid \mu_A(u) > \lambda\}$$

These two sets A_λ and $A_{\overline{\lambda}}$ are called the λ-cut and strong λ-cut of A, respectively. Sets formed by the λ-cut and strong λ-cut have the following properties:

1. $(A \cup B)_\lambda = A_\lambda \cup B_\lambda, \quad (A \cap B)_\lambda = A_\lambda \cap B_\lambda;$

2. $(A \cup B)_{\overline{\lambda}} = A_{\overline{\lambda}} \cup B_{\overline{\lambda}}, \quad (A \cap B)_{\overline{\lambda}} = A_{\overline{\lambda}} \cap B_{\overline{\lambda}};$

3. $\left(\bigcup_{t \in T} A^{(t)} \right)_\lambda \supset \bigcup_{t \in T} A^{(t)}_\lambda;$

4. $\left(\bigcup_{t \in T} A^{(t)} \right)_{\overline{\lambda}} = \bigcup_{t \in T} A^{(t)}_{\overline{\lambda}};$

5. $\left(\bigcap_{t \in T} A^{(t)} \right)_{\overline{\lambda}} \subset \bigcap_{t \in T} A^{(t)}_{\overline{\lambda}};$

6. $\left(\bigcap_{t \in T} A^{(t)} \right)_\lambda = \bigcap_{t \in T} A^{(t)}_\lambda;$

7. $A_{\overline{\lambda}} \subset A_\lambda;$

8. If $\lambda \leq \eta$ then $A_\lambda \subset A_\eta$ and $A_{\overline{\lambda}} \supset A_\eta \supset A_{\overline{\eta}};$

9. $A_{\bigvee_{t \in T} \lambda_t} = \bigcap_{t \in T} A_{\lambda_t};$

10. $A_{\overline{\bigvee_{t \in T} \lambda_t}} = \bigcup_{t \in T} A_{\lambda_t};$

11. $(A^c)_\lambda = A^c_{\overline{1-\lambda}};$

12. $(A^c)_{\overline{\lambda}} = A^c_{1-\lambda};$

13. $A_0 = U, \quad A_{\overline{1}} = \Phi.$

Let $A \in \mathcal{F}(U)$. We call A_1 (where $\lambda = 1$) the kernel of A, and it is denoted by $\ker A$. $A_{\overline{0}}$ (where $\lambda = 0$) is the **support** of A, and is denoted by $\operatorname{supp} A$. The difference set $A_{\overline{0}} \setminus A_1$ is called the **boundary** of A.

Example 5   Let $A = \{(a, 1), (b, 0.75), (c, 0.5), (d, 0.25), (e, 0)\}$. Then

$$A_1 = \{a\}, \ A_{0.7} = \{a, b\}, \ A_{0.5} = \{a, b, c\}, \ A_{0.2} = \{a, b, c, d\}, \ A_0 = U.$$

and

$$A_{\overline{1}} = \Phi, \ A_{\overline{0.7}} = \{a, b\}, \ A_{\overline{0.5}} = \{a, b\}, \ A_{\overline{0.2}} = \{a, b, c, d\}, \ A_0 = \{a, b, c, d\}.$$

For any real number $\lambda \in [0, 1]$ and fuzzy set $A \in \mathcal{F}(U)$, a new fuzzy set λA can be defined by the membership function

$$\mu_{\lambda A}(u) \overset{\triangle}{=} \lambda \wedge \mu_A(u), \quad u \in U.$$

The new fuzzy set λA defined in this way is called the scalar product of A. It is easy to show that the following properties hold for the scalar product of fuzzy sets.
1) $\lambda_1 \leq \lambda_2 \Longrightarrow \lambda_1 A \subset \lambda_2 A$.
2) $A \subset B \Longrightarrow \lambda A \subset \lambda B$.

The following resolution theorems reveal that a fuzzy set may be decomposed into λ-cut sets.

Theorem 1 (First Resolution Theorem)   If $A \in \mathcal{F}(U)$, then

$$A = \bigcup_{\lambda \in [0,1]} \lambda A_\lambda.$$

Proof.   We need only to prove that $\forall u \in U$.

$$\mu_A(u) = \bigvee_{\lambda \in [0,1]} (\lambda \wedge \chi_{A_\lambda}(u)).$$

Since

$$\chi_{A_\lambda}(u) = \begin{cases} 1, & 0 \leq \lambda \leq \mu_A(u) \\ 0, & \mu_A(u) \leq \lambda \leq 1 \end{cases},$$

we have that

$$\bigvee_{\lambda \in [0,1]} (\lambda \wedge \chi_{A_\lambda}(u)) = \left(\bigvee_{0 \leq \lambda \leq \mu_A(u)} (\lambda \wedge \chi_{A_\lambda}(u)) \right) \vee \left(\bigvee_{\mu_A(u) \leq \lambda \leq 1} (\lambda \wedge \chi_{A_\lambda}(u)) \right)$$

$$= \bigvee_{0 \leq \lambda \leq \mu_A(u)} \lambda = \mu_A(u). \quad \textbf{Q.E.D.}$$

Similarly, we can prove

Theorem 2 (Second Resolution Theorem)   If $A \in \mathcal{F}(U)$, then

$$A = \bigcup_{\lambda \in [0,1]} \lambda A_{\overline{\lambda}}.$$

Let $A \in \mathcal{F}(U)$. Define the mapping:

$$H : \ [0,1] \longrightarrow \mathcal{P}(U), \quad \lambda \longmapsto H(\lambda)$$

that $H(\lambda)$ satisfies

$$(\forall \lambda \in [0,1])(A_{\overline{\lambda}} \subset H(\lambda) \subset A_\lambda).$$

The next resolution theorem will show that A can be formed by $\{H(\lambda) \mid \lambda \in [0,1]\}$.

Theorem 3 (Third Resolution Theorem)   If $A \in \mathcal{F}(U)$, then

$$A = \bigcup_{\lambda \in [0,1]} \lambda H(\lambda).$$

Proof.   By definition of $H(\lambda)$, we have

$$A_{\overline{\lambda}} \subset H(\lambda) \subset A_\lambda \Longrightarrow \lambda A_{\overline{\lambda}} \subset \lambda H(\lambda) \subset \lambda A_\lambda$$

$$\Longrightarrow A = \bigcup_{\lambda \in [0,1]} \lambda A_{\overline{\lambda}} \subset \bigcup_{\lambda \in [0,1]} \lambda H(\lambda) \subset \bigcup_{\lambda \in [0,1]} \lambda A_\lambda = A$$

$$\Longrightarrow A = \bigcup_{\lambda \in [0,1]} \lambda H(\lambda). \quad \textbf{Q.E.D.}$$

Proposition 1   The mapping H defined previously has the following properties:
1) $\lambda < \eta \Longrightarrow H(\lambda) \supset H(\eta)$;
2) $A_\eta = \bigcup_{\lambda \in [0,\eta)} \lambda H(\lambda), \quad ; \eta \in (0,1]$
3) $A_{\overline{\eta}} = \bigcup_{\lambda \in (\eta,1]} \lambda H(\lambda), \quad \eta \in [0,1).$

Proof.   1) $\lambda < \eta \Longrightarrow H(\lambda) \supset A_{\overline{\lambda}} \supset A_\eta \supset H(\eta) \Longrightarrow H(\lambda) \supset H(\eta).$
2)   On one hand, for every $\lambda \in [0,\eta)$, we have

$$H(\lambda) \supset A_{\overline{\lambda}} \supset A_\eta \Longrightarrow \bigcup_{\lambda \in [0,\eta)} \lambda H(\lambda) \supset A_\eta.$$

On the other hand, $H(\lambda) \subset A_\lambda$; this implies that

$$\bigcap_{\lambda \in [0,\eta)} H(\lambda) \subset \bigcap_{\lambda \in [0,\eta]} A_\lambda = A \bigvee_{\lambda \in [0,\eta)} \lambda = A_\eta.$$

Therefore 2) is valid.
3)   This is similar to the proof for 2).   **Q.E.D.**

We now see that a fuzzy set can be represented by a family of classical sets $\{A_\lambda\}_{\lambda \in [0,1]}$ (or $\{A_{\overline{\lambda}}\}_{\lambda \in [0,1]}$ or $\{H(\lambda)\}_{\lambda \in [0,1]}$), and that these sets are revelations of A at different levels of λ. As λ increases or decreases, sets A_λ (or $A_{\overline{\lambda}}$, or $H(\lambda)$) will contract or expand; that is, A behaves like having an elastic boundary.

1.4   A Representation Theorem

Like resolution theorems, the representation theorem is also a basic theorem in fuzzy set theory. The theorem offers another viewpoint of linking fuzzy sets to classical sets. First, we define the mapping:

$$H : [0,1] \longrightarrow \mathcal{P}(U), \quad \lambda \longmapsto H(\lambda).$$

If the mapping H satisfies

$$\lambda < \eta \Longrightarrow H(\lambda) \supset H(\eta), \quad \forall \, \lambda, \eta \in [0,1]$$

then H is called a **nested set** on U.

The set of all nested sets on U is denoted by $\Psi(U)$. We define union ($\cup$), intersection ($\cap$), and complement (c) operations on $\Psi(U)$ as follows:

$$\cup : \Psi(U) \times \Psi(U) \longrightarrow \Psi(U)$$
$$(H_1, H_2) \longmapsto H_1 \cup H_2$$
$$(H_1 \cup H_2)(\lambda) = H_1(\lambda) \cup H_2(\lambda)$$
$$\cap : \Psi(U) \times \Psi(U) \longrightarrow \Psi(U)$$
$$(H_1, H_2) \longmapsto H_1 \cap H_2$$
$$(H_1 \cap H_2)(\lambda) = H_1(\lambda) \cap H_2(\lambda)$$
$$c : \Psi(U) \longrightarrow \Psi(U)$$
$$H \longmapsto H^c : \ H^c(\lambda) = (H(1-\lambda))^c$$

More generally, these operations can be extended to an arbitrary index set T:

$$\bigcup_{t \in T} H_t : \ \Big(\bigcup_{t \in T} H_t\Big)(\lambda) = \bigcup_{t \in T} H_t(\lambda),$$

and

$$\bigcap_{t \in T} H_t : \ \Big(\bigcap_{t \in T} H_t\Big)(\lambda) = \bigcap_{t \in T} H_t(\lambda).$$

Theorem 4 (Third Representation Theorem)   Let the mapping φ be defined as

$$\varphi : \Psi(U) \longrightarrow \mathcal{F}(U), \quad H \longmapsto \varphi(H) = \bigcup_{\lambda \in [0,1]} \lambda H(\lambda)$$

then φ is a homomorphic surjection from $(\Psi(U), \cap, \cup, c)$ to $(\Psi(U), \cap, \cup, c)$ and satisfies

1) $(\varphi(H))_\eta = \bigcap_{\lambda \in [0,\eta)} H(\lambda), \quad \eta \in (0,1];$

2) $(\varphi(H))_{\bar{\eta}} = \bigcup_{\lambda \in (\eta,1]} H(\lambda), \quad \eta \in [0,1);$

3) $(\varphi(H))_{\bar{\eta}} \subset H(\eta) \subset (\varphi(H))_{\eta}, \quad \eta \in [0,1]$.

Proof. For any $H \in \Psi(U)$,

$$\varphi(H) = \bigcup_{\lambda \in [0,1]} \lambda H(\lambda) \in \mathcal{F}(U)$$

is uniquely determined. Hence, φ is an injection. On the other hand, for any $A \in \mathcal{F}(U)$, define $H(\lambda) = A_{\lambda}$; then $H \in \Psi(U)$ and H satisfy

$$\varphi(H) = \bigcup_{\lambda \in [0,1]} \lambda A_{\lambda} = A.$$

Thus, φ is a surjection.

We now prove 3). First we show that $H(\eta) \subset (\varphi(H))_{\eta}$ holds.

$$\text{For } u \in H(\eta) \Longrightarrow \chi_{H(\eta)}(u) = 1$$
$$\Longrightarrow \mu_{\varphi(H)}(u) = \bigvee_{\lambda \in [0,1]} (\lambda \wedge \chi_{H(\lambda)}(u))$$
$$\geq \eta \wedge \chi_{H(\eta)}(u) = \eta \wedge 1$$
$$\Longrightarrow u \in (\varphi(H))_{\eta}.$$

Next, we show that $(\varphi(H))_{\bar{\eta}} \subset H(\eta)$ holds.

$$\text{For } u \notin H(\eta) \Longrightarrow \chi_{H(\eta)}(u) = 0$$
$$\Longrightarrow (\forall \lambda \in [\eta, 1])(\chi_{H(\lambda)}(u) = 0)$$
$$\Longrightarrow \mu_{\varphi(H)}(u) = \bigvee_{\lambda \in [0,\eta)} (\lambda \wedge \chi_{H(\lambda)}(u))$$
$$\leq \bigvee_{\lambda \in [0,\eta)} \lambda = \eta$$
$$\Longrightarrow u \notin (\varphi(H))_{\bar{\eta}}.$$

Based on the proposition of the last section, it is easy to show that 1) and 2) are true. We now prove in three steps that φ is a homomorphic function.

(a) Prove that φ preserves the set union operation.

Let $A^{(t)} = \varphi(H_t)$ and $A = \varphi(\bigcup_{t \in T} H_t)$. For any $\eta \in [0,1]$, we have

$$A_{\bar{\eta}} = \bigcup_{\lambda \in [0,\eta)} \left(\bigcup_{t \in T} H_t\right)(\lambda) = \bigcup_{\lambda \in (\eta,1)} \left(\bigcup_{t \in T} H_t(\lambda)\right)$$
$$= \bigcup_{t \in T} \left(\bigcup_{\lambda \in (\eta,1]} H_t(\lambda)\right) = \bigcup_{t \in T} A^{(t)}_{\bar{\eta}} = \left(\bigcup_{t \in T} A^{(t)}\right)_{\bar{\eta}}.$$

From the second resolution theorem, we know that $A = \bigcup_{t \in T} A^{(t)}$; therefore,

$$\varphi\left(\bigcup_{t \in T} H_t\right) = \bigcup_{t \in T} \varphi(H_t).$$

(b)   Prove that φ preserves the set intersection operation.

Let $A^{(t)} = \varphi(H_t)$ and $A = \varphi(\bigcap_{t \in T} H_t)$. For any $\eta \in [0,1]$, we have

$$
A_\eta = \bigcap_{\lambda \in [0,\eta)} \left(\bigcap_{t \in T} H_t\right)(\lambda) = \bigcap_{\lambda \in [0,\eta,)} \left(\bigcap_{t \in T} H_t(\lambda)\right)
$$

$$
= \bigcap_{t \in T} \left(\bigcap_{\lambda \in [0,\eta)} H_t(\lambda)\right) = \bigcup_{t \in T} A_\eta^{(t)} = \left(\bigcap_{t \in T} A^{(t)}\right)_{\overline{\eta}}.
$$

From the first resolution theorem, we know that $A = \bigcap_{t \in T} A^{(t)}$; therefore,

$$
\varphi\left(\bigcap_{t \in T} H_t\right) = \bigcap_{t \in T} \varphi(H_t).
$$

(c)   Prove that φ preserves the set complementary operation.

Let $A = \varphi(H)$, $B = \varphi(H^c)$ and $\lambda' = 1 - \lambda$. For any $\eta \in [0,1]$, we have

$$
B_\eta = \bigcap_{\lambda \in [0,\eta)} H^c(\lambda) = \bigcap_{\lambda \in [0,\eta)} (H(1-\lambda))^c = \left(\bigcup_{1-\lambda \in (1-\eta,1]} H(1-\lambda)\right)^c
$$

$$
= \left(\bigcup_{\lambda' \in (1-\eta,1]} H(\lambda')\right)^c = (A_{\overline{1-\eta}})^c = \{u \in U \mid \mu_A(u) > 1-\eta\}^c
$$

$$
= \{u \in U \mid \mu_A(u) \leq 1-\eta\} = \{u \in U \mid 1-\mu_A(u) \geq \eta\}
$$

$$
= \{u \in U \mid \mu_{A^c}(u) \geq \eta\} = (A^c)_\eta.
$$

From the first resolution theorem, we have $B = A^c$, therefore,

$$
\varphi(H^c) = (\varphi(H))^c.
$$

This completes the proof that φ is a homomorphic function from $\Psi(U)$ to $\mathcal{F}(U)$.
Q.E.D.

Resolution theorems illustrate that a fuzzy set can be expressed by a nested set, and the representation theorem shows that any nested set, indeed, corresponds to a fuzzy set.

Since φ is a surjection from $\Psi(U)$ to $\mathcal{F}(U)$, by means of φ, we can define an equivalence relation "$\sim$" as

$$
(\forall\, H_1,\ H_2 \in \Psi(U))(H_1 \sim H_2 \iff \varphi(H_1) = \varphi(H_2)).
$$

Let the equivalent class that H belongs to be denoted by $\overline{H}$. We get the quotient set induced by the equivalent relation as

$$
\Psi'(U) \overset{\triangle}{=} \Psi(U)/\!\sim\ \overset{\triangle}{=} \{\overline{H} \mid H \in \Psi(U)\}.
$$

In $\Psi'(U)$, we define the set operations of $\cup$, $\cap$, and "c" as follows.

$$\overline{H}_1 \cup \overline{H}_2 = \overline{H_1 \cup H_2}. \quad \bigcup_{t \in T} \overline{H}_t = \overline{\bigcup_{t \in T} H_t},$$

$$\overline{H}_1 \cap \overline{H}_2 = \overline{H_1 \cap H_2}. \quad \bigcap_{t \in T} \overline{H}_t = \overline{\bigcap_{t \in T} H_t},$$

$$(\overline{H})^c = \overline{H^c}.$$

We have the following:

Corollary 1   The mapping $\varphi' : \Psi'(U) \longrightarrow \mathcal{F}(U)$ defined by $\overline{H} \longmapsto \varphi'(\overline{H}) = \varphi(H)$ is an isomorphism from $(\Psi'(U), \cup, \cap, c)$ to $(\mathcal{F}(U), \cup, \cap, c)$.

From the corollary, a fuzzy set is exactly an equivalence class and, conversely, an equivalence class can be regarded as a fuzzy set.

Let $F \in \Psi(U)$. The nested set F is called a **dizzy set** on U if it satisfies

$$F(0) = U, \quad F\left(\bigvee_{t \in T} \lambda_t\right) = \bigcap_{t \in T} F(\lambda_t) \tag{1.1}$$

where $\lambda_t \in [0,1]$ and $t \in T$. The set of all dizzy sets is denoted by $\Gamma(U)$. Let $\overline{F} \in \Psi(U)$. A nested set $\overline{F}$ is called an **open dizzy set** if it satisfies

$$\overline{F}(1) = \emptyset, \quad \overline{F}\left(\bigwedge_{t \in T} \lambda_t\right) = \bigcup_{t \in T} \overline{F}(\lambda_t) \tag{1.2}$$

The set of all open dizzy sets on U is denoted by $\overline{\Gamma}(U)$.

Example 6   Let $A \in \mathcal{F}(U)$ and define two mappings:

$$F : [0,1] \longrightarrow \mathcal{P}(U), \quad \lambda \longmapsto F(\lambda) = A_\lambda$$

and

$$F : [0,1] \longrightarrow \mathcal{P}(U), \quad \lambda \longmapsto \overline{F}(\lambda) = A_{\overline{\lambda}}$$

It is obvious that $F \in \Gamma(U)$ and $\overline{F} \in \overline{\Gamma}(U)$.

Proposition 2   For any $\overline{H} \in \Psi'(U)$, we have
1)   there uniquely exists a dizzy set $F_H \in \overline{H}$ such that

$$F_H(\eta) = \bigcap_{\lambda \in [0,\eta)} H(\lambda), \quad \eta \in (0,1],$$

2)   there uniquely exists an open dizzy set $\overline{F_H} \in \overline{H}$ such that

$$\overline{F}_H(\eta) = \bigcup_{\lambda \in (\eta,1]} H(\lambda), \quad \eta \in [0,1).$$

Proof.   We only prove 1). The proof of 2) is similar.

From the third representation theorem, we know $\varphi(H) \in \mathcal{F}(U)$. The existence of $F_H \in \Gamma(U)$ that satisfies

$$F_H(\eta) = (\varphi(H))_\eta = \bigcap_{\lambda \in [0,\eta)} H(\lambda), \quad \eta \in (0,1]$$

and its uniqueness is obvious. We prove that $F_H \in \overline{H}$. In fact,

$$(\varphi(F_H))_\eta = \bigcap_{\lambda \in [0,\eta)} F_H(\lambda) = \bigcap_{\lambda \in [0,\eta)} \left(\bigcap_{\nu \in [0,\eta)} H(\nu) \right)$$

$$= \bigcap_{\lambda \in [0,\eta)} H(\lambda) = (\varphi(H))_\eta.$$

Hence $F_H \in \overline{H}$, and this proves 1).   **Q.E.D.**

Note 2   If $H, H' \in \Psi(U)$, then the following conditions are equivalent.
1) $H \sim H'$;

2) $(\forall\, \eta \in [0,1]) \left(\bigcap_{\lambda \in [0,\eta)} H(\lambda) = \bigcap_{\lambda \in [0,\eta)} H'(\lambda) \right)$;

3) $(\forall\, \eta \in [0,1]) \left(\bigcup_{\lambda \in (\eta,1]} H(\lambda) = \bigcup_{\lambda \in (\eta,1]} H'(\lambda) \right)$.

Theorem 5
(1) $(\Psi'(U), \cup, \cap, c) \cong (\Gamma(U), \cup, \cap, c)$;
(2) $(\Psi'(U), \cup, \cap, c) \cong (\overline{\Gamma}(U), \cup, \cap, c)$.
Proof.   Define the following mappings:

$$f : \ \Psi'(U) \longrightarrow \Gamma(U), \quad \overline{H} \longmapsto f(\overline{H}) = F_H$$

and

$$\overline{f} : \ \Psi'(U) \longrightarrow \overline{\Gamma}(U), \quad \overline{H} \longmapsto f(\overline{H}) = \overline{F}_H.$$

It is easy to verify that both f and $\overline{f}$ are isomorphic mappings.   **Q.E.D.**

Corollary 2 (First Representation Theorem)

$$(\Gamma(U), \cup, \cap, c) \cong (\mathcal{F}(U), \cup, \cap, c).$$

Proof.   Let $g = \varphi' \circ f^{-1}$, where φ' and f are defined as before. Then it is easy to show that

$$g : \ \Gamma(U) \longrightarrow \mathcal{F}(U), \quad F \longmapsto g(F) = \bigcup_{\lambda \in [0,1]} \lambda F(\lambda)$$

is an isomorphic mapping from $\Gamma(U)$ to $\mathcal{F}(U)$.   **Q.E.D.**

Corollary 3 (Second Representation Theorem)

$$(\overline{\Gamma}(U), \cup, \cap, c) \cong (\mathcal{F}(U), \cup, \cap, c).$$

Proof.   Let $\overline{g} = \varphi' \circ \overline{f}^{-1}$, where φ' and $\overline{f}$ are defined as before. Similar to the above proof

$$\overline{g} : \overline{\Gamma}(U) \longrightarrow \mathcal{F}(U), \quad \overline{F} \longmapsto \overline{g}(\overline{F}) = \bigcup_{\lambda \in [0,1]} \lambda \overline{F}(\lambda)$$

is an isomorphic mapping from $\overline{\Gamma}(U)$ to $\mathcal{F}(U)$.   **Q.E.D.**

Clearly, the first and second representation theorems are special cases of the third representation theorem.

1.5   Extension Principle

The extension principle is an important premise that has broad applications.

Let X and Y be two universes. Given a mapping $f : X \longrightarrow Y$, the extension principle tells us how to induce a mapping $f : \mathcal{F}(X) \longrightarrow \mathcal{F}(Y)$ by means of the mapping f.

First, we review the extension principle from the classical set theory. Let $f : X \longrightarrow Y$ be a mapping. The extension of f, still denoted by f, may be induced as

$$f : \mathcal{P}(X) \longrightarrow \mathcal{P}(Y), \quad A \longmapsto f(A) = \{f(x) \mid x \in A\}.$$

An inverse mapping f^{-1} can also be induced by f:

$$f^{-1} : \mathcal{P}(X) \longrightarrow \mathcal{P}(Y), \quad B \longmapsto f^{-1}(B) = \{x \in X \mid f(x) \in B\}.$$

We call $f(A)$ the image of A, and $f^{-1}(B)$ the inverse image of B (see Figure 5).

The characteristic functions of A, B, $f(A)$, and $f^{-1}(B)$ have the following properties:

1)

$$\chi_{f(A)}(y) = \bigvee_{f(x)=y} \chi_A(x), \quad y \in Y; \tag{1.3}$$

2)

$$\chi_{f^{-1}(B)}(x) = \chi_B(f(x)), \quad x \in X, \tag{1.4}$$

and images and their inverse images satisfy

a) $f^{-1}(f(A)) \supset A$ ("$\supset$" becomes "$=$" when f is an injection),

b) $f(f^{-1}(B)) \subset B$ ("$\subset$" becomes "$=$" when f is a surjection).

The classical extension principle is now generalized below.

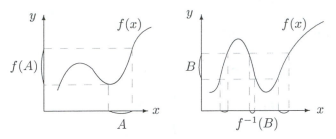

Figure 5   The classical extension principle

Extension Principle 1   The mapping $f : \mathcal{F}(X) \longrightarrow \mathcal{F}(Y)$ can be induced from the mapping $f : X \longrightarrow Y$ by defining

$$A \longmapsto f(A) = \bigcup_{\lambda \in [0,1]} \lambda f(A_\lambda), \quad \forall A \in \mathcal{F}(X). \tag{1.5}$$

The mapping f induced by f is called the fuzzy transformation from X to Y. Similarly, f^{-1} may be induced by f by defining

$$f^{-1} : \mathcal{F}(Y) \longrightarrow \mathcal{F}(X), \quad B \longmapsto f^{-1}(B) = \bigcup_{\lambda \in [0,1]} \lambda f^{-1}(B_\lambda). \tag{1.6}$$

The mapping f^{-1} induced by f is called the fuzzy inverse transformation.

The following theorem will show that A, B, $f(A)$, and $f^{-1}(B)$ have properties similar to those of a) and b).

Theorem 6   Given a mapping $f : X \longrightarrow Y$, we have
1)   if $A \in \mathcal{F}(X)$, then for any $y \in Y$.

$$\mu_{f(A)}(y) = \bigvee_{f(x)=y} \mu_A(x) \tag{1.7}$$

(when $\{x \in X \mid f(x) = y\} = \emptyset$, then set $\mu_{f(A)}(y) = 0$);
2)   if $B \in \mathcal{F}(Y)$, then for any $x \in X$,

$$\mu_{f^{-1}(B)}(x) = \mu_B(f(x)). \tag{1.8}$$

Proof.

$$\mu_{f(A)}(y) = \bigvee_{\lambda \in [0,1]} (\lambda \wedge \chi_{f(A_\lambda)}(y))$$

$$= \bigvee \{\lambda \in [0,1] \mid y \in f(A_\lambda)\}$$

$$= \bigvee \{\lambda \in [0,1] \mid (\exists x \in A_\lambda)(y = f(x))\}$$

$$= \bigvee_{f(x)=y} \{\lambda \in [0,1] \mid x \in A_\lambda\}$$

$$= \bigvee_{f(x)=y} \left(\bigvee_{\lambda \in [0,1]} (\lambda \wedge \chi_{A_\lambda}(x)) \right) = \bigvee_{f(x)=y} \mu_A(x).$$

This proves 1). For 2),

$$\mu_{f^{-1}(B)}(x) = \bigvee_{\lambda \in [0,1]} (\lambda \wedge \chi_{f^{-1}(B_\lambda)}(x))$$

$$= \bigvee \{\lambda \in [0,1] \mid x \in f^{-1}(B_\lambda)\}$$

$$= \bigvee \{\lambda \in [0,1] \mid f(x) \in B_\lambda\}$$

$$= \bigvee_{\lambda \in [0,1]} (\lambda \wedge \chi_{B_\lambda}(f(x))) = \mu_B(f(x)).$$

This shows 2) is valid.   **Q.E.D.**

Theorem 7 (Extension Principle 2)   Given a mapping $f : X \longrightarrow Y$,
1)   if $A \in \mathcal{F}(X)$, then
$$f(A) = \bigcup_{\lambda \in [0,1]} \lambda f(A_{\overline{\lambda}}); \tag{1.9}$$

2)   if $B \in \mathcal{F}(Y)$, then
$$f^{-1}(B) = \bigcup_{\lambda \in [0,1]} \lambda f^{-1}(B_{\overline{\lambda}}). \tag{1.10}$$

Proof.   Using a proof similar to the one for the last theorem, we can easily verify that

$$\mu_{\cup_{\lambda \in [0,1]} \lambda f(A_{\overline{\lambda}})}(y) = \bigvee_{f(x)=y} \mu_A(x) = \mu_{f(A)}(y),$$

$$\mu_{\cup_{\lambda \in [0,1]} \lambda f^{-1}(B_{\overline{\lambda}})}(x) = \mu_B(f(x)) = \mu_{f^{-1}(B)}(x).$$

Hence the theorem is proved.   **Q.E.D.**

Theorem 8 (Extension Principle 3)   Given a mapping $f : X \longrightarrow Y$,
1)   if $A \in \mathcal{F}(X)$, then
$$f(A) = \bigcup_{\lambda \in [0,1]} \lambda f(H_A(\lambda)), \tag{1.11}$$

where $H_A \in \Psi(X)$ and $A_{\bar{\lambda}} \subset H_A(\lambda) \subset A_\lambda$;

2)   if $B \in \mathcal{F}(Y)$, then

$$f^{-1}(B) = \bigcup_{\lambda \in [0,1]} \lambda f^{-1}(H_B(\lambda)), \qquad (1.12)$$

where $H_B \in \Psi(Y)$ and $B_{\bar{\lambda}} \subset H_B(\lambda) \subset B_\lambda$.

Proof.   The proof is omitted since it is not difficult.   **Q.E.D.**

Theorem 9 (The Extension Principle for Composite Mappings)   Let $f : X \longrightarrow Y$ and $g : Y \longrightarrow Z$. The composite mapping of f and g is defined by

$$g \circ f : X \longrightarrow Z, \text{ such that } (g \circ f)(x) = g(f(x)).$$

1)   If $A \in \mathcal{F}(X)$, then

$$(g \circ f)(A) = g(f(A)) = \bigcup_{\lambda \in [0,1]} \lambda g(f(A_\lambda)). \qquad (1.13)$$

2)   If $D \in \mathcal{F}(Z)$, then

$$(g \circ f)^{-1}(D) = f^{-1}(g^{-1}(D)) = \bigcup_{\lambda \in [0,1]} \lambda f^{-1}(g^{-1}(D_\lambda)). \qquad (1.14)$$

Proof.   We will only prove 1), because 2) can be shown in a similar manner. Since

$$\begin{aligned} (g \circ f)(A_\lambda) &= \{z \in Z \mid (\exists\, x \in A_\lambda)((g \circ f)(x) = g(f(x)))\} \\ &= \{z \in Z \mid (\exists\, x \in A_\lambda)(f(x) = y,\ g(y) = z)\} \\ &= \{z \in Z \mid (\exists\, y \in f(A_\lambda))(g(y) = z)\} \\ &= g(f(A_\lambda)), \end{aligned}$$

then

$$(g \circ f)(A) = \bigcup_{\lambda \in [0,1]} \lambda(g \circ f)(A_\lambda) = \bigcup_{\lambda \in [0,1]} \lambda g(f(A_\lambda)).$$

By the extension principle 1 and the representation theorem,

$$f(A) = \bigcup_{\lambda \in [0,1]} \lambda f(A_\lambda), \quad \text{and} \quad (f(A))_{\bar{\lambda}} \subset f(A_\lambda) \subset (f(A))_\lambda.$$

By the extension principle 3, we have

$$g(f(A)) = \bigcup_{\lambda \in [0,1]} \lambda g(f(A_\lambda)) = (g \circ f)(A).$$

This completes the proof.   **Q.E.D.**

Proposition 3   1) $f^{-1}(f(A)) \supset A$. The "$\supset$" becomes "$=$" when f is an injection,

2) $f(f^{-1}(B)) \subset B$, The "$\subset$" becomes "$=$" when f is a surjection.

Proof.   Since

$$\mu_{f^{-1}(f(A))}(x) = \mu_{f(A)}(f(x)) = \bigvee_{f(t)=f(x)} \mu_A(t) \geq \mu_A(x),$$

clearly, when f is an injection, the "$\geq$" becomes "$=$". Hence 1) is true. Next, since

$$\mu_{f(f^{-1}(B))}(y) = \bigvee_{f(x)=y} \mu_{f^{-1}(B)}(x) = \bigvee_{f(x)=y} \mu_B(f(x))$$

$$= \begin{cases} \mu_B(y), & (\exists x)(f(x)=y) \\ 0, & (\forall x)(f(x) \neq y) \end{cases}$$

$$\leq \mu_B(y).$$

Obviously, when f is a surjection "$\leq$" becomes "$=$". This proves 2).   **Q.E.D.**

References

1. L. A. Zadeh, Fuzzy sets, *Information and Control*, Vol. 8, pp. 338-353, 1965.

2. L. A. Zadeh, Fuzzy sets as a basis for a theory of possibility, *Fuzzy Sets and Systems,* Vol. 1, pp. 3-28, 1978.

3. R. Bellman and M. Giertz, On the analytic formalism of the theory of fuzzy sets, *Information Sciences,* Vol. 5, pp. 149-156, 1973.

4. D. Dubois and H. Prade, *Fuzzy Sets and Systems*, Academic Press, New York, 1980.

5. H. J. Zimmermann, *Fuzzy Sets Theory and Its Applications,* Kluwer Academic Publications, Hingham, 1984.

6. T. Terano, K. Asai, and M. Sugeno, *Fuzzy Systems Theory and its Applications,* Academic Press, San Diego, 1987.

7. G. J. Klir and T. A. Folger, *Fuzzy Sets, Uncertainty, and Information,* Prentice-Hall, Englewood Cliffs, 1988.

8. T. Terano, K. Asia, and M. Sugeno, *Fuzzy Systems Theory and its Applications,* Academic Press, San Diego, 1992.

9. R. R. Yager and D. P. Filev, *Essentials of Fuzzy Modeling and Control,* John Wiley and Sons, New York, 1994.

10. D. Dubois, H. Prade, and R. R. Yager, *Fuzzy Information Engineering*, John Wiley and Sons, New York, 1997.

11. C. T. Lin and C. S. G. Lee, *Neural Fuzzy Systems*, Prentice-Hall, Englewood Cliffs, 1996.

12. J. S. Jang, C. T. Sun, and E. Mizutani, *Neuro-Fuzzy and Soft Computing*, Prentice-Hall, Englewood Cliffs, 1997.

Chapter 2

Determination of Membership Functions

In real world application of fuzzy sets, an important task is to determine membership functions of the fuzzy sets in question. Like the estimation of probabilities in the probability theory, we can only obtain an approximate membership function of a fuzzy set because of our cognitive limitations. In this chapter we discuss how to determine a membership function of a fuzzy set.

2.1 A General Method for Determining Membership Functions

In our natural world and daily lives, we experience all kinds of phenomena; broadly speaking, we can divide them into two types: phenomena of certainty and phenomena of uncertainty.

The class of uncertain phenomena can further be subdivided into random (stochastic) phenomena and fuzzy phenomena. Therefore, we have three categories of phenomena and their associated mathematical models:

1. Deterministic mathematical models–This is a class of models where the relationships between objects are fixed or known with certainty.

2. Random (stochastic) mathematical models–This is a class of models where the relationships between objects are uncertain or random in nature.

3. Fuzzy mathematical models–This is a class of models where objects and relationships between objects are fuzzy.

The main distinction between random phenomena and fuzzy phenomena is that random events themselves have clear and well-defined meaning, whereas a fuzzy concept does not have a precise extension because it is hard to judge if an object belongs to the concept. We may say that randomness is a deficiency of the law of causality and that fuzziness is a deficiency of the law of the excluded middle. Probability theory applies the random concept to generalized laws of causality–laws of probability. Fuzzy set theory applies the fuzzy property to the generalized law of the excluded middle–the law of membership from fuzziness.

Probability reflects the internal relations and interactions of events under certain conditions. It could be very objective if a stable frequency is available from repeated experiments. Similarly, a stable frequency results from fuzzy statistical tests (discussed later) and can serve as the degree of membership in the objective sense. In many cases, the degree of membership can be determined by fuzzy statistical methods.

Before entering into fuzzy statistics, we first outline some basic notions of probability and statistics. A random experiment has four basic requirements:

1. A sample space Ω that is a high-dimensional direct product space formed by all related factors;

2. An event A that is a fixed and crisp subset of Ω;

3. A variable ω that is in Ω; once ω is determined, so are all the factors on their state spaces;

4. A condition S that sets forth the restrictions on variable ω.

The basic reason for the existence of randomness is that the condition S is too weak to restrict ω to a fixed point. Moreover, if $S \cap A \neq \Phi$ and $S \cap A^c \neq \Phi$, then $\omega \in S \cap A$ signifies that the event A has occurred; otherwise it has not. That means A is a random event under the condition S. On random experiments we have the following observations:

1. The purpose of a random experiment is to study the uncertainty through the use of certain or deterministic procedures.

2. A fundamental requirement of a random experiment is to unambiguously tell whether an event A has occurred or not.

3. A special characteristic of a random experiment is that A is fixed, while ω is not (see Figure 1(a)).

4. After n times of repeated trials (experiments), we compute the frequency of A by

$$\text{Frequency of } A = \frac{\text{The number of trials when } \omega \in A}{n}.$$

It is well known that as n becomes very large, the above frequency tends to stabilize or converge to a fixed number. This number is called the probability of A under S.

A *fuzzy statistical experiment* also has four basic requirements:

1. A universe U;

2. A fixed element $u_0 \in U$;

3. An alterable crisp set A_* in U that is related to a fuzzy set A (the extension of some fuzzy concept α). Every fixed instance of A_* is a partition of the concept α; it also represents an approximate extension of α.

4. A condition S that relates to all objective and psychological factors during the process of partitioning of the concept α, and limits the range of change that A_* may have.

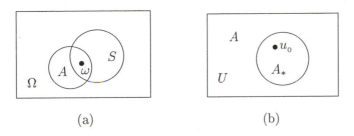

Figure 1
(a) A random experiment: A is fixed, but w varies
(b) A fuzzy statistical experiment: u_0 is fixed, but A_* varies

On the properties of fuzzy statistical experiments, we have the following observations:

1. The purpose of fuzzy statistical experiments is to study such uncertainties with certain or definite procedures.

2. A fundamental requirement of a fuzzy statistical experiment is to unambiguously tell whether $u_0 \in A_*$ or not.

3. A special characteristic of a fuzzy statistical experiment is that u_0 is fixed while X_* varies.

4. After collecting n sample observations from an experiment, we compute the frequency of u_0 in A_* by

$$\text{Frequency of } u_0 \text{ in } A_* = \frac{\text{The number of observations of } u_0 \in A_*}{n}.$$

As n increases, the frequency tends to stabilize or converge to a fixed number; this number is called the *degree of membership* of u_0 in A_*.

The question we consider is what is the acceptable range of age for the class "young men"? Results of the 129 represents are given in Table 1. Table 2 shows the frequency of membership $u_0 (= 27)$ in these age ranges.

Table 2 demonstrates that the normalized frequency of membership for age 27 stabilizes at 0.78 as the sample size increases (The sum of the relative frequencies).

We may now define

$$\mu_A(u_0) = \mu_A(27) = 0.78.$$

To find the membership function for the "young man" that is denoted by A, we partition the universe U into unit length intervals with integers as the midpoint of the intervals. Table 3 contains the normalized frequencies of each interval. We can draw a histogram based on the values of Table 3 (Figure 2). This way we obtain an empirical membership curve for $\mu_A(u)$.

Table 1   The Age Ranges for "Young Men" Specified by 129 Respondents

18–25	17–30	17–28	18–25	16–35	14–25	18–30	18–35	18–35	16–25
15–30	18–35	17–30	18–25	10–25	18–35	20–30	18–30	16–30	20–35
18–30	18–30	15–25	18–30	15–25	16–28	16–30	18–30	16–30	18–35
18–25	18–25	16–28	18–30	16–30	16–28	18–35	18–35	17–27	16–28
15–28	16–30	19–28	15–30	15–26	17–25	15–36	18–30	17–30	18–35
16–35	15–25	15–25	18–28	16–30	15–28	18–35	18–30	17–28	18–35
15–28	18–30	15–25	15–25	18–30	16–24	15–25	16–32	15–27	18–35
16–25	18–28	16–28	18–30	18–35	18–30	18–30	17–30	18–30	18–35
16–30	18–35	17–25	15–30	18–25	17–30	14–25	18–26	18–29	18–35
18–28	18–30	18–25	16–35	17–29	18–25	17–30	16–28	18–30	16–28
15–30	15–35	18–30	20–30	20–30	16–25	17–30	15–30	18–30	16–30
18–28	18–35	16–30	15–30	18–35	18–35	18–30	17–30	18–35	17–30
15–25	18–35	15–30	15–25	15–30	18–30	17–25	18–29	18–28	—

Example 1   Let $U = [0, 100]$ be the "age" universe and $A \in \mathcal{F}(U)$ be the extension of the fuzzy concept α: " a young man". Choose a fixed age, say 27, i.e., $u_0 = 27$; we wish to determine its degree of membership in A. Suppose we conducted a fuzzy statistical experiment. We picked 129 suitable people for our experiment.

Table 2   Cumulative Frequency of Membership for Age 27
(Continued from Table 1)

n	Count	Frequency
10	6	0.60
20	14	0.70
30	23	0.77
40	31	0.78
50	39	0.78
60	47	0.78
70	53	0.76
80	62	0.78
90	68	0.76
100	76	0.76
110	85	0.75
120	95	0.79
129	101	0.78

Table 3   Frequency of Membership

Order	Grouping	Frequency	Relative Frequency
1	13.5–14.5	2	0.0155
2	14.5–15.5	27	0.2093
3	15.5–16.5	51	0.0155
4	16.5–17.5	67	0.2093
5	17.5–18.5	124	0.9612
6	18.5–19.5	125	0.9690
7	19.5–20.5	129	1.0000
8	20.5–21.5	129	1.0000
9	21.5–22.5	129	1.0000
10	22.5–23.5	129	1.0000
11	23.5–24.5	129	1.0000
12	24.5–25.5	128	1.9922
13	25.5–26.5	103	0.7984
14	26.5–27.5	101	0.7829
15	27.5–28.5	99	0.9922
16	28.5–29.5	80	0.6202
17	29.5–30.5	77	0.5969
18	30.5–31.5	27	0.2093
19	31.5–32.5	27	0.2093
20	32.5–33.5	26	0.2016
21	33.5–34.5	26	0.2016
22	34.5–35.5	26	0.2016
23	35.5–36.5	1	0.0078

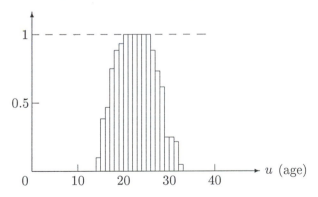

Figure 2   Histogram of the membership function for the "young men"

2.2   The Three-phase Method

The *three-phase method* is another fuzzy experimental procedure that uses random intervals. For example, suppose we want to establish a membership function for each type of man called: short, medium, and tall. Let the universe be $U = [0, 3]$ (in meters), A_1 be the short type , A_2 be the medium type, and A_3 then be the tall type. Every fuzzy statistical experiment determines a pair of numbers ζ and η, where ζ is a demarcation point of "short man" and "medium man", and η a demarcation point of "medium man" and "tall man". We can view (ζ, η) as a two-dimensional random variable. Then through sampling (Figure 3) we obtain its marginal probability distributions $p_\zeta(u)$ and $p_\eta(u)$. The membership function for each type of man, therefore, is

$$\mu_{A_1}(u) = \int_u^{+\infty} p_\zeta(u)\, du,$$

$$\mu_{A_2}(u) = \int_{-\infty}^u p_\eta(u)\, du,$$

and

$$\mu_{A_3}(u) = 1 - \mu_{A_1}(u) - \mu_{A_2}(u),$$

respectively.

In general, ζ and η follow a normal distribution. Assume

$$\zeta \sim N(\mu_1, \sigma_1) \qquad \text{and} \qquad \eta \sim N(\mu_2, \sigma_2),$$

then

$$\mu_{A_1}(u) = 1 - \varphi\Big(\frac{u - \mu_1}{\sigma_1}\Big)$$

$$\mu_{A_2}(u) = \varphi\Big(\frac{u - \mu_2}{\sigma_2}\Big)$$

and

$$\mu_{A_3}(u) = \varphi\Big(\frac{u - \mu_1}{\sigma_1}\Big) - \varphi\Big(\frac{u - \mu_2}{\sigma_2}\Big),$$

where

$$\varphi(u) = \int_{-\infty}^u \frac{1}{\sqrt{2\pi}} e^{-x^2/2}\, dx.$$

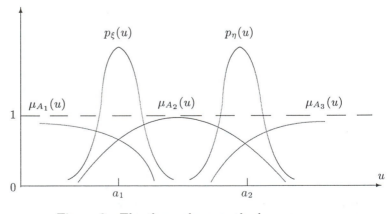

Figure 3   The three-phase method

2.3   The Incremental Method

We explain, through an example, how to apply the incremental method for calibrating membership functions. Let A be the extension of the fuzzy concept "old man" and $A \in \mathcal{F}([0, 100])$. For simplicity, we define $\mu(u) = \mu_A(u)$, $u \in [0, 100]$.

Given an arbitrary increment Δu on u, there will be a corresponding incremental change $\Delta \mu$ on μ. For simplicity, we assume that $\Delta \mu$ is directly proportional to Δu. Also, for the same increment Δu, the bigger the value of u is, the larger the $\Delta \mu$ will be. Since μ is bounded by 1, $\Delta \mu$ should become smaller as μ is getting close to 1. That is,

$$\Delta \mu = k \cdot \Delta u \cdot u(1 - \mu) \tag{2.1}$$

where k is a constant of proportionality. Dividing both sides by Δu, we obtain

$$\frac{\Delta \mu}{\Delta u} = k \cdot u(1 - \mu).$$

When $\Delta u \longrightarrow 0$, we get a differential equation:

$$\frac{d\mu}{du} = k \cdot u(1 - \mu) \tag{2.2}$$

whose solution is

$$\mu(u) = 1 - ce^{-kx^2/2} \tag{2.3}$$

where c is an integral constant. If we properly choose k and c, then $\mu(u)$ will be completely determined, and so will $\mu_A(u)$.

2.4   The Multiphase Fuzzy Statistical Method

Strictly speaking, the two fuzzy statistical methods mentioned above belong to the two-phase method. Let's analyze the way we establish the degree of membership for age 27 $(= u_0)$ in the concept of "young man". The steps taken are as follows:

1.   Select a set $P = \{p_1, p_2, \cdots, p_n\}$ of n subjects for sampling.

2.   Collect response of p_1, in terms of a subset of U on the concept of "young man". For example, to p_1, a young man means any man whose age is the range of $[18, 29]$. Let's denote it as

$$A(u_0, p_1) = 1, \qquad u_0 \in [18, 29].$$

Similarly, for p_2 we may have

$$A(u_0, p_2) = 1, \qquad u_0 \in [16, 25].$$

Thus, we collect n such responses. Each of them can affirmatively say yes or no to the question of whether a 27-year-old man is "young" or "not young", The "young" and "not young" are what we mean by phases; in this case we have two.

3. Compute the degree of membership by

$$\mu_A(u_0) = \frac{1}{n} \sum_{i=1}^{n} A(u_0, p_i).$$

The three-phase discussed earlier is a three-phase fuzzy statistical method.

In many cases, it is necessary to consider more than two fuzzy sets simultaneously. Therefore, we need to generalize the two-phase fuzzy statistical method to a multiphase fuzzy statistical method.

Let $A_1, A_2, \cdots, A_m \in \mathcal{F}(U)$. To see if $u \in U$ belongs to some A_i, $1 \le i \le m$, compute $\mu_{A_i}(u)$. Let $P = \{p_1, p_2, \cdots, p_n\}$ as defined earlier. We assume that for any $p_j \in P$, it satisfies

1) $A_i(u, p_j) = \begin{cases} 1, & p_j \text{ says } u \text{ belongs to } A_i \\ 0, & p_j \text{ says } u \text{ does not belong to } A_i \end{cases}$

2) $\sum_{i=1}^{m} A_i(u, p_j) = 1$, that is u belongs to one and only one of the A's.

Then we can compute $\mu_{A_i}(u)$ from the following formula:

$$\mu_{A_i}(u) = \frac{1}{n} \sum_{j=1}^{n} A_i(u, p_j), \quad i = 1, 2, \cdots, m.$$

Example 2 Find the degree of membership of John's height (u) in fuzzy sets of A_1 (short men), A_2 (medium size men), and A_3 (tall men).

We apply the three-phase fuzzy statistical method to this problem. Let $P = \{p_1, p_2, \cdots, p_{10}\}$ be a set of ten subjects in this experiment. Their responses from this experiment are in Table 4.

Table 4   Degree of Membership by Respondents

$A_i \backslash p$	p_1	p_2	p_3	p_4	p_5	p_6	p_7	p_8	p_9	p_{10}
A_1	0	0	1	1	1	0	0	0	1	0
A_2	1	1	0	0	0	1	1	0	1	1
A_3	0	0	0	0	0	0	0	1	0	0

From the table, we compute

$$\mu_{A_1}(u) = \frac{1}{10} \sum_{j=1}^{10} A_1(u, p_j) = \frac{4}{10} = 0.4.$$

Likewise, we find

$$\mu_{A_2}(u) = 0.5 \quad \text{and} \quad \mu_{A_3}(u) = 0.1.$$

Notice that

$$\mu_{A_1}(u) + \mu_{A_2}(u) + \mu_{A_3}(u) = 0.4 + 0.5 + 0.1 = 1.$$

In general, the multiphased fuzzy statistical method has the following two properties:

Property 1 $\sum_{i=1}^{m} \mu_{A_i}(u) = 1.$

Proof.

$$\sum_{i=1}^{m} \mu_{A_i}(u) = \sum_{i=1}^{m} \left(\frac{1}{n} \sum_{j=1}^{n} A_i(u, p_j) \right)$$

$$= \frac{1}{n} \sum_{i=1}^{m} \sum_{j=1}^{n} A_i(u, p_j)$$

$$= \frac{1}{n} \sum_{j=1}^{n} \left(\sum_{i=1}^{m} A_i(u, p_j) \right)$$

$$= \frac{1}{n} \sum_{j=1}^{n} 1 = 1. \quad \square$$

Property 2 Let $U = \{u_1, u_2, \cdots, u_k\}$ be a universe with k elements. Then,

$$\sum_{i=1}^{m} \sum_{j=1}^{k} \mu_{A_i}(u_j) = k.$$

Proof.

$$\sum_{i=1}^{m} \sum_{j=1}^{k} \mu_{A_i}(u_j) = \sum_{j=1}^{k} \left(\sum_{i=1}^{m} \mu_{A_i}(u_j) \right) = \sum_{j=1}^{k} 1 = k. \quad \square$$

2.5   The Method of Comparisons

2.5.1   Binary Comparisons

Very often in a multiphase fuzzy statistical experiment, participants have difficulty declaring the highest degree of membership of objects being evaluated. In practice, pairwise comparisons have been used to solve this problem. For example, "George is more enthusiastic than Martin". If we let the extension of the fuzzy concept "enthusiastic" be denoted by A, then

$$\mu_A(\text{George}) \geq \mu_A(\text{Martin}).$$

The comparison made between George and Martin on the basis of the degree of membership in the "enthusiastic" set is called the *binary comparison*.

In the process of learning and recognition, binary comparison is used routinely by people. In simple cases, a binary comparison can establish an order relation ($\geq$), and from it we can roughly determine a membership function. Unfortunately, the order relation generated by the binary comparison does not satisfy the transitivity property. For instance, if George is more enthusiastic than Martin, and Martin is more enthusiastic than Peter, we may not conclude that George is more enthusiastic than Peter. It could be that Peter is more enthusiastic than George–a feature of fuzzy concepts. Therefore, to overcome this problem, we need to modify the binary comparison method.

2.5.2   Preferred Comparisons

Let $P = \{p_1, p_2, \cdots, p_n\}$ be the set of subjects or participants in the fuzzy statistical experiment, U be the universe, and $A_1, A_2, \cdots, A_m \in \mathcal{F}(U)$ be m fuzzy sets. Our problem is this: for a given $u \in U$, find its degree of membership in each A_i, i.e., $\mu_{A_i}(u)$, for $i = 1, 2, \cdots, m$.

For every pair of fuzzy sets (A_s, A_k), $1 \leq s \leq k \leq m$, and any $p_j \in P$, we stipulate that

$$A_i(u, p_j) = \begin{cases} 1, & p_j \text{ says } u \text{ belongs to } A_i \\ 0, & p_j \text{ says } u \text{ does not belongs to } A_i \end{cases}$$

and

$$A_s(u, p_j) + A_k(u, p_j) = 1.$$

Thus, the comparison made by p_k between A_s and A_k, if $\mu_{A_k}(u) = 1$ (so $\mu_{A_s}(u) = 0$) is determined. Then we can write

$$f_{A_s}(A_k, p_j) = 1, \quad \text{and} \quad f_{A_k}(A_s, p_j) = 0.$$

We sum up all comparisons made by p_j in Table 5.

Table 5 $f_{A_k}(A_s, p_j)$ from p_j by Binary Comparisons

$A_s \setminus A_k$	A_1	A_2	A_3	$\cdots$	A_m
A_1	—	0	1	$\cdots$	0
A_2	1	—	0	$\cdots$	0
A_3	0	1	—	$\cdots$	1
$\cdots$	$\cdots$	$\cdots$	$\cdots$	$\cdots$	$\cdots$
A_m	1	1	0	$\cdots$	—

We present the results of the binary comparisons made by n participants in Table 6.

Table 6   Binary Comparisons Made by n Participants

$A_s \setminus A_k$	A_1	A_2	A_3	$\cdots$	A_m	$\sum$
A_1	—	$\sum_{j=1}^{n} f_{A_2}(A_1, p_j)$	$\sum_{j=1}^{n} f_{A_3}(A_1, p_j)$	$\cdots$	$\sum_{j=1}^{n} f_{A_m}(A_1, p_j)$	$\cdot$
A_2	$\sum_{j=1}^{n} f_{A_1}(A_2, p_j)$	—	$\sum_{j=1}^{n} f_{A_3}(A_2, p_j)$	$\cdots$	$\sum_{j=1}^{n} f_{A_m}(A_2, p_j)$	$\cdot$
A_3	$\sum_{j=1}^{n} f_{A_1}(A_3, p_j)$	$\sum_{j=1}^{n} f_{A_2}(A_3, p_j)$	—	$\cdots$	$\sum_{j=1}^{n} f_{A_m}(A_3, p_j)$	$\cdot$
$\cdots$	$\cdots$	$\cdots$	$\cdots$	$\cdots$	$\cdots$	$\cdot$
A_m	$\sum_{j=1}^{n} f_{A_1}(A_m, p_j)$	$\sum_{j=1}^{n} f_{A_2}(A_m, p_j)$	$\sum_{j=1}^{n} f_{A_3}(A_m, p_j)$	$\cdot$	—	$\cdot$

When $s \neq k$, we can state the following properties:

1) $\sum\limits_{j=1}^{n} f_{A_s}(A_k, p_j) + \sum\limits_{j=1}^{n} f_{A_k}(A_s, p_j) = \sum\limits_{j=1}^{n} (f_{A_s}(A_k, p_j) + (f_{A_k}(A_s, p_j)))$

$= \sum\limits_{j=1}^{n} 1 = n,$

2) $\sum\limits_{k=1}^{m} \sum\limits_{s=1}^{m} \left(\sum\limits_{j=1}^{n} f_{A_k}(A_s, p_j) \right) = \frac{n(m^2 - m)}{2} = \frac{nm(m-1)}{2}.$

Finally, the degree of membership $\mu_{A_i}(u)$, $i = 1, 2, \cdots, m$ can be computed by

$$\mu_{A_i}(u) = \frac{\sum\limits_{k=1}^{m} \left(\sum\limits_{j=1}^{n} f_{A_k}(A_i, p_j) \right)}{\frac{1}{2} nm(m-1)}.$$

Obviously, the above $\mu_{A_i}(u)$'s satisfies

$$\sum\limits_{i=1}^{m} \mu_{A_i}(u) = 1.$$

2.5.3   A Special Case of Preferred Comparisons

Let $U = \{u_1, u_2, \cdots, u_q\}$ be a finite universe and A be a fuzzy set on U. Find the membership function $\mu_A(u_i)$ for each i, $i = 1, 2, \cdots, q$. Let $P = \{p_1, p_2, \cdots, p_n\}$ be the set of subjects or participants in the fuzzy statistical experiment. For every pair (u_s, u_k), where $1 \leq s \leq k \leq m$, and any $p_j \in P$, we stipulate that

$$A(u_i, p_j) = \begin{cases} 1, & p_j \text{ says } u_i \text{ belongs to } A \\ 0, & p_j \text{ says } u_i \text{ does not belongs to } A \end{cases}$$

and

$$A(u_s, p_j) + A(u_k, p_j) = 1.$$

When p_j compares u_s to u_k, if $\mu_A(u_k) = 1$ (so that $\mu_A(u_s) = 0$) is determined, then we can write

$$f_{u_s}(u_k, p_j) = 1 \quad \text{and} \quad f_{u_k}(u_s, p_j) = 0.$$

This is summed up in Table 7.

Table 7 $f_{u_k}(u_s, p_j)$ from p_j by Binary Comparisons

$u_s \backslash u_k$	u_1	u_2	u_3	$\cdots$	u_q
u_1	—	0	1	$\cdots$	0
u_2	0	—	1	$\cdots$	1
u_3	1	0	—	$\cdots$	0
$\cdots$	$\cdots$	$\cdots$	$\cdots$	$\cdots$	$\cdots$
u_q	1	0	1	$\cdots$	—

The results of binary comparisons by n participants are in Table 8.
When $s \neq k$, the following properties hold:

1) $\sum\limits_{j=1}^{n} (f_{u_k}(u_s, p_j) + f_{u_s}(u_k, p_j)) = n$;

2) $\sum\limits_{k=1}^{q} \sum\limits_{s=1}^{q} \left(\sum\limits_{j=1}^{n} f_{u_k}(u_s, p_j) \right) = \frac{1}{2}nq(q-1)$.

The degree of membership of $\mu_A(u_i)$, $i = 1, 2, \cdots, q$, may be calculated from

$$\mu_A(u_i) = \frac{\sum\limits_{k=1}^{q} \left(\sum\limits_{j=1}^{n} f_{u_k}(u_s, p_j) \right)}{\frac{1}{2}nq(q-1)}.$$

Clearly, these $\mu_A(u_i)$'s satisfy

$$\sum\limits_{k=1}^{q} \mu_A(u_i) = 1.$$

Table 8   Binary Comparisons Made by n Participants

$u_s \setminus u_k$	u_1	u_2	u_3	$\cdots$	u_q	$\sum$
u_1	—	$\sum\limits_{j=1}^{n} f_{u_2}(u_1, p_j)$	$\sum\limits_{j=1}^{n} f_{u_3}(u_1, p_j)$	$\cdots$	$\sum\limits_{j=1}^{n} f_{u_q}(u_1, p_j)$	$\cdot$
u_2	$\sum\limits_{j=1}^{n} f_{u_1}(u_2, p_j)$	—	$\sum\limits_{j=1}^{n} f_{u_3}(u_2, p_j)$	$\cdot$	$\sum\limits_{j=1}^{n} f_{u_q}(u_2, p_j)$	$\cdots$
u_3	$\sum\limits_{j=1}^{n} f_{u_1}(u_3, p_j)$	$\sum\limits_{j=1}^{n} f_{u_2}(u_3, p_j)$	—	$\cdots$	$\sum\limits_{j=1}^{n} f_{u_q}(u_3, p_j)$	$\cdot$
$\cdots$	$\cdots$	$\cdots$	$\cdots$	$\cdots$	—	$\cdot$
u_q	$\sum\limits_{j=1}^{n} f_{u_1}(u_q, p_j)$	$\sum\limits_{j=1}^{n} f_{u_2}(u_q, p_j)$	$\sum\limits_{j=1}^{n} f_{u_3}(u_q, p_j)$	$\cdot$	$\cdot$	

2.5.4   An Example

In the production of ping-pong paddles, we would like to know which color is the most preferred by the players? Let the universe of colors be

$$U = \{\text{red, orange, yellow, green, blue}\}.$$

We now randomly select 500 players. Each player is asked to make comparisons, and each comparison proceeds according to the order as shown in Table 9. The player then selects his or her most preferred color. The results are given in Table 10. Let α be the concept "good color" and A is its extension. Then

$$\mu_A(\text{Red}) = \frac{2248}{1000 \times 5 \times 4 \times 0.5} = 0.2248,$$

$$\mu_A(\text{Orange}) = 0.2377, \quad \mu_A(\text{Yellow}) = 0.1782,$$

$$\mu_A(\text{Green}) = 0.2087, \quad \mu_A(\text{Blue}) = 0.1506.$$

Table 9   Order of Comparison

Color	Red	Orange	Yellow	Green	Blue
Red	—	—	—	—	—
Orange	1	—	—	—	—
Yellow	5	2	—	—	—
Green	8	6	3	—	—
Blue	10	9	7	4	—

Table 10   Frequency

$u_i \setminus u_j$	Red	Orange	Yellow	Green	Blue	$\sum$
Red	—	517	525	545	661	2248
Orange	483	—	841	477	576	2237
Yellow	475	159	—	534	614	1782
Green	455	523	466	—	643	2087
Blue	339	524	386	357	—	1506

2.6   The Absolute Comparison Method

The pairwise comparison used in the last section must satisfy

$$A_i(u, p_j) = 1 \text{ or } 0,$$
$$\mu_{A_s}(u) + \mu_{A_k}(u) = 1, \quad 1 \le s \le k \le 1.$$

Now we modify this condition. We allow the values of $\mu_{A_s}(u)$ and $\mu_{A_k}(u)$ from participants/subjects of an experiment to be any number in [0,1]. Especially, in certain problems, we can use a more convenient approach called the absolute comparison method now described.

Let $A_1, A_2, \cdots, A_m \in \mathcal{F}(U)$ be m fuzzy sets. For any $u \in U$, we want to determine the degree of membership of u in A_i, i.e., $\mu_{A_i}(u)$, $i = 1, 2, \cdots, m$.

Suppose there exists an index $r \in \{1, 2, \cdots, m\}$ such that u obviously belongs to A_r than others, i.e.,

$$\mu_{A_r}(u) = \max \{\mu_{A_i}(u) \mid 1 \le i \le m\}.$$

Then we do not need to perform a complete set of pairwise comparisons. Instead, we need only to perform pairwise comparisons with respect to A_r (a partial set of pairwise comparisons). That is, to find

$$f_{A_r}(A_i, p_j), \quad i = 1, 2, \cdots, m,$$

where $f_{A_r}(A_r, p_j)$ acts as a standard reference value.

Assume $P = \{p_1, p_2, \cdots, p_m\}$ to be the set of subjects in our experiment. Let

$$a_i = \frac{1}{n} \sum_{j=1}^{n} f_{A_r}(A_i, p_j). \tag{2.4}$$

Then we may define the degree of membership of u in A_i as

$$\mu_{A_i}(u) = a_i \left(\sum_{k=1}^{m} a_k \right)^{-1} \tag{2.5}$$

It can be readily verified that $\mu_{A_i}(u)$ satisfies this normalized condition:

$$\sum_{i=1}^{m} \mu_{A_i}(u) = \sum_{i=1}^{m} \left(a_i \left(\sum_{k=1}^{m} a_k \right)^{-1} \right) = \left(\sum_{k=1}^{m} a_k \right)^{-1} \left(\sum_{i=1}^{m} a_i \right) = 1.$$

Now suppose $U = \{u_1, u_2, \cdots, u_k\}$ to be a finite universe and A to be a fuzzy set on U. The problem is to find $\mu_A(u_i)$ for all $i \in \{1, 2, \cdots, k\}$. We may proceed as follows:

Step 1. Determine an index r such that

$$\mu_A(u_r) = \max\{\mu_A(u_1), \mu_A(u_2), \cdots, \mu_A(u_k)\},$$

i.e., u_r is perceived as best fit in A than other u's.

Step 2. Find
$$f_{u_r}(u_i, p_j), \qquad i = 1, 2, \cdots, k$$

by each p_j, $j = 1, 2, \cdots, n$.

Step 3. Compute a_i that sums up the results of binary comparisons from n participants as in

$$a_i = \frac{1}{n} \sum_{j=1}^{n} f_{u_r}(u_i, p_j), \qquad i = 1, 2, \cdots, k.$$

Step 4. Calculate the degree of membership according to

$$\mu_A(u_i) = a_i \left(\sum_{j=1}^{k} a_j \right), \qquad i = 1, 2, \cdots, k.$$

Example 3   In an engineering project evaluation, assume there are four basic (atomic) factors in a factor set V. Let $V = \{v_1, v_2, v_3, v_4\}$, where

$$v_1\text{---technology feasibility,} \quad v_2\text{---facility investment,}$$
$$v_3\text{---fixed operating costs,} \quad v_4\text{---cost of labor and material}$$

Find a weight vector for V. Stated in other words, find a fuzzy set A on V, where

$$A = (\mu_A(v_1), \mu_A(v_2), \mu_A(v_3), \mu_A(v_4)) \in \mathcal{F}(V).$$

Choose a group $P = \{p_1, p_2, \cdots, p_{10}\}$ of ten evaluators. They all agree that v_4 is the primary factor concerning economic benefits. This is recorded as $f_A(v_4)$, and a score of 10 is assigned to it. Moreover, we have

$$
\begin{aligned}
f_{v_4}(v_1, p_1) &= 4, & f_{v_4}(v_2, p_1) &= 9, & f_{v_4}(v_3, p_1) &= 6, \\
f_{v_4}(v_1, p_2) &= 4, & f_{v_4}(v_2, p_2) &= 8, & f_{v_4}(v_3, p_2) &= 7, \\
f_{v_4}(v_1, p_3) &= 4, & f_{v_4}(v_2, p_3) &= 8, & f_{v_4}(v_3, p_3) &= 7, \\
f_{v_4}(v_1, p_4) &= 5, & f_{v_4}(v_2, p_4) &= 8, & f_{v_4}(v_3, p_4) &= 5, \\
f_{v_4}(v_1, p_5) &= 5, & f_{v_4}(v_2, p_5) &= 8, & f_{v_4}(v_3, p_5) &= 5, \\
f_{v_4}(v_1, p_6) &= 4, & f_{v_4}(v_2, p_6) &= 8, & f_{v_4}(v_3, p_6) &= 5, \\
f_{v_4}(v_1, p_7) &= 5, & f_{v_4}(v_2, p_7) &= 9, & f_{v_4}(v_3, p_7) &= 5, \\
f_{v_4}(v_1, p_8) &= 4, & f_{v_4}(v_2, p_8) &= 9, & f_{v_4}(v_3, p_8) &= 5, \\
f_{v_4}(v_1, p_9) &= 4, & f_{v_4}(v_2, p_9) &= 9, & f_{v_4}(v_3, p_9) &= 6, \\
f_{v_4}(v_1, p_{10}) &= 4, & f_{v_4}(v_2, p_{10}) &= 9, & f_{v_4}(v_3, p_{10}) &= 6.
\end{aligned}
$$

The a_i's are found by computing the averages:

$$
a_1 = \frac{1}{10} \sum_{j=1}^{10} f_{v_4}(v_1, p_j) = 4.3,
$$

$$
a_2 = \frac{1}{10} \sum_{j=1}^{10} f_{v_4}(v_2, p_j) = 8.5,
$$

$$
a_3 = \frac{1}{10} \sum_{j=1}^{10} f_{v_4}(v_3, p_j) = 5.7.
$$

For a_4, due to the initial agreement, we have

$$
a_4 = \frac{1}{10} \sum_{j=1}^{10} f_{v_4}(v_4, p_j)
$$

$$
= \frac{1}{10} \sum_{j=1}^{10} 10 = 10.
$$

Normalizing a_1, a_2, a_3, a_4, we obtain

$$
\mu_A(v_1) = a_1 \left(\sum_{j=1}^{10} a_j \right)^{-1}
$$

$$
= \frac{4.3}{4.3 + 8.5 + 5.7 + 10} \approx 0.35.
$$

Similarly,

$$
\mu_A(v_2) \approx 0.30,\, \mu_A(v_3) \approx 0.20,\, \mu_A(v_4) \approx 0.15.
$$

Hence,

$$
A = (0.15, 0.30, 0.20, 0.35).
$$

2.7   The Set-valued Statistical Iteration Method

2.7.1   Statement of the Problem

Let $U = \{u_1, u_2, \cdots, u_k\}$ be a finite universe, A be a fuzzy set of interest in $\mathcal{F}(U)$, and $P = \{p_1, p_2, \cdots, p_m\}$ be the set of subjects in our experiment. The problem is to find the degree of membership $\mu_A(u_i)$ for $i = 1, 2, \cdots, k$.

When k is large, it is hard to determine the values of $f_{u_r}(u_i, p_j)$ via pairwise comparisons. The following method can overcome this difficulty.

2.7.2   Basic Steps of Set-valued Statistical Method

First we choose an initial number q such that $1 \le q \ll k$, and then a p_j in P so that we can carry out the statistical experiment according to the following steps:

(1)   Select $r_1 = q$ elements from U such that they are the first group of elements best fit to A by p_j. This generates a subset of U, and it is referred to as set-valued statistics denoted by

$$U_1^{(j)} = \{u_{i_1}^{(j)}, u_{i_2}^{(j)}, \cdots, u_{i_q}^{(j)}\} \subset U.$$

(2)   Select $r_2 = 2q$ elements from U (which includes the q elements already selected in step (1)) in such a way that all $2q$ elements are considered better fit to A than other elements of U by p_j. This generates another subset of U:

$$U_2^{(j)} = \{u_{i_1}^{(j)}, u_{i_2}^{(j)}, \cdots, u_{i_q}^{(j)}, u_{i_{q+1}}^{(j)}, \cdots, u_{i_{2q}}^{(j)}\} \supset U_1^{(j)}.$$

The reason that the earlier q elements must be a part of $2q$ elements is that the first q elements are considered better fit to A than the next q elements. So when selecting a total of $2q$ elements from U that are considered as best fit to A we must include the q elements already chosen.

(3)   In a similar manner, the s-th subset of U is established. This subset consists of $r_s = s \cdot q$ elements of U. That is,

$$U_s^{(j)} = \{u_{i_1}^{(j)}, u_{i_2}^{(j)}, \cdots, u_{i_{sq}}^{(j)}\} \supset U_{s-1}^{(j)}.$$

If t is a natural number that satisfies $k = t \cdot q + v$, where $1 \le v \le q$, then the iterative process stops at step $(t + 1)$, setting

$$U_{t+1}^{(j)} = U.$$

Next, we calculate $m(u_i)$, the average frequency of u_i, using

$$m(u_i) = \frac{1}{n(t+1)} \sum_{s=1}^{t+1} \sum_{j=1}^{n} \chi_{U_s^{(j)}}(u_i), \quad i = 1, 2, \cdots, k, \tag{2.6}$$

where $\chi_{U_s^{(j)}}$ is the characteristic function of the set $U_s^{(j)}$.

2.7   The Set-valued Statistical Iteration Method

We obtain the degree of membership of u_i in A by normalizing the values of $m(u_i)$, for each $i = 1, 2, \cdots, k$. Therefore,

$$\mu_A(u_i) = m(u_i)\left(\sum_{j=1}^{k} m(u_j)\right)^{-1}, \quad i = 1, 2, \cdots, k. \tag{2.7}$$

Example 4   Let $U = \{u_1, u_2, u_3, u_4\}$, A, and $P = \{p_1, p_2, \cdots, p_5\}$ be the same as in the last section.

Take $q = 1$. Each participant's selection process can be presented in a triangular form:

$$
\begin{array}{llllll}
p_1: & r_1 = 1 & u_1 & & & \\
     & r_2 = 2 & u_2 & u_1 & & \\
     & r_3 = 3 & u_2 & u_1 & u_4 & \\
     & r_4 = 4 & u_2 & u_1 & u_4 & u_3
\end{array}
$$

The presentation may be simplified as

$$
\begin{array}{lcccc}
p_1: & u_2 & u_1 & u_4 & u_3 \\
     & 4 & 3 & 2 & 1
\end{array}
$$

Likewise, the results of other participants are given below.

$$
\begin{array}{lcccc}
p_2: & u_2 & u_4 & u_1 & u_3 \\
     & 4 & 3 & 2 & 1
\end{array}
$$

$$
\begin{array}{lcccc}
p_3: & u_1 & u_2 & u_4 & u_3 \\
     & 4 & 3 & 2 & 1
\end{array}
$$

$$
\begin{array}{lcccc}
p_4: & u_2 & u_1 & u_3 & u_4 \\
     & 4 & 3 & 2 & 1
\end{array}
$$

$$
\begin{array}{lcccc}
p_5: & u_2 & u_3 & u_1 & u_4 \\
     & 4 & 3 & 2 & 1
\end{array}
$$

The average frequency counts of u_i's are

$$m(u_1) = \frac{14}{4 \times 5} = \frac{7}{10}, \quad m(u_2) = \frac{19}{20},$$

$$m(u_3) = \frac{8}{20} = \frac{2}{5}, \quad m(u_4) = \frac{9}{20}.$$

After normalization, we obtain

$$\mu_A(u_1) = \frac{7}{25} \approx 0.28, \quad \mu_A(u_2) = \frac{19}{50} \approx 0.38,$$

$$\mu_A(u_3) = \frac{4}{25} \approx 0.16, \quad \mu_A(u_4) = \frac{9}{50} \approx 0.18.$$

Therefore, $A = (0.28, 0.38, 0.16, 0.18)$.

2.8   Ordering by Precedence Relations

Among n objects $u_1, u_2, \cdots, u_n$, we wish to determine an order by a precedence criterion. Very often a precedence criterion is a fuzzy concept whose extension A is a fuzzy set on $U = \{u_1, u_2, \cdots, u_n\}$. If we have an "order" relationship, then we can use it to determine the membership function $\mu_A(u)$.

2.8.1   Precedence Relations

First we establish a precedence relation among n objects. Let $C = (c_{ij})_{n \times n}$ be a matrix where c_{ij} is a measure of precedence of u_i over u_j, and satisfies

1) $c_{ii} = 0, \ i = 1, 2, \cdots, n$.
2) $0 \le c_{ii} \le 1, \ i \ne j, \ i, j = 1, 2, \cdots, n$,
3) $c_{ij} + c_{ji} = 1, \ i \ne j, \ i, j = 1, 2, \cdots, n$.

Condition 1) says that the precedence measure is 0 when compared to the object itself. Condition 3) says that the sum of the precedence measure of u_i over u_j and the precedence measure of u_j over u_i is 1. When the precedence of u_i over u_j and the precedence of u_j over u_i are indistinguishable, we set $c_{ij} = c_{ji} = 0.5$. If u_i overwhelmingly precedes u_j, then c_{ij} closes to 1 and c_{ji} closes to 0.

The matrix $C = (c_{ij})_{n \times n}$ is called the precedence relation matrix. It is a fuzzy set or a fuzzy matrix on U^2, in other words, $C \in \mathcal{F}(U \times U)$.

2.8.2   Creating Order

Take a "threshold" $\lambda \in [0, 1]$ and form a cut matrix:

$$
C_\lambda = \begin{bmatrix}
c_{11}^{(\lambda)} & c_{12}^{(\lambda)} & \cdots & c_{1n}^{(\lambda)} \\
c_{21}^{(\lambda)} & c_{22}^{(\lambda)} & \cdots & c_{2n}^{(\lambda)} \\
\vdots & \vdots & & \vdots \\
c_{n1}^{(\lambda)} & c_{n2}^{(\lambda)} & \cdots & c_{nn}^{(\lambda)}
\end{bmatrix}
$$

where

$$
c_{ij}^{(\lambda)} = \begin{cases} 1, & c_{ij} \ge \lambda \\ 0, & c_{ij} < \lambda \end{cases}.
$$

When λ decreases from 1 to 0, and if λ_1 is the threshold where for the first time all of the entries except the diagonal one in the i_1-th row of C_{λ_1} are one, then u_{i_1} is taken as an element in the first batch of the precedence set (it may not be unique). We take away the first batch of elements from U and work out another batch in the same way from the remaining elements of U. Repeating this procedure, we can create a precedence order for all the elements of U.

2.8.3   An Example

Let $U = \{u_1, u_2, u_3\}$ and $A \in \mathcal{F}(U)$, a fuzzy set of interest. Suppose we know the precedence matrix, say

$$C = \begin{bmatrix} 0 & 0.9 & 0.2 \\ 0.1 & 0 & 0.7 \\ 0.8 & 0.3 & 0 \end{bmatrix}.$$

When λ decreases from 1 to 0, we have the corresponding cut matrices:

$$C_{0.9} = \begin{bmatrix} 0 & 1 & 0 \\ 0 & 0 & 0 \\ 0 & 0 & 0 \end{bmatrix}, \qquad C_{0.8} = \begin{bmatrix} 0 & 1 & 0 \\ 0 & 0 & 0 \\ 1 & 0 & 0 \end{bmatrix},$$

$$C_{0.7} = \begin{bmatrix} 0 & 1 & 0 \\ 0 & 0 & 1 \\ 0 & 0 & 0 \end{bmatrix}, \qquad C_{0.3} = \begin{bmatrix} 0 & 1 & 0 \\ 0 & 0 & 1 \\ 1 & 1 & 0 \end{bmatrix}.$$

When $\lambda = 0.3$, we find for the first time that there is a row, namely the third row in $C_{0.3}$, that has 1's in every entry except the diagonal one. This means that the precedence measure of u_3 over the other two is definitely greater than 0.3. So u_3 is an element of the first batch of the precedence set.

After deleting u_3, we obtain a new precedence matrix of u_1 and u_2 as follows:

$$C^{(1)} = \begin{bmatrix} 0 & 0.9 \\ 0.1 & 0 \end{bmatrix}.$$

Since

$$C_{0.9}^{(1)} = \begin{bmatrix} 0 & 1 \\ 1 & 0 \end{bmatrix},$$

we see that the elements in the first row have the aforementioned "property", so u_1 is an element of the second batch of the precedence set. In conclusion, u_3 has the highest precedence; next is u_1, and the last is u_2. Applying the set-valued statistical iteration method, we have

$$\mu_A(u_3) = \frac{3}{3} = 1, \quad \mu_A(u_1) = \frac{2}{3} \approx 1, \quad \mu_A(u_2) = \frac{1}{3} \approx 0.33.$$

2.8.4   Generalizations

We now generalize the method for establishing the degree of membership to a more general case.

Let $U_j = \{u_{i_1}, u_{i_2}, \cdots, u_{i_{k_j}}\}$ be the j-th batch of the precedence set, where

$$j = 1, 2, \cdots, s, \quad k_1 + k_2 + \cdots + k_s = n.$$

The degree of membership can be obtained by the following formula:

$$\mu_A(u) = \left(n - \sum_{p=1}^{j-1} k_p\right) n^{-1}, \tag{2.8}$$

where $u \in U_j$, $j = 1, 2, \cdots, s$, and define $\sum\limits_{p=1}^{0} k_p = 0$.

2.9   The Relative Comparison Method and the Mean Pairwise Comparison Method

2.9.1   The Relative Comparison Method

Let $A_i \in \mathcal{F}(U)$, $i = 1, 2, \cdots, m$, be m fuzzy sets of interest. We want to determine the degree of membership $\mu_{A_i}(u)$ of $u \in U$ for $i = 1, 2, \cdots, m$. Let the set $P = \{p_1, p_2, \cdots, p_n\}$ be interpreted as before, and let $f_{A_s}(A_k, p_j)$ be a value assigned by p_j for A_k that is a result of the comparison of A_k relative to A_s. See Table 11.

Table 11   Pairwise Comparisons by p_j

$A_k \setminus A_s$	A_1	A_2	A_3	$\cdots$	A_m
A_1	—	$f_{A_2}(A_1, p_j)$	$f_{A_3}(A_1, p_j)$	$\cdots$	$f_{A_m}(A_1, p_j)$
A_2	$f_{A_1}(A_2, p_j)$	—	$f_{A_3}(A_2, p_j)$	$\cdots$	$f_{A_m}(A_2, p_j)$
A_3	$f_{A_1}(A_3, p_j)$	$f_{A_2}(A_3, p_j)$	—	$\cdots$	$f_{A_m}(A_3, p_j)$
$\vdots$	$\vdots$	$\vdots$	$\vdots$	$\vdots$	$\vdots$
A_m	$f_{A_1}(A_m, p_j)$	$f_{A_2}(A_m, p_j)$	$f_{A_3}(A_m, p_j)$	$\cdots$	—

In the above terms, we agree that

$$0 \leq f_{A_s}(A_k, p_j) \leq 1 \quad \text{and} \quad f_{A_s}(A_s, p_j) = 1$$

but they may not need to satisfy

$$f_{A_s}(A_k, p_j) + f_{A_k}(A_s, p_j) = 1.$$

Define

$$f_{A_s}(A_k, p_j) = \frac{1}{n} \sum_{j=1}^{n} f_{A_s}(A_k, p_j).$$

Then we obtain results from the n participants as in Table 12.

Table 12   Values of $f_{A_s}(A_k)$ from the n Participants

$A_k \setminus A_s$	A_1	A_2	A_3	$\cdots$	A_m
A_1	—	$f_{A_s}(A_k)$	$f_{A_s}(A_k)$	$\cdots$	$f_{A_s}(A_k)$
A_2	$f_{A_s}(A_k)$	—	$f_{A_s}(A_k)$	$\cdots$	$f_{A_s}(A_k)$
A_3	$f_{A_s}(A_k)$	$f_{A_s}(A_k)$	—	$\cdots$	$f_{A_s}(A_k)$
$\vdots$	$\vdots$	$\vdots$	$\vdots$	$\vdots$	$\vdots$
A_m	$f_{A_s}(A_k)$	$f_{A_s}(A_k)$	$f_{A_s}(A_k)$	$\cdots$	—

By setting

$$f(A_s | A_k) = \frac{f_{A_k}(A_s)}{\max\{f_{A_s}(A_k), f_{A_k}(A_s)\}}$$

we get a matrix as shown in Table 13.

Table 13　A Comparison Matrix

$A_k \setminus A_s$	A_1	A_2	A_3	$\cdots$	A_m
A_1	—	$f(A_1\vert A_2)$	$f(A_1\vert A_3)$	$\cdots$	$f(A_1\vert A_m)$
A_2	$f(A_2\vert A_1)$	—	$f(A_2\vert A_3)$	$\cdots$	$f(A_2\vert A_m)$
A_3	$f(A_3\vert A_1)$	$f(A_3\vert A_2)$	—	$\cdots$	$f(A_3\vert A_m)$
$\vdots$	$\vdots$	$\vdots$	$\vdots$	$\vdots$	$\vdots$
A_m	$f(A_m\vert A_1)$	$f(A_m\vert A_2)$	$f(A_m\vert A_3)$	$\cdots$	—

Clearly, we have

$$f(A_s|A_k) = \begin{cases} \dfrac{f_{A_k}(A_s)}{f_{A_s}(A_k)}, & f_{A_k}(A_s) \le f_{A_s}(A_k) \\ 1, & f_{A_k}(A_s) > f_{A_s}(A_k) \text{ or } s = k \end{cases}.$$

By synthesizing the values of $f(A_s|A_k)$ we can obtain a degree of membership for u in A_i. For example,

$$\mu_{A_i}(u) = \bigwedge_{s=1}^{m} f(A_s|A_k), \qquad i = 1, 2, \cdots, m. \tag{2.9}$$

Note. In the next chapter we present a basic tool, a multifactorial function M_m, for "multifactorial analysis". It can be used for building membership functions. The way to do this is to determine an appropriate M_m from the context of the problem, and then compute

$$\mu_{A_i}(u) = M_m(f(A_1|A_i), f(A_2|A_i), \cdots, (A_m|A_i)). \tag{2.10}$$

2.9.2　The Mean Pairwise Comparison Method

From Table 12 and with an appropriate M_m (see Chapter 6), we can assign the degree of membership of u by

$$\mu_{A_i}(u) = M_m(f_{A_1}(A_i), f_{A_2}(A_i), \cdots, f_{A_m}(A_i),). \tag{2.11}$$

In particular, take M_m as the mean function. Then

$$\mu_{A_i}(u) = \frac{1}{m} \sum_{j=1}^{m} f_{A_j}(A_i). \tag{2.12}$$

This is a familiar formula.

Note. When U is a finite universe, and there is only one fuzzy set A $(\in \mathcal{F}(U))$ of interest, then we can follow the method in previous sections by replacing A_s with u_s as the basis of comparison. To illustrate, we give two examples below:

Example 5   A father has three sons named u_1, u_2, and u_3. Let $U = \{u_1, u_2, u_3\}$ and A be the fuzzy concept of "resemblance to their father". Suppose we obtained Table 13 by pairwise comparisons, and relative to the fuzzy set A we have

$$f_{u_1}(u_2) = 0.5, \quad f_{u_2}(u_1) = 0.8, \quad f_{u_1}(u_3) = 0.3,$$
$$f_{u_3}(u_1) = 0.5, \quad f_{u_2}(u_3) = 0.7, \quad f_{u_3}(u_2) = 0.4.$$

Table 14   Resemblance by Pairwise Comparisons

$u_j \setminus u_i$	u_1	u_2	u_3
u_1	1	0.8	0.5
u_2	0.5	1	0.4
u_3	0.3	0.7	1

Next, we calculate the following values:

$$f(u_1|u_2) = \frac{f_{u_2}(u_1)}{\max\{f_{u_1}(u_2), f_{u_2}(u_1)\}} = \frac{0.8}{\max\{0.5, 0.8\}} = 1.$$

Similarly,

$$f(u_2|u_1) \approx 0.62, \quad f(u_1|u_3) = 2, \quad f(u_3|u_1) = 0.6,$$
$$f(u_2|u_3) \approx 0.57, \quad f(u_3|u_2) = 1.$$

We summarize these values in Table 15.

Table 15   Values of $f(u_i|u_j)$

$u_j \setminus u_i$	u_1	u_2	u_3
u_1	1	1	1
u_2	0.62	1	0.57
u_3	0.6	1	1

We use the operator $\wedge$ on each column in Table 15 to derive

$$\mu_A(u_1) = 1 \wedge 0.62 \wedge 0.6 = 0.6.$$

Likewise, $\mu_A(u_2) = 1$ and $\mu_A(u_3) = 0.57$. That is, $A = (0.6, 1, 0.57)$.

Example 6   Let $U = \{u_1, u_2, u_3\}$ be a universe where u_1 represents an oriental cherry, u_2 represents a chrysanthemum, and u_3 represents a dandelion. Also, let A be a fuzzy concept "beautiful" on U. Find the membership function $\mu_A(u_i)$, $i = 1, 2, 3$.

Let the results of pairwise comparisons be

$$f_{u_i}(u_i) = 1, \qquad i = 1, 2, 3.$$

and

$$f_{u_2}(u_1) = 0.8, \quad f_{u_1}(u_2) = 0.7, \quad f_{u_3}(u_1) = 0.9,$$
$$f_{u_1}(u_3) = 0.5, \quad f_{u_2}(u_3) = 0.4, \quad f_{u_3}(u_2) = 0.8.$$

The mean values of each row are

$$\mu_A(u_1) = \frac{2.8}{3} \approx 0.93, \quad \mu_A(u_2) \approx 0.83, \quad \text{and} \quad \mu_A(u_3) \approx 0.63.$$

Hence, $A = (0.93, 0.83, 0.63)$.

References

1. Q. Z. Duan and H. X. Li, n-dimensional t-norm, *Journal of Tianjin College of Textile Engineering*, Vol. 7, No. 2, pp. 9-16, 1988.

2. H. X. Li, Multifactorial functions in fuzzy sets theory, *Fuzzy Sets and Systems*, Vol. 35, No. 1, pp. 69-84, 1990.

3. H. X. Li, C. Z. Luo and P. Z. Wang, The cardinality of fuzzy sets and the continuum hypothesis, *Fuzzy Sets and Systems*, Vol. 55, No. 1, pp. 61-77, 1993.

4. H. X. Li and V. C. Yen, Factor spaces and fuzzy decision making, *Journal of Beijing Normal University*, Vol. 30, No. 1, pp. 41-46, 1994.

5. H. X. Li and P. Z. Wang, Falling shadow representation of fuzzy concepts on factor spaces, *Journal of Yantai University*, No. 2, pp. 15-21, 1994.

6. P. Z. Wang, A factor space approach to knowledge representation, *Fuzzy Sets and Systems*, Vol. 36, pp. 113-124, 1990.

Chapter 3

Mathematical Essence and Structures of Feedforward Artificial Neural Networks

In this chapter, we first introduce the mathematical model and structure of artificial neurons. After that, we consider several artificial neural networks that assemble the neurons. This chapter does not intend to explain details of biological neurons. Instead, we only focus on artificial neurons that simply extract the abstract operation of biological neurons and their mathematical models. We start with an introduction in Section 1, followed by the discussion of neuron models in Section 2. We then discuss the interpolation mechanism of feedforward neural networks in Section 3 and a generalized model in Section 4. We prove that for any continuous functions, there exists a three-layer feedforward neural network such that the network can approximate the function satisfying any given precision. Finally, we discuss the mathematical interpolation of the learning algorithm in the rest of the sections.

3.1 Introduction

Artificial neural networks are intended to model the behavior of biological neural networks. The original hope for the development of artificial neural networks is intended to take advantage of parallel processors computing than traditional serial computation. Several acronyms such as parallel distributed processing (PDP) systems [1] or connectionist systems are used. Over the years, several models of artificial neural networks and the learning algorithms that associate with networks have been developed. In terms of structure, there are non-recurrent [2] and recurrent networks [3]. Depending upon functionality of the networks, there are associative memory networks [4], feedforward neural networks [2], self-organizing neural networks [5], adaptive resonance theory networks [6,7], and optimization networks [8]. Each network has its own "prototype" learning algorithm that achieves the function of storage and recall, learning and mapping, clustering and association, or optimizing a performance index. These prototype learning algorithms have demonstrated the superiority of the design that fits to both network architectures and applications.

In this chapter, we will focus mainly on feedforward neural networks. We intend to explain some phenomenon mathematically.

3.2   Mathematical Neurons and Mathematical Neural Networks

A feedforward artificial neuron that has multi-input and single output, for example, the McCulloch-Pitts (MP) model [9], is shown in Figure 1. Based on the forms of the outputs, a discrete output and a continuous output models are given below.

3.2.1 MP Model with Discrete Outputs

Let $\mathbf{X} = (x_1, x_2, \cdots, x_n)$ be an input vector, $\mathbf{W} = (w_1, w_2, \cdots, w_n)^T$ be a weight vector, θ be the threshold value of a neuron , and φ be the input-output function (or called an activation function) of the neuron. If z is used to indicate the outputs of the neuron, the equation that describes the model is:

$$z = \varphi\left(\sum_{i=1}^{n} w_i x_i - \theta\right) = \varphi(\mathbf{X} \cdot \mathbf{W} - \theta). \tag{3.1}$$

For discrete outputs, φ can be a step-up function:

$$\varphi(u) = \begin{cases} 1, & u \geq 0, \\ 0, & u < 0. \end{cases} \tag{3.2}$$

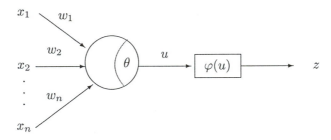

Figure 1   MP model of artificial neurons

3.2.2 MP Model with Continuous-valued Outputs

For continuous-valued output, φ can be a trapezoidal function, a sigmoidal function, a Gaussian function, or a displacement function (it will be an identity function when $c = 0$):

$$\varphi(u) = \begin{cases} 0, & u < u_0 \\ (u - u_0)/(u_1 - u_0), & u_0 \leq u \leq u_1 \\ 1, & u > u_1 \end{cases} \tag{3.3}$$

$$\varphi(u) = [1 + \exp(-u + c)]^{-1}, \quad c = \text{constant} \tag{3.4}$$

$$\varphi(u) = u + c, \quad c = \text{constant} \tag{3.5}$$

Remark 1. When $(\forall u)(\varphi(u) \in [0,1])$, the neuron can be called a fuzzy neuron. A neural network that consists of fuzzy neurons is called a fuzzy neural network. The operator, $(+, \cdot)$, in Equation (3.1), "$\mathbf{X} \cdot \mathbf{W}$," i.e., $\sum_{i=1}^{n} w_i x_i$ can be replaced by the pair of operators $(\vee, \wedge)$ meaning $\bigvee_{i=1}^{n} (w_i \wedge x_i)$, then the neuron can also be called a "fuzzy neuron." We will discuss this subject in Chapter 6.

We give a formal definition of "mathematical neurons" as follows.

Definition 1. Let R be the real number field. For any $D \subset R^n$ and $E \subset R$, a function as the following,

$$f: \ D \longrightarrow E, \ \ (x_1, x_2, \cdots, x_n) \longmapsto z = f(x_1, x_2, \cdots, x_n)$$

is called a mathematical neuron. When $f(D) = \{0,1\}$ or $\{-1,1\}$, f is called a discrete neuron.

Clearly, the neuron defined by Expression (3.1) is a special example of mathematical neurons:

$$z = f(x_1, x_2, \cdots, x_n) \triangleq \varphi\left(\sum_{i=1}^{n} w_i x_i - \theta \right).$$

A network is called a mathematical neural network, if it is formed by mathematical neurons.

Now we turn to consider the structure of the activation function of a neuron. For Expression (3.2), if put

$$\varphi'(u) = \begin{cases} 1, & u \geq \theta \\ 0, & u < \theta, \end{cases} \tag{3.6}$$

$$z = \varphi'\left(\sum_{i=1}^{n} w_i x_i \right) = \varphi'(\mathbf{W} \cdot \mathbf{X}), \tag{3.7}$$

then $\varphi'(\mathbf{W} \cdot \mathbf{X}) = \varphi(\mathbf{W} \cdot \mathbf{X} - \theta)$, which means that Expressions (3.1) and (3.7) are identical. In other words, φ and φ' are equivalent, for the threshold value θ has been absorbed into φ'. After change of a variable, $y = \mathbf{W} \cdot \mathbf{X} - \theta$, $\varphi'(y)$ has the following form:

$$\varphi'(y) = \begin{cases} 1, & y \geq \theta \\ 0, & y < \theta, \end{cases} \tag{3.8}$$

Moreover, the activation functions defined by Expressions (3.3), (3.4) and (3.5) have taken effect of absorbing threshold value.

If Expression (3.2) is rewritten as the following:

$$\varphi(u) = \begin{cases} 1, & u \geq \theta, \\ 0, & u < \theta, \end{cases} \tag{3.9}$$

then the relation between the input and output represented by Expression (3.1) can be simply written as follows:

$$z = \varphi\left(\sum_{i=1}^{n} w_i x_i\right) = \varphi(\mathbf{W} \cdot \mathbf{X}). \tag{3.10}$$

Figure 2 gives the graph of the activation function defined by Expression (3.9).

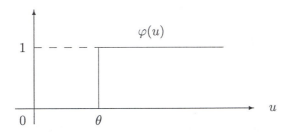

Figure 2   The graph of the activation function

From Figure 2, we know that the neuron will be excited when $u \geq \theta$, i.e., $\varphi(u) = 1$. In other words, however large a value of u may be when $u \geq \theta$, $\varphi(u) = 1$. Then we may find a problem: will there exist a bound u_0 $(> \theta)$ such that $\varphi(u) = 0$ when $u \geq u_0$?

As a matter of fact, for any neuron, u can not increase infinitely. Considering from biology, after θ there should be a bound $\theta' > \theta$ such that when $u \geq \theta'$, the neuron will turn to paralysis (i.e., inhibition) from excitation. This means that there should be $\theta' > \theta$ such that $\varphi(u) = 0$ when $u \geq \theta'$. So Expression (9) should be rewritten as follows:

$$\varphi(u) = \begin{cases} 1, & \theta \leq u \leq \theta' \\ 0, & \text{otherwise} \end{cases}. \tag{3.11}$$

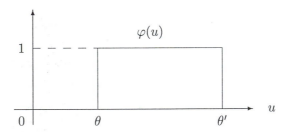

Figure 3   The activation function with bounded excitation

Figure 3 gives the graph of the activation function with bounded excitation defined by Expression (3.11). In fact, the excitation and inhibition should not be sudden changes or catastrophes. Generally speaking, there are transition or buffer zones both from excitation to inhibition and from inhibition to excitation. Thus, we can design the following activation function as an example where its graph is a trapezoid (see Figure 4):

$$\varphi(u) = \begin{cases} (u - \theta_1)/(\theta_2 - \theta_1), & \theta_1 \leq u \leq \theta_2 \\ 1, & \theta_2 < u < \theta_3 \\ (\theta_4 - u)/(\theta_4 - \theta_3), & \theta_3 \leq u \leq \theta_4 \\ 0, & \text{otherwise} \end{cases} . \tag{3.12}$$

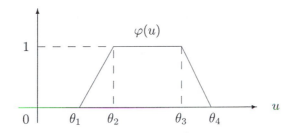

Figure 4 The trapezoidal activation function

Especially, when $\theta_2 = \theta_3$, Expression (3.12) will degenerate into a "triangle wave" activation function (see Figure 5):

$$\varphi(u) = \begin{cases} (u - \theta_1)/(\theta_2 - \theta_1), & \theta_1 \leq u \leq \theta_2 \\ (\theta_3 - u)/(\theta_3 - \theta_2), & \theta_2 < u \leq \theta_3 \\ 0, & \text{otherwise} \end{cases} \tag{3.13}$$

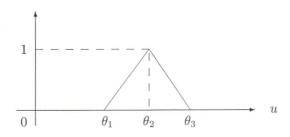

Figure 5 The triangle wave activation function

For a given n-dimensional vector $\mathbf{W} = (w_1, w_2, \cdots, w_n)^T$, if make a mapping $\Sigma : R^n \longrightarrow R$ such that

$$(x_1, x_2, \cdots, x_n) \longmapsto \Sigma(x_1, x_2, \cdots, x_n) \triangleq \sum_{i=1}^{n} w_i x_i = \mathbf{X} \cdot \mathbf{W}$$

then the neurons mentioned above can all expressed as $f = \varphi \circ \Sigma$, i.e.,

$$z = f(x_1, x_2, \cdots, x_n) = (\varphi \circ \Sigma)(x_1, x_2, \cdots, x_n)$$

$$= \varphi(\Sigma(x_1, x_2, \cdots, x_n)) = \varphi\left(\sum_{i=1}^{n} w_i x_i\right) = \varphi(\mathbf{X} \cdot \mathbf{W}). \qquad (3.14)$$

This reflects that the neurons have a basic feature of integrating by Σ first and then activating by φ.

3.3   The Interpolation Mechanism of Feedforward Neural Networks

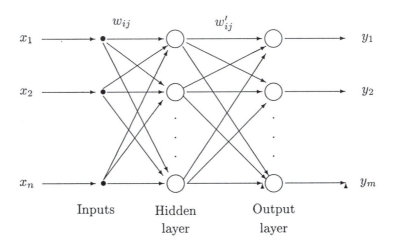

Figure 6   Feedforward neural networks

The neural network shown in Figure 6 is a typical kind of feedforward neural network. As indicated in Figure 6, neurons in the hidden layer have the weight connection between the previous layer (in this case, is inputs) and the next layer. From the graph theory, a complicated network can be always regarded as forming by some simpler networks. Therefore, we should study the mathematical essence and structure of neural networks from some simpler neural networks.

As a matter of fact, in most cases, a feedforward neural network can be regarded as a mapping (cf. Figure 6):

$$F: \ R^n \longrightarrow R^m, \quad (x_1, \cdots, x_n) \longmapsto F(x_1, \cdots, x_n) = (y_1, \cdots, y_m)$$

When $n = m = 1$, F is just a function of a single variable: $y = F(x)$, which can be represented as a two-layer feedforward neural network with single input single output shown as Figure 7.

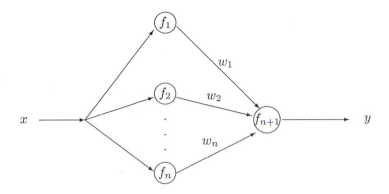

Figure 7   The two-layer feedforward neural network with single input single output

Now let the activation function of the neuron f_i be φ_i $(i = 1, 2, \cdots, n + 1)$, the connection weight from f_i to f_{n+1} be w_i $(i = 1, 2, \cdots, n)$, and φ_{n+1} be taken as an identity function: $\varphi_{n+1}(u) = u$. From Expression (3.10), the output of the network shown as the Figure 7 is the following:

$$F(x) = y = f_{n+1}(f_1(x), f_2(x), \cdots, f_n(x))$$
$$= f_{n+1}(\varphi_1(x), \varphi_2(x), \cdots, \varphi_n(x)) = \varphi_{n+1}\left(\sum_{i=1}^{n} \varphi_i(x)w_i\right)$$
$$= \sum_{i=1}^{n} \varphi_i(x)w_i. \tag{3.15}$$

Clearly, Expression (3.15) is just the general formula of interpolation functions, where the activation functions $\varphi_1, \varphi_2, \cdots, \varphi_n$ do be the basic functions of interpolation. For any system of the basic functions $\varphi_1, \varphi_2, \cdots, \varphi_n$ given adequately, Expression (3.15) will be some kind of interpolation function.

Example 1. In Expression (3.15), if the activation function φ_i $(i = 1, 2, \cdots, n)$ are taken as "rectangular waves" (see Figure 8):

$$\varphi_i(u) = \begin{cases} 1, & \theta_i \le u \le \theta_{i+1} \\ 0, & \text{otherwise, } i = 1, 2, \cdots, n, \end{cases} \tag{3.16}$$

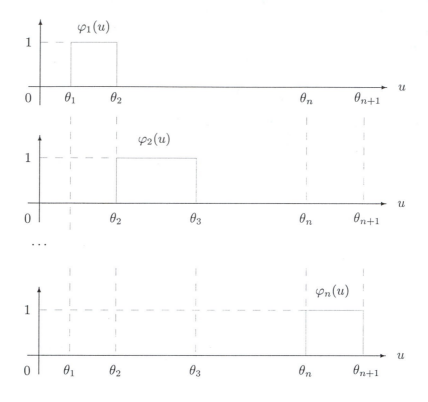

Figure 8 The rectangular wave activation functions
as the basic functions of interpolation

then $F(x) = \sum\limits_{i=1}^{n} \varphi_i(x)w_i$ is a piecewise zero-order interpolation function, which the graph of the interpolation function presents as shape of steps (see Figure 9).

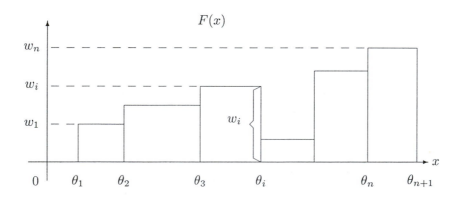

Figure 9 The piecewise zero-order interpolation function

From Figure 9, it is easy to comprehend that a two-layer feedforward neural network is an interpolation function $F(x) = \sum\limits_{i=1}^{n} \varphi_i(x)w_i$, for a given group of data $\{(\theta_i, w_i)\}_{(1 \leq i \leq n)}$. It is interesting that, for such a network, the threshold values θ_i $(i = 1, 2, \cdots, n)$ are the basic points of interpolation, the weights w_i $(i = 1, 2, \cdots, n)$ are the coefficients, and the activation functions φ_i $(i = 1, 2, \cdots, n)$ are the basic functions of interpolation, which means that the neurons of the hidden layer are the basic functions of interpolation.

Example 2. In Expression (3.15), if the activation function φ_i $(i = 1, 2, \cdots, n)$ is taken as "triangle waves" (see Figure 10):

$$\varphi_1(u) = \begin{cases} (\theta_2 - u)/(\theta_2 - \theta_1), & \theta_1 \leq u \leq \theta_2 \\ 0, & \text{otherwise,} \end{cases} \tag{3.17}$$

$$\varphi_i(u) = \begin{cases} (u - \theta_{i-1})/(\theta_i - \theta_{i-1}), & \theta_{i-1} \leq u \leq \theta_i \\ (\theta_{i+1} - u)/(\theta_{i+1} - \theta_i), & \theta_i < u \leq \theta_{i+1} \\ 0, & \text{otherwise,} \end{cases} \tag{3.18}$$

$$i = 2, 3, \cdots, n - 1,$$

$$\varphi_n(u) = \begin{cases} (u - \theta_{n-1})/(\theta_n - \theta_{n-1}), & \theta_{n-1} \leq u \leq \theta_n \\ 0, & \text{otherwise,} \end{cases} \tag{3.19}$$

then $F(x) = \sum\limits_{i=1}^{n} \varphi_i(x)w_i$ is a piecewise linear interpolation function, which the graph of the interpolation function presents as shape of polygonal line (see Figure 11).

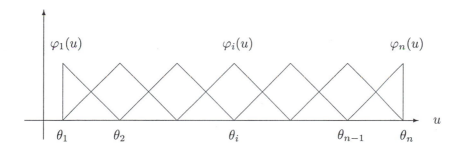

Figure 10   The triangle wave activation functions
as the basic functions of interpolation

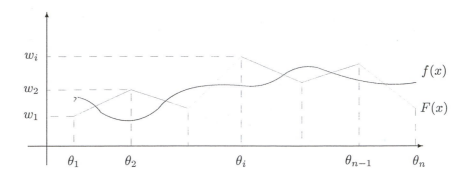

Figure 11 The piecewise linear interpolation function

From the application point of view, a neural network is considered to achieve some objectives or tasks. For the purpose of interpolation, a feedforward neural network can be regarded to interpolate a function, defined by $f(x)$ and called a task function (see Figure 11).

It is well known that the learning in the neural networks is to adjust the connection weights and to achieve some performance indices. As for interpolation (cf. Figure 11), the adjustment of weights is just to try to adjust w_i to approximate $f(\theta_i)$ sufficiently ($i = 1, 2, \cdots, n$), such that the interpolation function $F(x)$ can sufficiently approximate the task function $f(x)$. Moreover, sometimes the threshold values θ_i ($i = 1, 2, \cdots, n$) can contribute the learning, because θ_i ($i = 1, 2, \cdots, n$) can be adjusted for them to be situated on optimal places such that $F(x)$ can approximate $f(x)$ more sufficiently.

From the discussion above we have a clear consequence: **For any continuous function $y = f(x)$, there exists a two-layer neural network shown as Figure 7 such that the network can approximate $f(x)$ satisfying any given precision.** Detailed proof of a two-layer neural network as a universal function approximator can be found in reference [10].

3.4   A Three-layer Feedforward Neural Network with Two Inputs One Output

Figure 12 gives a three-layer feedforward neural network with two inputs one output, where the neurons h_{ij} are taken as hyperbolic functions, i.e., $h_{ij}(u, v) = uv$ ($i = 1, \cdots, n, j = 1, \cdots, m$), and the neuron h is taken as Σ, that is,

$$h(u_{11}, u_{12}, \cdots, u_{nm}) = \Sigma(u_{11}, u_{12}, \cdots, u_{nm}) = \sum_{i=1}^{n}\sum_{j=1}^{m} w_{ij}u_{ij}.$$

We can still consider this network as a fully connected layer-network as shown in Figure 6, with the exception that some weights are set to 0.

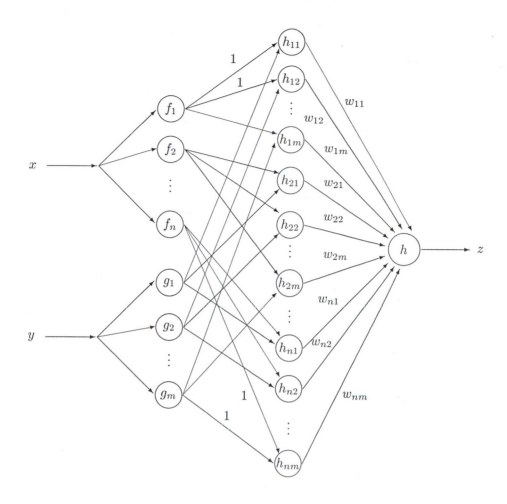

Figure 12   The three-layer feedforward neural network
with two inputs one output

Let the connection weights from f_i and g_j to h_{ij} $(i = 1, \cdots, n, \; j = 1, \cdots, m)$ be all 1. Then the output of the network is as follows:

$$
\begin{aligned}
F(x, y) = z &= h(h_{11}(f_1(x), g_1(y)), \cdots, h_{1m}(f_1(x), g_m(y)), \cdots, \\
&\quad h_{n1}(f_n(x), g_1(y)), \cdots, h_{nm}(f_n(x), g_m(y)) \\
&= h(f_1(x)g_1(y), \cdots, f_1(x)g_m(y), \cdots, f_n(x)g_1(y), \\
&\quad \cdots, f_n(x)g_m(y)) \\
&= \sum_{i=1}^{n} \sum_{j=1}^{m} w_{ij} f_i(x) g_j(y).
\end{aligned}
\tag{3.20}
$$

This is a typical formula of piecewise bivariate interpolation function.

Example 3. In Expression (3.20), $f_i(x)$ are defined as Expression (3.16), and $g_j(y)$ as the following:

$$g_j(y) = \begin{cases} 1, & \delta_j \leq y \leq \delta_{j+1} \\ 0, & \text{otherwise} \end{cases} \quad j = 1, 2, \cdots, m, \tag{3.21}$$

where δ_j and δ_{j+1} (cf. Section 2) are the thresholds of the neuron g_j ($j = 1, 2, \cdots, m$). Then the function $F(x, y)$ determined by Expression (3.20) is a piecewise bivariate zero-order interpolation function which can approximate the task function $f(x, y)$ by adjusting the weights w_{ij}.

Example 4. In Expression (3.20), $f_i(x)$ and $g_j(y)$ are all taken as triangle waves activation functions, i.e., $f_i(x)$ are defined by Expressions (3.17)–(3.19), and $g_j(y)$ similar to the expressions; however, θ_i are replaced by δ_j, and n by m. Then $F(x, y)$ is a piecewise bivariate linear interpolation function.

Like Section 3, we have a second consequence: **For any bivariate continuous function $z = f(x, y)$, there exists a three-layer neural network shown as Figure 12 such that the network can approximate $f(x, y)$ satisfying any given precision.**

3.5   Analysis of Steepest Descent Learning Algorithms of Feedforward Neural Networks

We begin with a simple case, considering the learning algorithm of the two-layer feedforward neural network with one input and one output shown as in Figure 7. We have known that the relation between the input and output of the network is:

$$F(x) = \sum_{i=1}^{n} \varphi_i(x) w_i \tag{3.22}$$

Referring to supervised learning, a group of training samples should be given, for example,

$$\{(x_i, y_i) \mid i = 1, 2, \cdots, p\} \tag{3.23}$$

Upon the substitution of the training samples in Expression (3.22), we get a system of linear equations regarding w_j ($j = 1, 2, \cdots, m$) as unknowns:

$$\left. \begin{array}{l} \varphi_1(x_1)w_1 + \varphi_2(x_1)w_2 + \cdots + \varphi_n(x_1)w_n = y_1 \\ \varphi_1(x_2)w_1 + \varphi_2(x_2)w_2 + \cdots + \varphi_n(x_2)w_n = y_2 \\ \qquad\qquad \cdots\cdots \\ \varphi_1(x_p)w_1 + \varphi_2(x_p)w_2 + \cdots + \varphi_n(x_p)w_n = y_p \end{array} \right\}. \tag{3.24}$$

Let $a_{ij} = \varphi_j(x_i)$, $\mathbf{A} = (a_{ij})_{p \times n}$, $\mathbf{Y} = (y_1, y_2, \cdots, y_p)^T$, and $\mathbf{W} = (w_1, w_2, \cdots, w_n)^T$. Then Expression (3.24) can be written as follows:

$$\mathbf{AW} = \mathbf{Y}. \tag{3.25}$$

Generally speaking, the system of linear equations (3.24) (or (3.25)) is a contradictory system; in other words, if denoting

$$\mathbf{AW} = \hat{\mathbf{Y}} = (\hat{y}_1, \hat{y}_2, \cdots, \hat{y}_p)^T, \tag{3.26}$$

then $\hat{\mathbf{Y}} \neq \mathbf{Y}$ generally. In this case, the learning is to find $\mathbf{W}^* = (w_1^*, w_2^*, \cdots, w_n^*)^T$ such that $\mathbf{AW}$ sufficiently approximates $\mathbf{Y}$ according to some sort of norm $\| \cdot \|$, i.e.,

$$\min \| \mathbf{AW} - \mathbf{Y} \|. \tag{3.27}$$

Therefore, it becomes an optimization problem.

In most neural network researching and applications, the norm $\| \cdot \|$ is often taken as the following form:

$$E(w_1, \cdots, w_n) = \frac{1}{2} \sum_{i=1}^{p} (\hat{y}_i - y_i)^2 = \frac{1}{2} \sum_{i=1}^{p} \left(\sum_{j=1}^{n} a_{ij} w_j - y_j \right)^2. \tag{3.28}$$

Clearly, it is a problem of optimization of quadratic function without constraints that can be usually solved by gradient descent algorithm. In fact, denote

$$\nabla E(\mathbf{W}) = \left(\frac{\partial E(\mathbf{W})}{\partial w_1}, \frac{\partial E(\mathbf{W})}{\partial w_2}, \cdots, \frac{\partial E(\mathbf{W})}{\partial w_n} \right)^T.$$

In order to solve Equation (3.27), assume k interactions to have been done and $\mathbf{W}_k$ has been found. The $(k+1)$th interaction point $\mathbf{W}_{k+1}$ can be obtained by a search from $\mathbf{W}_k$ along the direction of $-\nabla E(\mathbf{W}_k)$, i.e.,

$$\mathbf{W}_{k+1} = \mathbf{W}_k - \mu_k \nabla E(\mathbf{W}_k), \tag{3.29}$$

where the step size factor μ_k should satisfy the following optimal condition:

$$E(\mathbf{W}_k - \mu_k \nabla E(\mathbf{W}_k)) = \min_{\mu} E(\mathbf{W}_k - \mu \nabla E(\mathbf{W}_k)). \tag{3.30}$$

Expressions (3.29) and (3.30) can be written into one expression:

$$\mathbf{W}_{k+1} = ls(\mathbf{W}_k, -\nabla E(\mathbf{W}_k)) \tag{3.31}$$

which means that we will move to $\mathbf{W}_{k+1}$ by performing a linear search from $\mathbf{W}_k$ along $-\nabla E(\mathbf{W}_k)$, where "l" and "s" represent the linear search.

When $k = 0, 1, 2, \cdots$, we have the sequence of points $\mathbf{W}_0, \mathbf{W}_1, \mathbf{W}_2, \cdots$, where $\mathbf{W}_0$ is the initial point that can be chosen arbitrarily. The sequence of points $\{\mathbf{W}_k\}$ can converge on a minimal point $\mathbf{W}^*$ of $E(\mathbf{W})$ in a quadratic case.

Remark 2. The learning algorithms are often simplified. For example, denoting $\frac{\partial E}{\partial \mathbf{W}} = \nabla E(\mathbf{W})$, then the following expression,

$$\mathbf{W}_{k+1} = \mathbf{W}_k + \mu \left(-\frac{\partial E}{\partial \mathbf{W}} \right)_{\mathbf{W} = \mathbf{W}_k} \tag{3.32}$$

is just one of the algorithms commonly used. It is said to be the simplifying form of Expression (3.29), because here μ is chosen in advance instead of changing μ_k based on each interaction step in Expression (3.29). But μ is chosen depending on the experience of users so that the interactive process can not be ensured to convergence. Sometimes the convergence speed is very low though Expression (3.32) is of convergence.

Remark 3. It is easy to show that $(-\nabla E(\mathbf{W}_{k+1}))(-\nabla E(\mathbf{W}_k)) = 0$ which means that $-\nabla E(\mathbf{W}_{k+1})$ is perpendicular to $-\nabla E(\mathbf{W}_k)$, so that there is a "sawtooth phenomenon". This is one of the main reasons why neural networks usually have a low speed of convergence.

Remark 4. Noticing Expression (3.28) is a quadratic function, we rewrite it as follows:

$$
\begin{aligned}
E(\mathbf{W}) &= \frac{1}{2}\sum_{i=1}^{p}\left(\sum_{j=1}^{n} a_{ij}w_j - y_i\right)^2 \\
&= \frac{1}{2}\sum_{i=1}^{p}\left[\left(\sum_{j=1}^{n} a_{ij}w_j\right)^2 - 2y_i\sum_{j=1}^{n} a_{ij}w_j + y_i^2\right] \\
&= \frac{1}{2}\left[\left(\sum_{i=1}^{p} a_{i1}^2\right)w_1^2 + \left(\sum_{i=1}^{p} a_{i2}^2\right)w_2^2 + \cdots + \left(\sum_{i=1}^{p} a_{in}^2\right)w_n^2\right. \\
&\quad + 2\left(\sum_{i=1}^{p} a_{i1}a_{i2}\right)w_1w_2 + 2\left(\sum_{i=1}^{p} a_{i1}a_{i3}\right)w_1w_3 + \cdots + 2\left(\sum_{i=1}^{p} a_{i1}a_{in}\right)w_1w_n \\
&\quad + 2\left(\sum_{i=1}^{p} a_{i2}a_{i3}\right)w_2w_3 + 2\left(\sum_{i=1}^{p} a_{i2}a_{i4}\right)w_2w_4 + \cdots + 2\left(\sum_{i=1}^{p} a_{i2}a_{in}\right)w_2w_n \\
&\quad + \cdots + 2(\sum_{i=1}^{p} a_{i,n-1}a_{in})w_{n-1}w_n\Bigg] + \left(-\sum_{i=1}^{p} a_{i1}y_i\right)w_1 \\
&\quad + \left(-\sum_{i=1}^{p} a_{i2}y_i\right)w_2 + \cdots + \left(-\sum_{i=1}^{p} a_{in}y_i\right)w_n + \frac{1}{2}\sum_{i=1}^{p} y_i^2 \\
&= \frac{1}{2}\mathbf{W}^T\mathbf{Q}\mathbf{W} + \mathbf{B}^T\mathbf{W} + C,
\end{aligned}
\tag{3.33}
$$

where $\mathbf{Q}=(q_{ij})_{n\times n}$, $\mathbf{B}=(b_1, b_2, \cdots, b_n)^T$, $C = \frac{1}{2}\sum_{i=1}^{p} y_i^2$, $q_{ij} = \sum_{k=1}^{p} a_{ki}a_{kj}$ with $q_{ij} = q_{ji}$, and $b_i = -\sum_{k=1}^{p} a_{ki}y_k$. Clearly, we have

$$
\nabla E(\mathbf{W}) = \mathbf{Q}\mathbf{W} + \mathbf{B}
\tag{3.34}
$$

In order to find minimal points of $E(\mathbf{W})$, we should have $\nabla E(\mathbf{W}) = 0 \triangleq (0, 0, \cdots, 0)^T$. When $\mathbf{Q}$ is a positive definite matrix, the unique minimal point can be found as follows:

$$\mathbf{W}^* = -\mathbf{Q}^{-1}\mathbf{B} \tag{3.35}$$

The Hessian matrix at this point is the following:

$$\nabla^2 E(\mathbf{W}^*) = \nabla(\nabla E(\mathbf{W}^*)) = \begin{pmatrix} \frac{\partial^2 E(\mathbf{W})}{\partial w_1^2} & \frac{\partial^2 E(\mathbf{W})}{\partial w_2 \partial w_1} & \cdots & \frac{\partial^2 E(\mathbf{W})}{\partial w_n \partial w_1} \\ \frac{\partial^2 E(\mathbf{W})}{\partial w_1 \partial w_2} & \frac{\partial^2 E(\mathbf{W})}{\partial w_2^2} & \cdots & \frac{\partial^2 E(\mathbf{W})}{\partial w_n \partial w_2} \\ \cdots & \cdots & \cdots & \cdots \\ \frac{\partial^2 E(\mathbf{W})}{\partial w_1 \partial w_n} & \frac{\partial^2 E(\mathbf{W})}{\partial w_2 \partial w_n} & \cdots & \frac{\partial^2 E(\mathbf{W})}{\partial w_n^2} \end{pmatrix}_{\mathbf{W}=\mathbf{W}^*} = \mathbf{Q}$$

which is a positive definite matrix because $\mathbf{Q}$ is positive definite and $\nabla^2 E(\mathbf{W}^*)$ is just equal to $\mathbf{Q}$. Based on the consequences in optimization theory, $\mathbf{W}^* = -\mathbf{Q}^{-1}\mathbf{B}$ is the unique minimal point of the problem.

Remark 5. The learning algorithm (3.31) (i.e., (3.29) and (3.30)) is an implicit scheme and it is not very convenient to find the direction of the search. Now we consider its explicit scheme. In fact, since $\nabla E(\mathbf{W}) = \mathbf{Q}\mathbf{W}+\mathbf{B}$, $\nabla E(\mathbf{W}_k) = \mathbf{Q}\mathbf{W}_k + \mathbf{B}$. For we have known that $(\nabla E(\mathbf{W}_{k+1}))^T \nabla E(\mathbf{W}_k) = 0$, from Expression (3.29) we have

$$(\mathbf{Q}(\mathbf{W}_k - \mu_k \nabla E(\mathbf{W}_k)) + \mathbf{B})^T (\mathbf{Q}\mathbf{W}_k + \mathbf{B}) = 0.$$

Rearrange the equation, we have the following:

$$(\nabla E(\mathbf{W}_k) - \mu_k \mathbf{Q}\nabla E(\mathbf{W}_k))^T \nabla E(\mathbf{W}_k) = 0.$$

From the expression, we can easily get an expression of μ_k:

$$\mu_k = (\nabla E(\mathbf{W}_k))^T \nabla E(\mathbf{W}_k) / (\nabla E(\mathbf{W}_k))^T \mathbf{Q}\nabla E(\mathbf{W}_k) \tag{3.36}$$

After the expression is substituted into Expression (3.29), we have the explicit scheme:

$$\mathbf{W}_{k+1} = \mathbf{W}_k - \frac{(\nabla E(\mathbf{W}_k))^T \nabla E(\mathbf{W}_k)}{(\nabla E(\mathbf{W}_k))^T \mathbf{Q}\nabla E(\mathbf{W}_k)} \nabla E(\mathbf{W}_k) \tag{3.37}$$

Now we consider the learning algorithm of feedforward neural networks with two-input one-put. From Section 4, the relation between the input and output of the networks is as follows:

$$F(x,y) = \sum_{i=1}^{n} \sum_{j=1}^{m} f_i(x)g_j(y)w_{ij} \tag{3.38}$$

For a given group of training samples:

$$\{((x_k, y_k), z_k) \mid k = 1, 2, \cdots, p\}; \tag{3.39}$$

substituting it into Expression (3.38), we get a system of linear equations regarding w_{ij} $(i = 1, 2, \cdots, n, \ j = 1, 2, \cdots, m)$ as unknowns:

$$\sum_{i=1}^{n} \sum_{j=1}^{m} f_i(x_k)g_j(y_k)w_{ij} = z_k, \quad k = 1, 2, \cdots, p. \tag{3.40}$$

Let $a_{kij} = f_i(x_k)g_j(y_k)$. Then the above expression can be written as the following:

$$\sum_{i=1}^{n}\sum_{j=1}^{m} a_{kij}w_{ij} = z_k, \quad k = 1, 2, \cdots, p \,. \tag{3.41}$$

In order to simplify the triple subscripts of the above expression, let

$$l = (i-1)m + j. \tag{3.42}$$

Then Expression (3.41) can be written as follows:

$$\sum_{l=1}^{q} a_{kl}w_l = z_k, \quad k = 1, 2, \cdots, p \tag{3.43}$$

where $q = nm$. Furthermore, if we let $\mathbf{A} = (a_{kl})_{p \times q}$, $\mathbf{Z}=(z_1, z_2, \cdots, z_p)^T$ and $\mathbf{W}=(w_1, w_2, \cdots, w_q)^T$, then Expression (3.43) becomes a matrix equation:

$$\mathbf{AW} = \mathbf{X} \tag{3.44}$$

which is similar to Expression (3.25). So the learning algorithm should be similar with Expression (3.37).

3.6   Feedforward Neural Networks with Multi-input One Output and Their Learning Algorithm

Referring to feedforward neural networks with multi-input one output, we can draw a graph similar to Figure 12. For example, Figure 13 shows the graph of the network with three inputs one output.

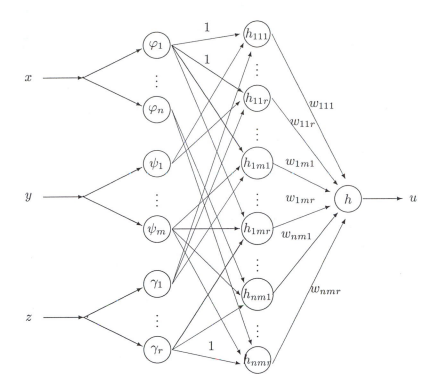

Figure 13   The three-layer feedforward neural network
with three-input one-output

The neurons h_{ijs} are taken as $h_{ijs}(u, v, w) = uvw$. In imitation of expression (3.20), the relation between the input and output of the network shown as in Figure 13 is as follows:

$$F(x, y, z) = \sum_{i=1}^{n} \sum_{j=1}^{m} \sum_{s=1}^{r} \varphi_i(x)\psi_j(y)\gamma_s(z)w_{ijs}. \qquad (3.45)$$

For a given group of training samples:

$$\{((x_k, y_k, z_k), u_k) \mid k = 1, 2, \cdots, p\}; \qquad (3.46)$$

substituting it into the above expression, we get a system of linear equations regarding w_{ijs} $(i = 1, 2, \cdots, n, \ j = 1, 2, \cdots, m, \ s = 1, 2, \cdots, r)$ as unknowns:

$$\sum_{i=1}^{n} \sum_{j=1}^{m} \sum_{s=1}^{r} \varphi_i(x_k)\psi_j(y_k)\gamma_s(z_k)w_{ijs} = u_k, \quad k = 1, 2, \cdots, p. \qquad (3.47)$$

Let $a_{kijs} = \varphi_i(x_k)\psi_j(y_k)\gamma_s(z_k)$. Expression (3.47) is written as the following:

$$\sum_{i=1}^{n} \sum_{j=1}^{m} \sum_{s=1}^{r} a_{kijs}w_{ijs} = u_k, \quad k = 1, \cdots, p \qquad (3.48)$$

We also can simplify the above subscripts, being similar to Expression (3.42). In fact, let

$$l = (i-1)mr + (j-1)r + s \qquad (3.49\ )$$

then Expression (3.48) can be rewritten as follows:

$$\sum_{l=1}^{q} a_{kl}w_l = u_k, \quad k = 1, \cdots, p, \qquad (3.50)$$

where $q = nmr$. Furthermore, if we let $\mathbf{A}=(a_{kl})_{p \times q}$, $\mathbf{U}=(u_1, u_2, \cdots, u_p)^T$, and $\mathbf{W}=(w_1, w_2, \cdots, w_q)^T$, then Expression (3.50) becomes a matrix equation similar to Expression (3.44):

$$\mathbf{AW} = \mathbf{U}. \qquad (3.51)$$

Of course, the learning algorithm of Expression (3.50) is very similar to Expression (3.44).

Now we consider the feedforward neural network with n-input one-output. We can also draw a graph of the network similar to Figure 13, and here omit it. Clearly the relation between the input and output of the network is as the following interpolation function:

$$F(x_1, x_2, \cdots, x_n) = \sum_{i_1=1}^{m_1} \sum_{i_2=1}^{m_2} \cdots \sum_{i_n=1}^{m_n} \varphi_{i_1}(x_1)\varphi_{i_2}(x_2) \cdots \varphi_{i_n}(x_n)w_{i_1 i_2 \cdots i_n}. \qquad (3.52)$$

Like Section 4, we have this third consequence: **For any n-variate continuous function $y = f(x_1, x_2, \cdots, x_n)$, there exists a three-layer feedforward neural network similar to Figure 13 such that the network can approximate $f(x_1, x_2, \cdots, x_n)$ satisfying any given precision.**

Let us consider a group of training samples:

$$\{((x_1^{(k)}, x_2^{(k)}, \cdots, x_n^{(k)}), y_k) \mid k = 1, 2, \cdots, p\}. \qquad (3.53)$$

After substituting it into Expression (3.52), we get a system of linear equations regarding $w_{i_1 i_2 \cdots i_n}$ $(i_1 = 1, \cdots, m_1; i_2 = 1, \cdots, m_2; \cdots; i_n = 1, \cdots, m_n)$ as unknowns:

$$\sum_{i_1=1}^{m_1} \cdots \sum_{i_n=1}^{m_n} \varphi_{i_1}(x_1^{(k)}) \cdots \varphi_{i_n}(x_n^{(k)})w_{i_1 \cdots i_n} = y_k, \quad k = 1, \cdots, p. \qquad (3.54)$$

Let $a_{ki_1 i_2 \cdots i_n} = \varphi_{i_1}(x_1^{(k)})\varphi_{i_2}(x_2^{(k)}) \cdots \varphi_{i_n}(x_n^{(k)})$. Then the above expression becomes the following:

$$\sum_{i_1=1}^{m_1} \cdots \sum_{i_n=1}^{m_n} a_{ki_1 i_2 \cdots i_n} w_{i_1 i_2 \cdots i_n} = y_k, \quad k = 1, \cdots, p. \qquad (3.55)$$

In imitation of Expression (3.49), let

$$l = (i_1-1)m_2 m_3 \cdots m_n + (i_2-1)m_3 m_4 \cdots m_n + \cdots + (i_n-1)m_n + i_n \qquad (3.56)$$

and $q = i_1 i_2 \cdots i_n$. Then the expression can be written as follows:

$$\sum_{l=1}^{q} a_{kl} w_l = y_k, \quad k = 1, 2, \cdots, p. \tag{3.57}$$

Furthermore, if denoting $\mathbf{A} = (a_{kl})_{p \times q}$, $\mathbf{Y} = (y_1, \cdots, y_p)^T$ and $\mathbf{W} = (w_1, \cdots, w_q)^T$, the above expression comes back to Expression (3.25). As for the learning algorithm, there is no difference with Expression (3.37).

Remark 6. All the neural networks defined so far have a unit-value weight that connects the hidden layers. The only unknowns that need to be found are the weights, w_{ij} or w_{ijk}. If the weights of the hidden nodes are not fixed values, i.e., these weights are to be found through the learning, and the neurons are not linear neurons, then the system equation is no longer a linear function of weights and inputs. Furthermore, the performance index in Equation (3.33) is not a quadratic form of w_{ij} or w_{ijk}. Hence, the abovementioned approach cannot be applied directly to train the neural network. Instead, the backpropagation algorithm [11] and other derivative-based optimization techniques [12, 13] that calculate a gradient vector for search direction are used. In order to avoid the complicated calculation of gradient vector and search direction, we will introduce some "flat" neural networks architecture in Chapter 5.

3.7 Feedforward Neural Networks with One Input Multi-output and Their Learning Algorithm

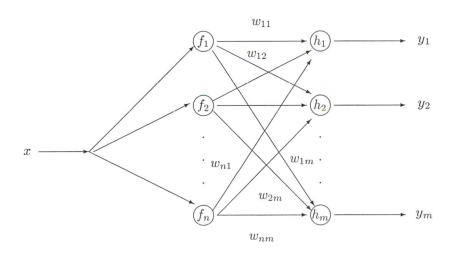

Figure 14   The two-layer feedforward neural network
with one-input m-output

The graph in Figure 14 offers a sort of two-layer feedforward neural network with one input m-output. Let the neurons h_j $(j = 1, 2, \cdots, m)$ be all taken as Σ. Then jth component of the network output is as follows:

$$y_j = h_j(f_1(x), f_2(x), \cdots, f_n(x)) = \sum_{i=1}^{n} f_i(x) w_{ij}, \tag{3.58}$$

which is an interpolation function. Thus, the vector-value output of the network is:

$$F(x) = (y_1, \cdots, y_m) = \left(\sum_{i=1}^{n} f_i(x) w_{i1}, \cdots, \sum_{i=1}^{n} f_i(x) w_{im} \right). \tag{3.59}$$

Now we consider the learning algorithm of the network. For a group of training samples:

$$\{(x_k, (y_1^{(k)}, y_2^{(k)}, \cdots, y_m^{(k)})) \mid k = 1, 2, \cdots, p\}; \tag{3.60}$$

after substituting it into Expression (3.58), we can get m systems of linear equations:

$$\sum_{i=1}^{n} f_i(x_k) w_{i1} = y_1^{(k)}, \quad k = 1, \cdots, p \tag{3.61.1}$$

$$\sum_{i=1}^{n} f_i(x_k) w_{i2} = y_2^{(k)}, \quad k = 1, \cdots, p \tag{3.61.2}$$

$$\cdots$$

$$\sum_{i=1}^{n} f_i(x_k) w_{im} = y_m^{(k)}, \quad k = 1, \cdots, p. \tag{3.61.m}$$

If let $a_{ij} = f_j(x_i)$, $\mathbf{A} = (a_{ij})_{p \times n}$, $\mathbf{Y}_s = (y_s^{(1)}, y_s^{(2)}, \cdots, y_s^{(p)})^T$ and $\mathbf{W}_s = (w_{1s}, w_{2s}, \cdots, w_{ns})$ then Expressions (3.61.s) $(s = 1, 2, \cdots, m)$ become m matrix equations:

$$\mathbf{AW}_s = \mathbf{Y}_s, \quad s = 1, 2, \cdots, m. \tag{3.62}$$

Clearly, the learning algorithms of the matrix equations are similar to the above algorithm with the exception of the number of the matrix equations.

Naturally we have a fourth consequence: **For any one-variate vector-value continuous function $f(x) = (y_1, y_2, \cdots, y_m)$, there exists a two-layer feedforward neural network shown as Figure 14 such that the network can approximate $f(x)$, satisfying any given precision.**

3.8 Feedforward Neural Networks with Multi-input Multi-output and Their Learning Algorithm

Without loss of generality, we now consider a three-layer feedforward neural network with two inputs two outputs shown as Figure 15, and the cases of the others are similar to this discussion.

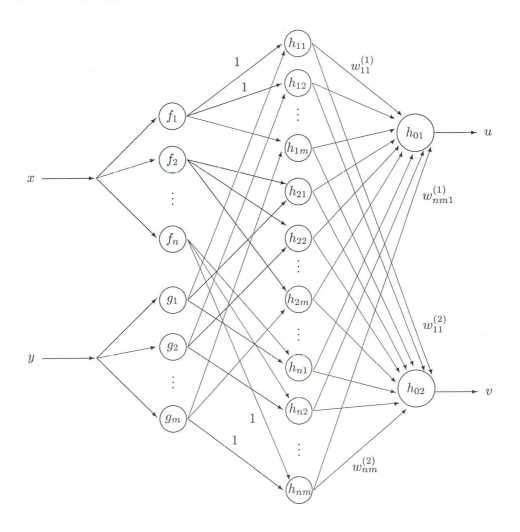

Figure 15   A three-layer feedforward neural network
with two inputs and two outputs

Based on the result of the previous section, the relation between the input and output of the network should be as follows:

$$F(x,y) = (u,v) = \left(\sum_{i=1}^{n} \sum_{j=1}^{m} f_i(x) g_j(y) w_{ij}^{(1)}, \ \sum_{i=1}^{n} \sum_{j=1}^{m} f_i(x) g_j(y) w_{ij}^{(2)} \right) \qquad (3.63)$$

which is a vector-value interpolation function.

Clearly, we can give a fifth consequence: **For any multi-variate vector-value continuous function** $f(x_1, x_2, \cdots, x_n) = (y_1, y_2, \cdots, y_m)$, **there exists a three-layer feedforward neural network like Figure 15 such that the network can approximate** $f(x_1, x_2, \cdots, x_n)$ **satisfying any given precision.**

It is easy to consider the learning algorithm of the network. For a given group of samples:

$$\{((x_k, y_k), (u_k, v_k)) \mid k = 1, 2, \cdots, p\}; \tag{3.64}$$

substituting it into Expression (3.63), we get two systems of linear equations:

$$\sum_{i=1}^{n} \sum_{j=1}^{m} f_i(x_k) g_j(y_k) w_{ij}^{(1)} = u_k, \quad k = 1, 2, \cdots, p \tag{3.65}$$

$$\sum_{i=1}^{n} \sum_{j=1}^{m} f_i(x_k) g_j(y_k) w_{ij}^{(2)} = u_k, \quad k = 1, 2, \cdots, p. \tag{3.66}$$

Let $a_{kij} = f_i(x_k) g_j(y_k)$, $l = (i-1)m+j$ and $q = nm$. Then the above two expressions become the following:

$$\sum_{l=1}^{q} a_{kl} w_l^{(1)} = u_k, \quad k = 1, 2, \cdots, p \tag{3.67}$$

$$\sum_{l=1}^{q} a_{kl} w_l^{(2)} = v_k, \quad k = 1, 2, \cdots, p. \tag{3.68}$$

If denoting $\mathbf{A} = (a_{kl})_{p \times q}$, $\mathbf{W}_1 = (w_1^{(1)}, w_2^{(1)}, \cdots, w_q^{(1)})^T$, $\mathbf{W}_2 = (w_1^{(2)}, w_2^{(2)}, \cdots, w_q^{(2)})^T$, $\mathbf{U} = (u_1, u_2, \cdots, u_p)^T$, and $\mathbf{V} = (v_1, v_2, \cdots, v_p)^T$, then Expressions (3.67) and (3.68) can be written as matrix equations:

$$\mathbf{A}\mathbf{W}_1 = \mathbf{U}, \quad \mathbf{A}\mathbf{W}_2 = \mathbf{V} \tag{3.69}$$

So the learning algorithm for them is similar to the ones mentioned above.

3.9   A Note on the Learning Algorithm of Feedforward Neural Networks

We have discussed the learning algorithms of feedforward neural networks in the above sections. However, there is an assumption that the activation functions at the output layer are identity functions and the threshold values are absorbed into the activation functions.

Taking the network with one input one output for an example (see Figure 7), the output of the network should be as follows:

$$y = \varphi \left(\sum_{i=1}^{n} \varphi_i(x) w_i - \theta \right), \tag{3.70}$$

where φ is the activation function of the neuron at output layer. If assuming φ to be an identity function, then

$$y = \sum_{i=1}^{n} \varphi_i(x)w_i - \theta. \tag{3.71}$$

To move θ to the left and put $y' = y + \theta$, then the above expression becomes an expression without θ:

$$y' = \sum_{i=1}^{n} \varphi_i(x)w_i \tag{3.72}$$

which is just like Expression (3.15). So its learning algorithm is very similar to the one in Expression (3.15). Without loss of generality, we can suppose $\theta = 0$. Thus Expressions (3.70) and (3.72) are written as the following simple forms:

$$y = \varphi\left(\sum_{i=1}^{n} \varphi_i(x)w_i\right) \tag{3.73}$$

$$y = \sum_{i=1}^{n} \varphi_i(x)w_i \tag{3.74}$$

In most cases, for activations functions in common use, there is hardly a difference between the learning algorithms with respect to Expression (3.73) and (3.74).

It is well known that activation functions at the output layer in common use have three types:

Type 1: Step-up function, i.e., $\varphi(u) = 1$, $u \geq 0$; $\varphi(u) = 0$, $u < 0$.

Type 2: Identity function, $\varphi(u) = u$.

Type 3: Sigmoid function, $\varphi(u) = \left(1 + e^{-(\alpha u + \beta)}\right)^{-1}$, where α and β are parameters.

When φ is an identity function, Expression (3.73) is just (3.74). When $\beta = 0$, after α is chosen properly, a sigmoid function can approximate a step-up function. So we may only discuss the case for φ being sigmoid functions. Noticing that sigmoid functions are monotone increasing continuous functions, for more generality, $\varphi(u)$ is supposed to be a monotone continuous function. Thus $\varphi(u)$ has its inverse function $\varphi^{-1}(u)$. Therefore, Expression (3.73) can be rewritten as follows:

$$\varphi^{-1}(y) = \sum_{i=1}^{n} \varphi_i(x)w_i \tag{3.75}$$

For a given group of training samples $\{(x_k, y_k) \mid k = 1, 2, \cdots, p\}$, substituting it into the above expression and letting $b_k = \varphi^{-1}(y_k)$ and $a_{ki} = \varphi_i(x_k)$, we have the following system of linear equations:

$$\sum_{i=1}^{n} a_{ki} w_i = b_k, \quad k = 1, 2, \cdots, p \tag{3.76}$$

Clearly, its learning algorithm is very similar to the ones in previous sections.

3.10 Conclusions

We have discussed the mathematical model of artificial neural networks in this chapter. Most of the models that we defined here have a unit value weights connection in hidden nodes. Based on the assumption, we can form a quadratic equation of the weights. Thus, the search of the weights becomes the search of the gradient direction of the performance index. However, practically, the weights are not simply the unit values in hidden nodes for most multi-layer neural networks. Thus, the backpropagation algorithm [11] or other derivative-based optimization techniques [12, 13] are used to find weights in different layers. However, we will not focus on the analysis of multi-layer neural networks here. Instead, we will use the concept that we derived here to introduce some "flat" neural networks. The system equations of the flat neural networks are similar to the equations discussed here. The learning algorithm will be much easier and straightforward compared to the backpropagation algorithm.

References

1. T. L. McClelland and D. E. Rumelhart, et al., *Parallel Distributed Processing*, The MIT Press, Cambridge, 1986.

2. P. J. Werbo, Beyond Regression: New Tools for Prediction and Analysis in the Behavioral Sciences, *Ph.D. Thesis*, Harvard University, Cambridge, 1974.

3. S. Haykin and L. Li, Nonlinear and adaptive prediction of nonstationary signals, *IEEE Transactions on Signal Processing*, Vol. 43, No. 2, pp. 526-535, 1995.

4. A. N. Michel and J. A. Farrell, Associative memories via artificial neural networks, *IEEE Control Systems Magazine*, April, pp. 6-17, 1990.

5. T. Kohonen, *Self-Organization and Associative Memory*, 2nd Ed., Springer-Verlag, Berlin, 1987.

6. G. A. Carpenter and S. Grossberg, ART2 Self-organization of state category recognition codes for analog input patterns, *Applied Optics*, Vol. 26, No. 23, pp. 4919-4930, 1987.

7. G. A. Carpenter, S. Grossberg, and D. S. Rosen, Fuzzy ART: Fast stable learning and categorization of analog patterns by an adaptive resonance systems, *Neural Networks*, Vol. 4, pp. 759-771, 1991.

8. J. J. Hopfield and D. W. Tank, Neural computation of decisions in optimization problems, *Biological Cybernetics*, Vol. 52, pp. 141-154, 1985.

9. W. S. McCulloc and W. H. Pitts, A logical calculus of the ideas imminent in nervous activity, *Bulletin of Mathematical Biophysics*, Vol. 5, pp. 115-133, 1943.

10. K. M. Honik, M. Stinchcombe, and H. White, Multilayer feedforward networks are universal approximators, *Neural Networks*, Vol. 2, No. 5. pp. 359-366, 1989.

11. D. E. Rumelhart, G. E. Hinton, and R. J. Williams, Learning in representations by back-propagation errors, *Nature*, Vol. 323, pp. 533-536, 1986.

12. C. Charalambous, Conjugate gradient algorithm for efficient training of artificial neural networks, *IEEE Proceedings*, Vol. 139, No. 3, pp. 301-310, 1992.

13. M. T. Hagan, H. B. Demuth, and M. Beale, *Neural Network Design*, PWS Pub. Boston, 1995.

Chapter 4

Functional-link Neural Networks and Visualization Means of Some Mathematical Methods

This chapter focuses on functional-link neural networks. Beginning with the XOR problem, we discuss the mathematical essence and the structures of functional-link neural networks. Extending this idea, we give the visualization means of mathematical methods. We also give neural network representations of linear programming and fuzzy linear programming.

4.1   Discussion of the XOR Problem

A single-layer neural network, first studied by Minsky and Papert, was named perceptron in 1969 [1]. It is well known that a single-layer perceptron network cannot solve a nonlinear problem. A typical problem is the Exclusive-OR (XOR) problem.

Generally, there are two approaches to solve this nonlinear problem by modifying the architecture of this single-layer perceptron. The first one is to *increase number of the hidden layers*, and the second one is to *add higher order input terms*. There are numerous applications using either of these approaches [2-4]. Here we will illustrate that these two approaches, in fact, are essentially mathematical equivalence.

Figure 1 shows a simple neuron with two inputs. Figure 2 is the same neuron with a higher order term, $x_1 \cdot x_2$

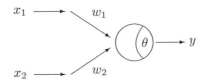

Figure 1   A simple one-layer perceptron

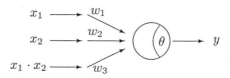

Figure 2   Add a higher order term $x_1 \cdot x_2$

Pao [5] showed that the XOR problem can be solved easily by adding this extra higher order term. As Pao indicated this extra term is an interaction of function of x_1 and x_2. By adding this extra term, we increase the input dimension by one, thus transforming the linear unseparable problem to linear separable. Here, we take a different explanation. The output of the this network in Figure 2 is given as follows.

$$y = \varphi(x_1 w_1 + x_2 w_2 + x_1 x_2 w_3 - \theta) .\tag{4.1}$$

By taking $w_1 = w_2 = 1$, $w_3 = 2$, and $\theta = 1$, then Expression (4.1) becomes the following:

$$y = \varphi(x_1 + x_2 - 2x_1 x_2 - 1) .\tag{4.2}$$

If we take a 45° rotation of axes and the new coordinates are denoted by x_1' and x_2', then the curve $x_1 + x_2 - 2x_1 x_2 = 1$ can be expressed as follows:

$$-\frac{\left(x_2' - \frac{\sqrt{2}}{2}\right)^2}{\left(\frac{\sqrt{2}}{2}\right)^2} + \frac{(x_1')^2}{\left(\frac{\sqrt{2}}{2}\right)^2} = 1 \tag{4.3}$$

which is a hyperbola. It can classify the two patterns (as inputs) into two classes, i.e., 0-class and 1-class (as outputs) (see Figure 3).

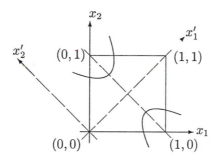

Figure 3   The hyperbola for solving XOR problem

In fact, by simply rearranging the input, the single neuron network with a higher order term can be considered as a two-layer network, shown in Figure 4, where the activation function of neuron f_i is φ_i $(i = 1, 2, 3)$, and the activation function of neuron h is φ. We define φ_1 and φ_2 to be the identity functions (i.e., $\varphi_1(u) = u$, $\varphi_2(u) = u$), φ_3 to be a hyperbolic function (i.e., $\varphi_3(u, v) = uv$), and φ to be a step-up function (i.e., $\varphi(u) = 1$, $u \geqslant 0$; $\varphi(u) = 0, u < 0$). Thus, the output of the network is given as follows:

$$y = \varphi(w_1\varphi_1(x_1) + w_2\varphi_2(x_2) + w_3\varphi_3(x_1, x_2) - \theta)$$
$$= \varphi(w_1x_1 + w_2x_2 + w_3x_1x_2 - \theta).$$

This is the same as Equation (4.1). In other words, the network shown in Figure 4 is equivalent to the one shown in Figure 2.

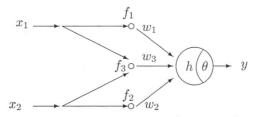

Figure 4   A z-lager perceptron for solving XOR problem

Remark 1.   Clearly the neurons f_1 and f_2 in Figure 4 play only roles of "passageways", since φ_1 and φ_2 are identity functions. So we can remove the two "stations" f_1 and f_2 . That gives us Figure 5.

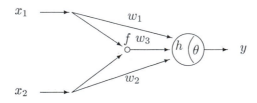

Figure 5   The simplifying of Figure 4

The activation functions of the neurons f and h are denoted by ψ and φ, and taken as hyperbolic function and step-up function. So we also have

$$y = \varphi(w_1x_1 + w_2x_2 + w_3x_1x_2 - \theta).$$

Remark 2.   If we add more higher order terms to Figure 2 (see Figure 6), then the network output is given below:

$$y = \varphi(w_1x_1 + w_2x_2 + w_3x_1x_2 + w_4x_1^2 + w_5x_2^2 - \theta) . \tag{4.4}$$

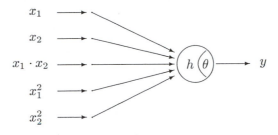

Figure 6   Add some more higher terms

Similarly, the single-layer network in Figure 6 can be accomplished by a two-layer network shown in Figure 7, where the activation function of the neurons f_1, f_2, f_3 and h are, respectively, $\varphi_1, \varphi_2, \varphi_3$ and φ. We can consider φ_2 as a hyperbolic function, φ_1 and φ_3 are square functions (i.e., $\varphi_1(u) = u^2$, $\varphi_2(u) = u^2$), and φ as a step-up function. Then the output of the network is given as follows.

$$y = \varphi(w_1\varphi_1(x_1) + w_2\varphi_2(x_1, x_2) + w_3\varphi_3(x_2) + w_4x_1 + w_5x_2 - \theta)$$
$$= \varphi(w_1x_1^2 + w_2x_1x_2 + w_3x_2^2 + w_4x_1 + w_5x_2 - \theta).$$

Again, this equation is the same as (4.4). Here we show by examples a one-layer network with higher order terms can be accomplished by a multi-layer network, or vice versa.

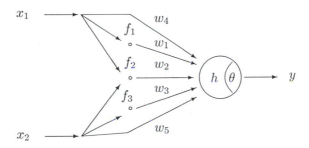

Figure 7   A network for accomplishing the network of Figure 6

4.2   Mathematical Essence of Functional-link Neural Networks

Pao first gave an interesting functional expansion functional-link network [5]. As shown in Figure 8, the output of the network is:

$$y = \varphi(w_1x + w_2\sin\pi x + w_3\cos\pi x + w_4\sin 2\pi x + w_5\cos 2\pi x + w_6\sin 4\pi x - \theta) \quad (4.5)$$

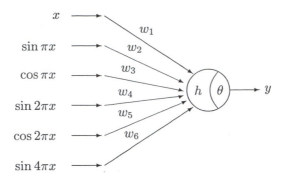

Figure 8   A function expansion network

In fact, the network is formed by adding five other inputs to an one-input network, where x is regarded as a generating input and the other five are function transformations of the original one (for instance, $x \mapsto \sin \pi x$). As a matter of fact, if we let the activation function of the neuron h be an identity function, $w_1 = 0$, $w_0' \triangleq -\theta$, $w_i' \triangleq w_{i+1}$ $(i = 1, \cdots, 5)$, then Expression (4.5) becomes the following:

$$y = w_0' + w_1' \sin \pi x + w_2' \cos \pi x + w_3' \sin \pi x + w_4' \cos 2\pi x + w_5' \sin 4\pi x . \qquad (4.6)$$

Clearly, it is a triangular interpolation function.

Figure 9 gives a different functional-link (also function expansion) network whose output is:

$$y = \varphi(w_1 x + w_2 x^2 + w_3 x^3 + \cdots + w_n x^n - \theta) . \qquad (4.7)$$

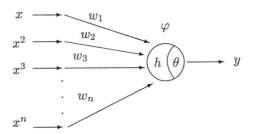

Figure 9   Another function expansion network

Now we if let $w_i = \frac{1}{i!}$ $(i = 1, 2, \cdots, n)$, $\theta = -1$ and φ be an identity function, then according to Maclaurin's expansion of e^x, we have

$$y = \varphi(w_1 x + w_2 x^2 + w_3 x^3 + \cdots + w_n x^n - \theta)$$
$$= 1 + \frac{x}{1!} + \frac{x^2}{2!} + \frac{x^3}{3!} + \cdots + \frac{x^n}{n!} \approx e^x.$$

This example illustrates that a network can express a function; conversely, a function can be also expressed by a network.

Let us consider a general form of functional-link networks (see Figure 10) with only one output.

The output of the network is:

$$y = \varphi(w_1 x_1 + \cdots + w_n x_n + w_{n+1} g_1(x_1, \cdots, x_n) + \cdots + w_{n+m} g_m(x_1, \cdots, x_n) - \theta) . \quad (4.8)$$

Clearly, all the functional-link networks discussed above are special cases of this network.

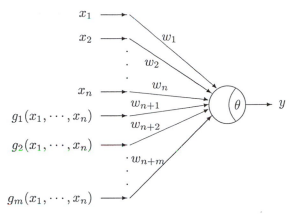

Figure 10   A network for representing Taylor expansion

Although the output function has non-linear terms $g_j(x_1, \cdots, x_n)$, we can make them linear by redefining the variables. In fact, let $z_i = x_i$ $(i = 1, \cdots, n)$, and

$$z_j = g_j(x_1, \cdots, x_n), \quad j = n+1, \cdots, n+m,$$

then the above output function becomes:

$$y = \varphi(w_1 z_1 + w_2 z_2 + \cdots + w_{n+m} z_{n+m} - \theta) . \quad (4.9)$$

This is a linear expression.

For a given group of training samples:

$$\left\{ ((x_1^{(k)}, \cdots, x_n^{(k)}), y_k) \mid k = 1, 2, \cdots, p \right\} , \quad (4.10)$$

assuming φ to be an identity function and

$$z_j^{(k)} = g_j(x_1^{(k)}, \cdots, x_n^{(k)}), \quad j = n+1, \cdots, n+m, \ k = 1, \cdots, p,$$

we have a system of linear equations regarding w_i $(i = 1, 2, \cdots, n+m)$ as unknowns:

$$a_{k1} w_1 + a_{k2} w_2 + \cdots + a_{kq} w_q = y_k, \ k = 1, 2, \cdots, p \quad (4.11)$$

where $a_{kj} = z_j^{(k)}$ $(j = 1, 2, \cdots, n+m)$ and $q = n+m$.

Remark 3.   The functional-link network shown as Figure 10 can also be expressed by a two-layer network shown as Figure 11.

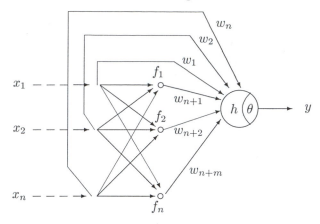

Figure 11   A two-layer network expressing Figure 10

The activation functions φ_j of the neurons f_j are taken as the following:

$$\varphi_j(x_1, \cdots, x_n) \overset{\triangle}{=} g_j(x_1, \cdots, x_n), \; j = 1, 2, \cdots, m.$$

Then the output of the network is given as follows:

$$y = \varphi\left(\sum_{i=1}^{n} w_i x_i + \sum_{j=1}^{m} w_{n+j} g_j(x_1, \cdots, x_n) - \theta \right).$$

This is the same as Equation (4.8). In other words, the function forms in a functional-link network can be perfectly expressed by the activation functions of the neurons in the network.

Remark 4.   In Figure 11, the input signals $x_1, \cdots, x_n$ flow into the neuron h "directly". Of course we can set up "relay stations" to avoid these "direct" connections (see Figure 12).

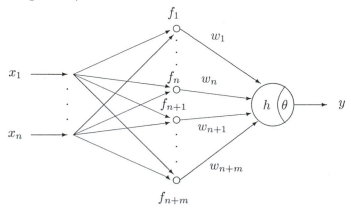

Figure 12   A network without "through trains"

Let the activation functions of the neurons $f_1, \cdots, f_m$ be $\varphi_j\ (x_1, \cdots, x_n)\ (j = 1, \cdots, n+m)$, where

$$\varphi_i(x_1, \cdots, x_n) \overset{\triangle}{=} x_i, \ i = 1, \cdots, n.$$

Then the output of the network is the following:

$$y = \varphi\left(\sum_{i=1}^{n+m} w_j \varphi_j(x_1, \cdots, x_n) - \theta \right)$$

$$= \varphi\left(\sum_{i=1}^{n} w_i x_i + \sum_{j=n+1}^{n+m} w_j \varphi_j(x_1, \cdots, x_n) - \theta \right)$$

which is the same with Expression (4.8).

4.3   As Visualization Means of Some Mathematical Methods

From the above discussion, we realize that neural networks can be used for representing mathematical methods, mathematical forms, or mathematical structures. In other words, neural networks can be regarded as a visualization means of mathematics. We consider the following examples.

Example 1.   *The neural network representation of Taylor expansion*
Given a function $f(x)$ satisfying the condition: $f(x), f'(x), f''(x), \cdots, f^{(n)}(x)$ are continuous in closed interval $[a, b]$ and $f^{(n+1)}(x)$ is existential in open interval (a, b). We design a two-layer forward functional-link neural network with one-input one-output (see Figure 13).

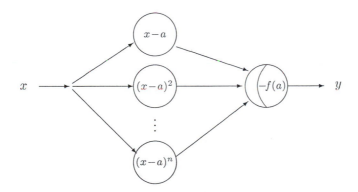

Figure 13   A network for representing Taylor expansion

The activation functions of the neurons in the network are all taken as identity functions, and the threshold values of the neurons at the first layer are all taken as zero and the threshold value of the neuron at the second layer is taken as $-f(a)$. If we let $w_1 = f'(a)$, $w_2 = \frac{f''(a)}{2!}, \cdots, w_n = \frac{f^{(n)}(a)}{n!}$, then the output of the network is:

$$y = \sum_{i=1}^{n} w_i(x-a)^i - (-f(a))$$

$$= f(a) + f'(a)(x-a) + \frac{f''(a)}{2!}(x-a)^2 + \cdots + \frac{f^{(n)}(a)}{n!}(x-a)^n \approx f(x) \ (4.12)$$

This is clearly a Taylor expansion neglecting the remainder term, and equal approximately to $f(x)$. Especially, when $a = 0$, it is just a Maclaurin's expansion neglecting the remainder term.

Remark 5. A Taylor expansion can be expressed as in Figure 14, where the activation functions of the neurons in the network are defined as $\varphi_0(x) \equiv f(a)$, $\varphi_1(x) = x - a$, $\varphi_2(x) = (x-a)^2, \cdots, \varphi_n(x) = (x-a)^n$, $\varphi_{n+1}(u) = u$; the weight values are taken as $w_0 = 1, w_1 = f'(a), w_2 = \frac{f''(a)}{2!}, \cdots, w_n = \frac{f^{(n)}(a)}{n!}$. Then the output of the network is:

$$y = \sum_{i=0}^{n} w_i\varphi_i(x) = f(a) + f'(a)(x-a) + \frac{f''(a)}{2!}(x-a)^2 + \cdots + \frac{f^{(n)}(a)}{n!}(x-a)^n$$

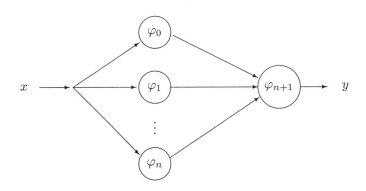

Figure 14   A Taylor expansion network designed by using activation functions

Example 2. *The neural network representation of Weierstrass's first approximation theorem* (see Figure 15).

The activation functions of the neurons in the network are defined as $\varphi_1(x) = \cos x$, $\varphi_2(x) = \cos x2x, \cdots, \varphi_n(x) = \cos nx$, $\psi_1(x) = \sin x$, $\psi_2(x) = \sin 2x, \cdots, \psi_n(x) = \sin nx$; h_1, h_2, g are taken as identity functions. The threshold values of the neurons at the second layer are zero, and the threshold value at the third layer is $-\frac{a_0}{2}$. And

a_k, b_k $(k = 1, 2, \cdots, n)$ are all weight values. Then the output of the network is:

$$T_n(x) = \sum_{k=1}^{n} a_k \varphi_k(x) + \sum_{k=1}^{n} b_k \psi_k(x) - \left(-\frac{a_0}{2} \right)$$

$$= \frac{a_0}{2} + \sum_{k=1}^{n} (a_k \cos kx + b_k \sin kx) \qquad (4.13)$$

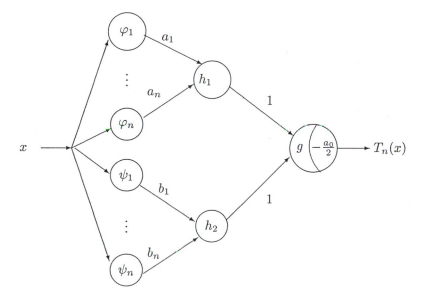

Figure 15   A network representing Weierstrass's first approximation theorem

This is the trigonometric polynomial in Weierstrass's first approximation theorem. In other words, for any a continuous function $f(x)$ with period 2π, and for any a positive real number $\varepsilon > 0$, there exists a network shown in Figure 15 such that $|f(x) - T_n(x)| < \varepsilon$ holds uniformly on the whole number axis.

4.4   Neural Network Representation of Linear Programming

A one-layer forward neural network is given in Figure 16, where the activation functions φ_i of the neurons f_i are all taken as step-up functions:

$$\varphi_i(u) = \begin{cases} 1, & u \geqslant 0 \\ 0, & u < 0, \end{cases} \quad i = 0, 1, \cdots, m \qquad (4.14)$$

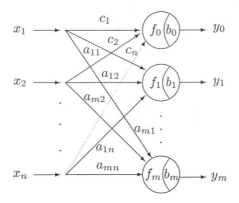

Figure 16   A network representing linear programming

and their threshold values are b_0, b_1, $\cdots$, b_m; the weight values in the network are c_j and $a_{ij}(i = 1, 2, \cdots, m,\ j = 1, 2, \cdots, n)$. For given input $(x_1, x_2, \cdots, x_n)$, the components of the output $(y_0, y_1, \cdots, y_m)$ are:

$$y_0 = \varphi_0 \left(\sum_{j=1}^{n} c_j x_j - b_0 \right), \quad y_i = \varphi_i \left(\sum_{j=1}^{n} a_{ij} x_j - b_i \right), \quad i = 1, \cdots, m \ . \qquad (4.15)$$

Problem 1. Let all weight values be known and the threshold values b_1, b_2, $\cdots$, b_m be given. We want to find the threshold value b_0 such that the neuron f_0 is inhibitory when the neurons $f_1, f_2, \cdots, f_m$ are all excited.

From Expressions (4.14) and (4.15), it is easy to realize that the problem is a linear programming problem described as follows.

$$\max \sum_{j=1}^{n} c_j x_j$$

$$\text{s.t.} \ \sum_{j=1}^{n} a_{ij} x_j \geqslant b_i, \quad i = 1, 2, \cdots, m \qquad (4.16)$$

$$x_j \geqslant 0, \quad j = 1, 2, \cdots, n \ .$$

If the set of feasible solutions is not empty, we should take $b_0 > \max \sum_{j=1}^{n} c_j x_j$ for the threshold value.

Problem 2. Under the same condition as in problem 1, we want to find the threshold value b_0 such that the neuron f_0 is inhibitory when the neurons f_1, f_2, $\cdots$, f_m are inhibitory.

Clearly, the problem is equivalent to the following linear programming problem:

$$\max \sum_{j=1}^{n} c_j x_j$$

$$\text{s.t.} \ \sum_{j-1}^{n} a_{ij} x_j \leqslant b_i, \quad i = 1, 2, \cdots, m \qquad (4.17)$$

$$x_j \geqslant 0, \quad j = 1, 2, \cdots, n \ .$$

Nevertheless, the activation functions φ_i (see Expression (4.14)) have a little difference except φ_0:

$$\varphi_i(u) = \begin{cases} 1, & u > 0 \\ 0, & u \leqslant 0, \end{cases} \quad i = 1, 2, \cdots, m \ . \tag{4.18}$$

Problem 3.   Under the same condition, we want to find b_0 such that f_0 is excited when $f_1, f_2, \cdots, f_m$ are excited.

The problem can be expressed as the following linear programming problem:

$$\min \sum_{j=1}^{n} c_j x_j$$
$$\text{s.t.} \sum_{j=1}^{n} a_{ij} x_j \geqslant b_i, \quad i = 1, 2, \cdots, m \tag{4.19}$$
$$x_j \geqslant 0, \quad j = 1, 2, \cdots, n \ .$$

If the set of feasible solutions is not empty, then we should take $b_0 \leqslant \min \sum_{j=1}^{n} c_j x$ as the threshold value.

Problem 4.   Under the same condition, we want to find b_0 such that f_0 is excited when $f_1, f_2, \cdots, f_m$ are inhibitory.

Of course, the problem is corresponding to the following linear programming problem:

$$\min \sum_{j=1}^{n} c_j x_j$$
$$\text{s.t.} \sum_{j=1}^{n} a_{ij} x_j < b_i, \quad i = 1, 2, \cdots, m \tag{4.20}$$
$$x_j \geqslant 0, \quad j = 1, 2, \cdots, n \ .$$

Similar to Problem 2, the activation functions $\varphi_i (i = 1, 2, \cdots, m)$ should be taken as Expression (4.18).

Problem 5.   *Inverse problem of linear programming*

As an example, we only consider the inverse problem of Problem 3 (other cases are similar). Given the threshold values $b_i (i = 0, 1, \cdots, m)$ and a group of training samples:

$$\left\{ X_k = (x_1^{(k)}, x_2^{(k)}, \cdots, x_n^{(k)})^T \mid k = 1, 2, \cdots, p \right\} \tag{4.21}$$

satisfying, for every k,

$$\min \sum_{j=1}^{n} c_j x_j = \sum_{j=1}^{n} c_j x_j^{(k)} = b_0$$

$$\text{s.t.} \sum_{j=1}^{n} a_{ij} x_j^{(k)} \geqslant b_i, \quad i = 1, 2, \cdots, m \qquad (4.22)$$

$$x_j(k) \geqslant 0, \quad j = 1, 2, \cdots, n$$

we want to find the weight values c_j and $a_{ij} (i = 1, 2, \cdots, m, \; j = 1, 2, \cdots, n)$.

In fact, if we use "surplus weight values" $a_{in+1} \geqslant 0 \, (i = 1, 2, \cdots, m)$ to force that

$$\sum_{j=1}^{n} a_{ij} x_j^{(k)} - a_{in+1} = b_i, \quad i = 1, 2, \cdots, m \qquad (4.23)$$

then Expression (4.22) becomes a "normal form" :

$$\min \sum_{j=1}^{n} c_j x_j = \sum_{j=1}^{n} c_j x_j^{(k)} = b_0$$

$$\text{s.t.} \sum_{j=1}^{n} a_{ij} x_j^{(k)} - a_{in+1} = b_i, \quad i = 1, 2, \cdots, m \qquad (4.24)$$

$$x_j^{(k)} \geqslant 0, \quad j = 1, 2, \cdots, n$$

This is the inverse problem of linear programming and can be solved by means of learning algorithms of neural networks.

Problem 6.   *The neural network representation of linear programming with several objects*

In Figure 17, activation functions of all neurons are taken as step-up functions. Given weight values c_{sj} and $a_{ij} (s = 1, 2, \cdots, t, \; i = 1, 2, \cdots, m, \; j = 1, 2, \cdots, n)$, and threshold values $b_i (i = 1, 2, \cdots, m)$, we want to find threshold values $d_s (s = 1, 2, \cdots, t)$ such that the neurons $g_1, g_2, \cdots, g_t$ are inhibitory when the neurons $f_1, f_2, \cdots, f_m$ are excited.

The problem can be expressed by the following linear programming with several objects:

$$\max \sum_{j=1}^{n} c_{sj} x_j, \quad s = 1, 2, \cdots, t$$

$$\text{s.t.} \sum_{j=1}^{n} a_{ij} x_j \geqslant b_i, \quad i = 1, 2, \cdots, m \qquad (4.25)$$

$$x_j \geqslant 0, \quad j = 1, 2, \cdots, n \; .$$

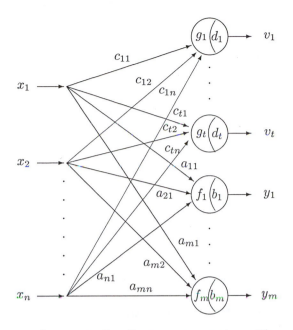

Figure 17   A network representing linear programming with several objects

If the set of feasible solutions is not empty, we should take $d_s > \max \sum_{j=1}^{n} c_{sj} x_j$ ($s = 1, 2, \cdots, t$).

Under the same condition as problem 6, the following problems can be formulated similarly. We will not discuss here.

Problem 7. Find threshold values $d_s (s = 1, 2, \cdots, t)$ such that $g_1, \cdots, g_t$ are inhibitory when $f_1, \cdots, f_m$ are inhibitory.

Problem 8. Find threshold $d_s (s = 1, 2, \cdots, t)$ such that $g_1, \cdots, g_t$ are excited when $f_1, \cdots, f_m$ are excited.

Problem 9. Find threshold $d_s (s = 1, 2, \cdots, t)$ such that $g_1, \cdots, g_t$ are excited when $f_1, \cdots, f_m$ are inhibitory.

Moreover, the following problem gives the "mixed" cases. We will not discuss in detail here.

Problem 10. Find $d_s (s = 1, 2, \cdots, t)$ such that $g_1, \cdots, g_q$ are excited but $f_{r+1}, \cdots, f_m$ inhibitory ($1 \leqslant r < m$).

Problem 11. *Inverse problem of linear programming with several objects*
As an example, we only consider the case in Problem 6 (other cases are similiar). Given threshold values $d_s (s = 1, 2, \cdots, t)$ and $b_i (i = 1, 2, \cdots, m)$ and a group of

training samples like Expression (6.21) satisfying, for every $k(k = 1, 2, \cdots, p)$,

$$\max \sum_{j=1}^{n} c_{sj}x_j = \sum_{j=1}^{n} c_{sj}x_j^{(k)} = d_s, \quad s = 1, 2, \cdots, t$$

$$\text{s.t.} \sum_{j=1}^{n} a_{ij}x_j^{(k)} \geqslant b_i, \quad i = 1, 2, \cdots, m \qquad (4.26)$$

$$x_j^{(k)} \geqslant 0, \quad j = 1, 2, \cdots, n$$

we want to find weight values c_{sj} and $a_{ij}(s = 1, 2, \cdots, t, \; i = 1, 2, \cdots, m, \; j = 1, 2, \cdots, n)$.

4.5   Neural Network Representation of Fuzzy Linear Programming

Based on the network shown in Figure 16 and Problem 1 (other cases are similar), we introduce neural network representation of fuzzy linear programming. As a matter of fact, if the activation functions $\varphi_1, \varphi_2, \cdots, \varphi_m$ of the neurons $f_1, f_2, \cdots, f_m$ are taken as membership functions, i.e., $\varphi_i : R \to [0, 1]$, $i = 1, 2, \cdots, m$, (for convenience, φ_i themselves can be regarded as fuzzy sets, i.e., $\varphi_i \in \mathcal{F}(R)$, $i = 1, 2, \cdots, m$), then the network shown in Figure 16 is a fuzzy neural network, where R is the real number field. Similar to linear programming networks, the m fuzzy activation functions $\varphi_i(i = 1, 2, \cdots, m)$ define m "elastic constraints" (i.e., m fuzzy constraints). Actually, we can consider the problem by such a way that, when $\varphi_1, \cdots, \varphi_m$ are "crisp" functions (i.e., step-up functions), the m constraints are just m inequalities expressed as the following:

$$\varphi_i\left(\sum_{j=1}^{n} a_{ij}x_j - b_i\right) = \begin{cases} 1, & \sum_{j=1}^{n} a_{ij}x_j \geqslant b_i \\ 0, & \sum_{j=1}^{n} a_{ij}x_j < b_i \end{cases} \qquad (4.27)$$
$$i = 1, 2, \cdots, m$$

which are characteristic functions to reflect the law of excluded middle with respect to "$\sum_{j=1}^{n} a_{ij}x_j \geqslant b_i (i = 1, 2, \cdots, m)$."

When $\varphi_1, \cdots, \varphi_m$ are fuzzy sets, the law of excluded middle reflected by Expression (4.27) no longer holds, thus "$\geqslant$" becomes "fuzzy $\geqslant$" denoted by "$\underset{\sim}{\geqslant}$". So

Expression (4.16) should be written as follows:

$$\max \sum_{j=1}^{n} c_j x_j$$

$$\text{s.t.} \sum_{j=1}^{n} a_{ij} x_j \underset{\sim}{\geqslant} b_i, \quad i = 1, 2, \cdots, m \tag{4.28}$$

$$x_j \geqslant 0, \quad j = 1, 2, \cdots, n \ .$$

This is a fuzzy linear programming with one object.

Example 3. In Expression (4.28), the activation functions $\varphi_1, \cdots, \varphi_m$ can be taken as the following:

$$\varphi_i(u) = (1 + e^{-(\alpha_i u + \beta_i)})^{-1}, \quad i = 1, 2, \cdots, m \tag{4.29}$$

where the parameters α_i and β_i can control the slope and parallel translation of the curve of $\varphi_i(u)$, $i = 1, 2, \cdots, m$.

Example 4. An alternate $\varphi_i(u)$ can be taken as the following simple form (see Figure 18):

$$\varphi_i(u) = \begin{cases} 1, & u \geqslant 0 \\ \dfrac{u + e_i}{e_i}, & -e_i \leqslant u < 0, \\ 0, & u < -e_i \end{cases} \quad (e_i > 0; \ i = 1, 2, \cdots, m) \ .$$

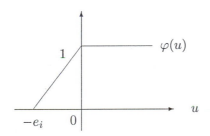

Figure 18   The polygonal form of $\varphi_i(u)$

4.6   Conclusions

In this chapter, we studied the modeling of diverse mathematical problems using neural networks. We first studied the functional-link neural network and extended the idea to model other mathematical functions. This modeling gives us a visualization means of mathematical functions. In this way, hardware realization is possible. We started with the well-known XOR problem, followed by a discussion of Taylor series expansion and Weierstrass's first approximation theorem. The neural representations of linear programming and fuzzy linear programming were also discussed. Using such representations, it is possible for us to overcome existing limitations. It also enables us to find new solutions through alternatives and to achieve synergistic effects through hybridization.

References

1. M. L. Minsky and S. A. Papert, *Perceptron* (expanded ed.), MIT Press, Cambridge, 1998.

2. C. T. Lin and C. S. G. Lee, *Neural Fuzzy Systems*, Prentice-Hall, Englewood Cliffs, 1997.

3. J. Hertz, A. Krogh, and R. G. Palmer, *Introduction to the Theory of Neural Computation*, Addison-Wesley, New York, 1991.

4. B. Kosko, *Neural Networks and Fuzzy Systems*, Prentice-Hall, Englewood Cliffs, 1992.

5. Y. H. Pao, *Adaptive Pattern Recognition and Neural Networks*, Addison-Wesley, New York, 1989.

Chapter 5

Flat Neural Networks and Rapid Learning Algorithms

In this chapter, we will introduce flat neural networks architecture. The system equations of flat neural networks can be formulated as a linear system. In this way, the performance index is a quadratic form of the weights, and the weights of the networks can be solved easily using a linear least-square method. Even though they have a linear-system-equations-like equation, the flat neural networks are also perfect for approximating non-linear functions. A fast learning algorithm is given to find an optimal weight of the flat neural networks. This formulation makes it easier to update the weights instantly for both a newly added input and a newly added node. A dynamic stepwise updating algorithm is given to update the weights of the system instantly. Finally, we give several examples of applications of the flat neural networks, such as an infrared laser data set, a chaotic time-series, a monthly flour price data set, and a non-linear system identification problem. The simulation results are compared to existing models in which more complex architectures and more costly training are needed. The results indicate that the flat neural networks are very attractive to real-time processes.

5.1 Introduction

Feedforward artificial neural networks have been a popular research subject recently. The research topics vary from the theoretical view of learning algorithms such as learning and generalization properties of the networks to a variety of applications in control, classification, biomedical, manufacturing, and business forecasting, etc. The backpropagation (BP) supervised learning algorithm is one of the most popular learning algorithms being developed for layered networks [1-2]. Improving the learning speed of BP and increasing the generalization capability of the networks have played a center role in neural network research [3-9]. Apart from multi-layer network architectures and the BP algorithm, various simplified architectures or different non-linear activation functions have been devised. Among those, so-called flat networks including functional-link neural networks and radial basis function networks have been proposed [10-15]. These flat networks remove the drawback of a long learning process with the advantage of learning only one set of weights. Most importantly, the literature has reported satisfactory generalization capability in function approximation [14-16].

This chapter discusses the flat networks along with an one-step fast learning algorithm and a stepwise update algorithm for training the flat networks. Although only the functional-link network is used as a prototype here, the algorithms can also be applied to the radial basis function network. The algorithms are developed based on the formulation of the functional-link network that has a set of linear system equations. Because the system equations of the radial basis function network have a similar form to the functional-link network and both networks share similar "flat" architecture, the update algorithm can be applied to the radial basis function network as well. The most significant advantage of the stepwise approach is that the weight connections of the network can be updated easily, when a new input is given later after the network has been trained. The weights can be updated easily based on the original weights and the new inputs. The stepwise approach is also able to update weights instantly when a new neuron is added to the existing network if the desired error criterion cannot be met. With this learning algorithm, the flat networks become very attractive in terms of learning speed.

Finally, the flat networks are used for several applications. These include an infrared laser data set, a chaotic time-series, a monthly flour price data set, and a non-linear system identification. The time-series is modeled by the AR(p) (Auto-Regression with p delay) model. During the training stage, a different number of nodes may be added as necessary. The update of weights is carried by the given algorithm. Contrary to the traditional BP learning and multi-layer models, the training of this network is fast because of an one-step learning procedure and the dynamic updating algorithm. We also applied the networks to non-linear system identification problems involving discrete-time single-input, single-output (SISO), and multiple-input, multiple-output (MIMO) plants which can be described by the difference equations [16]. With this learning algorithm, the training is easy and fast. The result is also very promising.

This chapter is organized as follows, wherein Section 2 briefly discusses the concept of the functional-link and its linear formulation. Sections 3 and 4 introduce the dynamic stepwise update algorithm followed by the refinement of the model in section 5. Section 6 discusses the procedures of the training. Finally, several examples and conclusions are given.

5.2 The Linear System Equation of the Functional-link Network

Figure 1 illustrates the characteristic flatness feature of the functional-link network. The network consists of a number of "enhancement" nodes. These enhancement nodes are used as extra inputs to the network. The weights from input nodes to the enhancement nodes are randomly generated and fixed thereafter. To be more precise, an enhancement node is constructed by first taking a linear combination of the input nodes, and then applying a nonlinear activation function $\xi(.)$ to it. This model has been discussed elsewhere by Pao [10]. A rigorous mathematical proof has also been given by Igelnik and Pao [12]. The literature has also discussed the advantage of the functional-link network in terms of training speed and its generalization property over the general feedforward networks [11]. In general, the functional-link network with k enhancement nodes can be represented as an equation of the form:

$$\mathbf{Y} = [\mathbf{x}|\xi(\mathbf{x}\mathbf{W}_h + \beta_h)]\mathbf{W} \qquad (5.1)$$

where W_h is the enhancement weight matrix, which is randomly generated, $\mathbf{W}$ is the weight matrix that needs to be trained, β_h is the bias function, $\mathbf{Y}$ is the output matrix, and $\xi(.)$ is a non-linear activation function. The activation function can be either a sigmoid or a *tanh* function. If the β_h term is not included, an additional constant bias node with -1 or +1 is needed. This will cover even function terms for function approximation applications, which have been explained using Taylor series expansion in Chen [17].

Denoting by $\mathbf{A}$ the matrix $[\mathbf{x}|\xi(\mathbf{x}\mathbf{W}_h + \beta_h)]$, where $\mathbf{A}$ is the expanded input matrix consisting of all input vectors combined with enhancement components, yields:

$$\mathbf{Y} = \mathbf{A}\mathbf{W} . \qquad (5.2)$$

The structure is illustrated in Figure 2.

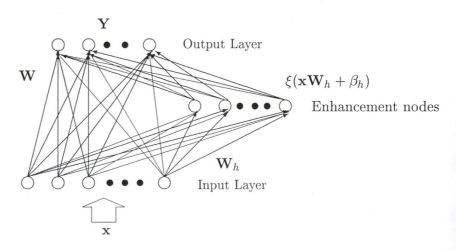

Figure 1   A flat Functional-link neural network

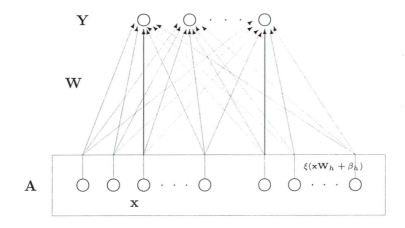

Figure 2   A Linear formulation of Functional-link network

5.3   Pseudoinverse and Stepwise Updating

Pao implemented a conjugate gradient search method that finds the weight matrix, $\mathbf{W}$ [11]. This chapter discusses a rapid method of finding the weight matrix. To learn the optimal weight connections for the flat network, it is essential to find the least-square solution of the equation, $\mathbf{AW} = \mathbf{Y}$. Recall that the least-square solution to the equation, $\mathbf{Y} = \mathbf{AW}$, is $\mathbf{W} = \mathbf{A}^{+}\mathbf{Y}$, where $\mathbf{A}^{+}$ is the pseudoinverse of matrix $\mathbf{A}$. To find the best weight matrix $\mathbf{W}$, the REIL (Rank Expansion with Instant Learning) algorithm is described in the following [17].

Algorithm Rank-Expansion with Instant Learning (REIL).

Input: The extended input pattern matrix, $\mathbf{A}$, and the output matrix, $\mathbf{Y}$, where N is the number of the input patterns.

Output: The weight matrix, $\mathbf{W}$, and the neural network.

Step 1. Add k hidden nodes and assign random weights, $k \leq N - r$, where r is the rank of $\mathbf{A}$.

Step 2. Solve weight, $\mathbf{W}$, by minimizing $||\mathbf{AW} - \mathbf{Y}||_2$.

Step 3. If mean-squared error criterion is not met, add additional nodes and go to Step 2; otherwise, stop.

End of Algorithm REIL

The computation complexity of this algorithm comes mostly from the time spent in Step 2. There are several methods for solving least-squares problems [18]. The complexity of FLOP count is the order of $O(Nq^2 + q^3)$, where N is the number of rows in the training matrix, and q is the number of columns. The singular value decomposition is the most common approach. Compared with gradient descent search, the least-squares method is time efficient [19].

The above algorithm is a batch algorithm in which we assume that all the input data are available at the time of training. However, in a real-time application, as a new input pattern is given to the network, the $\mathbf{A}$ matrix must be updated. It is not efficient at all if we continue using the REIL algorithm. We must pursue an alternative approach. Here we take advantage of the flat structure in which extra nodes can be added and the weights can be found very easily if necessary. In addition, weights can be easily updated without running a complete training cycle when either one or more new enhancement nodes are added, or more observations are available. The stepwise updating of the weight matrix can be achieved by taking the pseudoinverse of a partitioned matrix described below [20, 21]. Let us denote prime ($'$) as the transpose of a matrix. Let $\mathbf{A}_n$ be the $n \times m$ pattern matrix defined above, and $\mathbf{a}'$ be the $m \times 1$ new pattern entered to the neural network. Here the subscript denotes the discrete time instance. Denote $\mathbf{A}_{n+1}$ as the following.

$$\mathbf{A}_{n+1} \triangleq \begin{bmatrix} \mathbf{A}_n \\ \mathbf{a}' \end{bmatrix},$$

then the theorem states that the pseudoinverse of the new matrix $\mathbf{A}_{n+1}$ is

$$\mathbf{A}_{n+1}^+ = [\mathbf{A}_n^+ - \mathbf{b}\mathbf{d}'|\mathbf{b}],$$

where

$$\mathbf{d}' = \mathbf{a}'\mathbf{A}_n^+ \text{ and } \mathbf{b} = \begin{cases} (\mathbf{c}')^+, & if \ \mathbf{c} \neq 0, \\ (1 + \mathbf{d}'\mathbf{d})^{-1}\mathbf{A}_n^+\mathbf{d}, & if \ \mathbf{c} = 0, \end{cases}$$

where

$$\mathbf{c}' = \mathbf{a}' - \mathbf{d}'\mathbf{A}_n.$$

In other words, the pseudoinverse of $\mathbf{A}_{n+1}$ can be obtained through $\mathbf{A}_n^+$ and the added row vector $\mathbf{a}'$. A noteworthy fact is that, if $n > m$ and $\mathbf{A}_n$ is of full rank, then $\mathbf{c} = 0$. This can be shown as follows. If $\mathbf{A}_n$ is of full rank and $n > m$, then

$$\mathbf{A}_n^+ = (\mathbf{A}_n'\mathbf{A}_n)^{-1}\mathbf{A}_n',$$

therefore

$$\begin{aligned} \mathbf{c}' &= \mathbf{a}' - \mathbf{d}'\mathbf{A}_n \\ &= \mathbf{a}' - \mathbf{a}'(\mathbf{A}_n'\mathbf{A}_n)^{-1}\mathbf{A}_n'\mathbf{A}_n \\ &= \mathbf{a}'(\mathbf{I} - (\mathbf{A}_n'\mathbf{A}_n)^{-1}\mathbf{A}_n'\mathbf{A}_n) \\ &= 0. \end{aligned}$$

So the pseudoinverse of $\mathbf{A}_{n+1}$ can be updated based only on $\mathbf{A}_n^+$, and the new added row vector $\mathbf{a}'$ without recomputing the entire new pseudoinverse.

Let the output vector $\mathbf{Y}_{n+1}$ be partitioned as $\mathbf{Y}_n$,

$$\mathbf{Y}_{n+1} \triangleq \begin{bmatrix} \mathbf{Y}_n \\ \mathbf{y}' \end{bmatrix},$$

where $\mathbf{y}'$ is the new output corresponding to the new input $\mathbf{a}'$ and let

$$\mathbf{W}_{n+1} = \mathbf{A}_{n+1}^+ \mathbf{Y}_{n+1}, \quad \mathbf{W}_n = \mathbf{A}_n^+ \mathbf{Y}_n.$$

Then according to the above equations, the new weight, $\mathbf{W}_{n+1}$, can be found as below.

$$\mathbf{W}_{n+1} = \mathbf{W}_n - (\mathbf{y}' - \mathbf{a}'\mathbf{W}_n)\mathbf{b} . \tag{5.3}$$

Equation (5.3) has the same form with the recursive least-square solution, if $\mathbf{c} = 0$. However, Equation (5.3) considers the case if $\mathbf{A}_n$ is not the full-rank (i.e., $\mathbf{c} \neq 0$). Compared to the Least Mean Square (LMS) learning rule [22], Equation (5.3) has the optimal learning rate, $\mathbf{b}$, which leads the learning in one-step update, rather than iterative update. The stepwise updating in flat networks is also perfect for adding a new enhancement node to the network. In this case, it is equivalent to add a new column to the input matrix $\mathbf{A}_n$. Denote $\mathbf{A}_{n+1} \overset{\triangle}{=} [\mathbf{A}_n|\mathbf{a}]$. Then the pseudoinverse of the new $\mathbf{A}_{n+1}^+$ equals

$$\begin{bmatrix} \mathbf{A}_n - \mathbf{d}\mathbf{b}' \\ \mathbf{b}' \end{bmatrix},$$

where $\mathbf{d} = \mathbf{A}_n^+\mathbf{a}$,

$$\mathbf{b}' = \begin{cases} (\mathbf{c})^+, & if \ \mathbf{c} \neq 0, \\ (1 + \mathbf{d}'\mathbf{d})^{-1}\mathbf{A}_n^+, & if \ \mathbf{c} = 0, \end{cases}$$

and $\mathbf{c} = \mathbf{a} - \mathbf{A}_n\mathbf{d}$. Again the new weights are

$$\mathbf{W}_{n+1} = \begin{bmatrix} \mathbf{W}_n - \mathbf{d}\mathbf{b}'\mathbf{Y}_n \\ \mathbf{b}'\mathbf{Y}_n \end{bmatrix}, \tag{5.4}$$

where $\mathbf{W}_{n+1}$ and $\mathbf{W}_n$ are the weights after and before a new neuron is added, respectively. Since a new neuron is added to the existing network, the weights, $\mathbf{W}_{n+1}$, have one more dimension than $\mathbf{W}_n$. Also note again that, if $\mathbf{A}_n$ is of the full rank, then $\mathbf{c} = 0$ and no computation of pseudoinverse is involved in updating the pseudoinverse $\mathbf{A}_n^+$ or weight matrix $\mathbf{W}_n$.

The one-step dynamic learning is shown in Figure 3. This raises the question of the rank of input matrix $\mathbf{A}_n$. As can be seen from the above discussion, it is desirable to maintain the full rank condition of $\mathbf{A}_n$ when adding rows and columns. The rows consist of training patterns. In other words, it is practically impossible to observe any rank deficient matrix. Thus, during the training of the network, it is our advantage to make sure that the added nodes will increase the rank of input matrix. Also if the matrix becomes numerically rank deficient based on the adjustable tolerance on the singular values, we should consider removing the redundant input nodes. This is discussed in more detail in Section 5 on principal component analysis (PCA) related topics.

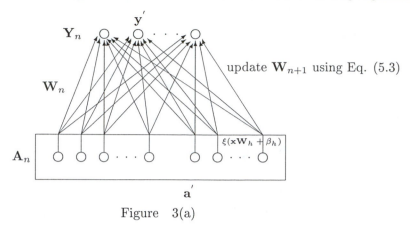

Figure   3(a)

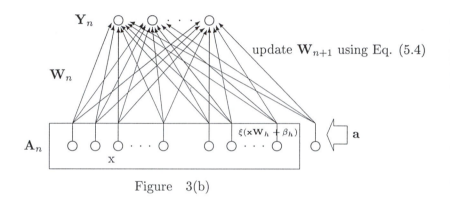

Figure   3(b)

Figure 3   Illustration of Stepwise Update algorithm

Another approach to achieve stepwise updating is to maintain the Q-R decomposition of the input matrix $\mathbf{A}_n$. The updating of the pseudoinverse (and therefore the weight matrix) involves only a multiplication of finitely sparse matrices and backward substitutions. Suppose we have the Q-R decomposition of $\mathbf{A}_n$ and denote $\mathbf{A}_n \stackrel{\triangle}{=} \mathbf{QR}$, where $\mathbf{Q}$ is an orthogonal matrix and $\mathbf{R}$ is an upper triangular matrix. When a new row or a new column is added, the Q-R decomposition can be updated based on a finite number of Givens rotations [18]. Denote $\mathbf{A}_{n+1} \stackrel{\triangle}{=} \hat{\mathbf{Q}}\hat{\mathbf{R}}$ where, $\hat{\mathbf{Q}}$ remains orthogonal, $\hat{\mathbf{R}}$ is an upper triangular matrix, and both are obtained through finitely many Givens rotations. The pseudoinverse of $\mathbf{A}_{n+1}$ is

$$\mathbf{A}_{n+1}^{+} \stackrel{\triangle}{=} \hat{\mathbf{R}}^{+} \hat{\mathbf{Q}}',$$

where $\hat{\mathbf{R}}^{+}$ (eventually, $\mathbf{W}_{n+1}$) can be computed by backward substitution. This stepwise weight update using Q-R and Givens rotation matrix is summarized in the following algorithm.

Q-R Implementation of Weights Matrix Updating

Input: $\mathbf{A}_n \overset{\triangle}{=} \mathbf{QR}$, vector a, and weight matrix $\mathbf{W}_n$, where $\mathbf{A}_n$ is a $n \times m$ matrix, $\mathbf{Q}$ is a $n \times n$ orthogonal matrix, $\mathbf{R}$ is a $n \times m$ upper triangular matrix, and $\mathbf{a}'$ is a $m \times 1$ row vector.

Output: $\mathbf{A}_{n+1} \overset{\triangle}{=} \begin{bmatrix} \mathbf{A}_n \\ \mathbf{a}' \end{bmatrix} \overset{\triangle}{=} \hat{\mathbf{Q}}\hat{\mathbf{R}}$ and weight matrix $\mathbf{W}_{n+1}$, where $\mathbf{A}_{n+1}$ is a $(n+1) \times m$ matrix, $\hat{\mathbf{Q}}$ is a $(n+1) \times (n+1)$ orthogonal matrix, and $\hat{\mathbf{R}}$ is a $(n+1) \times m$ upper triangular matrix.

Step 1. Expand $\hat{\mathbf{Q}}$ and $\hat{\mathbf{R}}$, i.e.,

$$\hat{\mathbf{Q}} \leftarrow \mathrm{diag}(\mathbf{Q}, 1) \overset{\triangle}{=} \begin{bmatrix} \mathbf{Q} & 0 \\ 0 & 1 \end{bmatrix},$$

$$\hat{\mathbf{R}} \leftarrow \begin{bmatrix} \mathbf{R} \\ \mathbf{a}' \end{bmatrix}.$$

Step 2. For $i = 1$ to m, do

$$\mathbf{J}_i = Rot(\mathbf{r}_i, \mathbf{e}_i),$$
$$\hat{\mathbf{Q}} \leftarrow \hat{\mathbf{Q}} * \mathbf{J}_i,$$
$$\hat{\mathbf{R}} \leftarrow \mathbf{J}_i * \hat{\mathbf{R}}.$$

Step 3. Since $\hat{\mathbf{R}}$ is an upper triangular matrix, the new $\mathbf{W}_{n+1}$ can be easily obtained by solving $\hat{\mathbf{R}} \, \mathbf{W}_{n+1} = \hat{\mathbf{Q}}' \begin{bmatrix} \mathbf{Y}_n \\ \mathbf{y}' \end{bmatrix}$ using backward substitution.

End of the Q-R Weight-Updating Algorithm

In Step 2, $\mathbf{J}_i$ is the Givens rotation matrix, $\mathbf{r}_i$ is the i-th column of $\hat{\mathbf{R}}$ and $\mathbf{e}_i$ is a column vector identical to $\mathbf{r}_i$ except the i-th and $(n+1)$-th components. The $n+1$ component is 0. *Rot* performs a plane rotation from vector $\mathbf{r}_i$ to $\mathbf{e}_i$. An example should make this clear. If $\mathbf{r}_2 = (1, 3, 0, 0, 4)'$ then $\mathbf{J}_2$ rotates $\mathbf{r}_2$ to $\mathbf{e}_2 = (1, 5, 0, 0, 0)'$. In fact, $\mathbf{J}_2$ transforms the plane vector (3,4) to (5,0) and keeps other components unchanged. The resulting $\hat{\mathbf{R}}$ is an upper triangular matrix while $\hat{\mathbf{Q}}$ remains orthogonal.

Similarly, with a few modifications, the above algorithm can be used to update the new weight matrix if a new column (a new node) is added to the network.

5.4   Training with Weighted Least Squares

In training and testing the fitness of a model, error is minimized in the sense of least mean-squares, that is in general,

$$E = \frac{1}{N}\|\mathbf{Y} - \mathbf{AW}\|_2 \ , \tag{5.5}$$

where N is the number of patterns. In other words, the average difference between network output and actual output is minimized over the span of the whole training data. If an overall fit is hard to achieve, it might be reasonable to train the network so that it achieves a better fit for most recent data. This leads to the so-called weighted least-squares problem. The stepwise updating of weight matrix based on weighted least-squares is derived as follows.

Let $\mathbf{K}_n = \mathrm{diag}(\Theta^{n-1}, \Theta^{n-2}, \cdots, \Theta, 1)$ be the weight factor matrix. Also, let $\mathbf{A}_n$ represent input matrix with n patterns and $\mathbf{A}_{n+1}$ is $\mathbf{A}_n$ with an added new row, that is,

$$\mathbf{A}_{n+1} \overset{\triangle}{=} \begin{bmatrix} \mathbf{A}_n \\ \mathbf{a}' \end{bmatrix}.$$

Then the weighted least-squares error for the equation $\mathbf{A}_n \mathbf{W}_n = \mathbf{Y}_n$ is:

$$E = \frac{1}{N}\|\mathbf{K}_n(\mathbf{Y}_n - \mathbf{A}_n \mathbf{W}_n)\|_2 \ . \tag{5.6}$$

With

$$\mathbf{K}_{n+1} \overset{\triangle}{=} \mathrm{diag}(\Theta^n, \Theta^{n-1}, \cdots, \Theta, 1) \overset{\triangle}{=} \mathrm{diag}(\Theta \mathbf{K}_n, 1) \ , \tag{5.7}$$

we have

$$\mathbf{K}_{n+1}^{1/2}\mathbf{A}_{n+1} = \begin{bmatrix} \Theta^{1/2}\mathbf{K}_n^{1/2} & 0 \\ 0 & 1 \end{bmatrix}\begin{bmatrix} \mathbf{A}_n \\ \mathbf{a}' \end{bmatrix} = \begin{bmatrix} \Theta^{1/2}\mathbf{K}_n^{1/2}\mathbf{A}_n \\ \mathbf{a}' \end{bmatrix} \ . \tag{5.8}$$

The weighted least-squares solution can be represented as

$$\mathbf{W}_n = (\mathbf{K}_n^{1/2}\mathbf{A}_n)^{+}\mathbf{K}_n^{1/2}\mathbf{Y}_n \ . \tag{5.9}$$

If $\mathbf{S}_n \overset{\triangle}{=} (\mathbf{K}_n^{1/2}\mathbf{A}_n)^{+}$ is known and a new pattern (i.e., a new row $\mathbf{a}'$) is imported to the network, then the weighted pseudoinverse of matrix $\mathbf{S}_{n+1} \overset{\triangle}{=} (\mathbf{K}_{n+1}^{1/2}\mathbf{A}_{n+1})^{+}$ can be updated by

$$\begin{aligned} \mathbf{S}_{n+1} &= [(\Theta^{1/2}\mathbf{K}_n^{1/2}\mathbf{A}_n)^{+} - \mathbf{bd}'|\mathbf{b}] \\ &= [\Theta^{-1/2}\mathbf{S}_n - \mathbf{bd}'|\mathbf{b}] \ , \end{aligned} \tag{5.10}$$

where,

$$\begin{aligned} \mathbf{d}' &= \mathbf{a}'(\Theta^{1/2}\mathbf{K}_n^{1/2}\mathbf{A}_n)^{+} \\ &= \Theta^{-1/2}\mathbf{a}'\mathbf{S}_n, \end{aligned}$$

$$
\mathbf{b} = \begin{cases}
(\mathbf{c}')^{+}, & \textit{if } \mathbf{c} \neq 0 \\
(1 + \mathbf{d}'\mathbf{d})^{-1}(\Theta^{1/2}\mathbf{K}_n^{1/2}\mathbf{A}_n)^{+}\mathbf{d} & \textit{if } \mathbf{c} = 0 \\
= (1 + \Theta^{-1}\mathbf{a}'\mathbf{S}_n\mathbf{S}_n'\mathbf{a})^{-1}\Theta^{-1/2}\mathbf{S}_n\Theta^{-1/2}\mathbf{S}_n'\mathbf{a} \\
= (\Theta + \mathbf{a}'\mathbf{S}_n\mathbf{S}_n'\mathbf{a})^{-1}\mathbf{S}_n\mathbf{S}_n'\mathbf{a},
\end{cases}
$$

$$
\mathbf{c}' = \mathbf{a}' - \mathbf{d}'\mathbf{A}_n.
$$

Similar to Equation (5.3), the updating rule for the weight matrix is

$$
\mathbf{W}_{n+1} = \mathbf{W}_n + (\mathbf{y}' - \mathbf{a}'\mathbf{W}_n)\mathbf{b} , \tag{5.11}
$$

the updating rule for the weight matrix, if $\mathbf{A}_n$ is of full rank (i.e., $\mathbf{c} = 0$), is

$$
\mathbf{W}_{n+1} = \mathbf{W}_n + (\Theta + \mathbf{a}'\mathbf{S}_n\mathbf{S}_n'\mathbf{a})^{-1}(\mathbf{y}' - \mathbf{a}'\mathbf{W})\mathbf{S}_n\mathbf{S}_n'\mathbf{a} . \tag{5.12}
$$

Equation (5.12) is exactly the same as the weighted recursive least-squares method [23] in which only the full-rank condition is discussed. However, Equation (5.11) is more complete because it covers both $\mathbf{c} = 0$ and $\mathbf{c} \neq 0$ cases. Thus, the weighted weight matrix $\mathbf{W}$ can be easily updated based on the current weights and new observations, without running a complete training cycle, as long as the weighted pseudoinverse is maintained. Similar derivation can be applied to the network with an added neuron.

5.5   Refine the Model

Let us take a look again at an input matrix $\mathbf{A}$ of size $n \times m$, which represents n observations of m variables, The singular value decomposition of $\mathbf{A}$ is:

$$
\mathbf{A} = \mathbf{U}\sum\mathbf{V}',
$$

where $\mathbf{U}$ is a $n \times n$ orthogonal matrix of the eigenvectors of $\mathbf{A}\mathbf{A}'$ and $\mathbf{V}$ a $m \times m$ orthogonal matrix of eigenvectors of $\mathbf{A}'\mathbf{A}$. $\sum$ is a $n \times m$ 'diagonal' matrix whose diagonals are singular values of $\mathbf{A}$. That is,

$$
\sum = \begin{bmatrix}
\sigma_1 & 0 & \cdots & 0 & \cdots & 0 \\
0 & \sigma_2 & \cdots & 0 & \cdots & 0 \\
0 & 0 & \cdots & 0 & \cdots & 0 \\
0 & 0 & \cdots & 0 & \cdots & 0 \\
0 & 0 & \cdots & \sigma_r & \cdots & 0 \\
0 & 0 & \cdots & 0 & \cdots & 0 \\
0 & 0 & \cdots & 0 & \cdots & 0
\end{bmatrix} \triangleq \begin{bmatrix} \sigma & 0 \\ 0 & 0 \end{bmatrix},
$$

where r is the rank of matrix $\mathbf{A}$. $\mathbf{A}\mathbf{A}'$ is the so-called correlation matrix, whose eigenvalues are squares of the singular values. Small singular values might be the result of noise in the data or due to round off errors in computations. This can lead to very large values of weights because the pseudoinverse of $\mathbf{A}$ is given by $\mathbf{A}^+ = \mathbf{V}\Sigma^+\mathbf{U}'$, where

$$\Sigma^+ = \begin{bmatrix} \frac{1}{\sigma} & 0 \\ 0 & 0 \end{bmatrix}.$$

Clearly, small singular values of A will result in a very large value of weights which will, in turn, amplify any noise in the new data. The same question arises as more and more enhancement nodes are added to the model during the training.

A possible solution is to round off small singular values to zeros and therefore avoid large values of weights. If there is a gap among all the singular values, it is easy to cutoff at the gap. Otherwise, one of the following approaches may work:

i) Set an upper bound on the norm of weights. This will provide a criterion to cutoff small singular values. The result is an optimal solution within a bounded region.

ii) Investigate the relation between the cutoff values and the performance of the network in terms of prediction error. If there is a point where the performance is not improved when small singular values are included, it is then reasonable to set a cutoff value corresponding to that point.

The orthogonal least squares learning approach is another way to generate a set of weights that can avoid an ill-conditioning problem [13]. Furthermore, regularization and cross-validation methods are the techniques to avoid both overfitting and generalization problems [24].

5.6   Time-series Applications

The literature has discussed time-series forecasting using different neural network models [25, 26]. Here the algorithm proposed above is applied to the forecasting model. Represent the time-series by the AR(p) (Auto-Regression with p delay) model. Suppose $\mathbf{X}$ is a stationary time-series. The AR(p) model can be represented as the following equation:

$$\mathbf{X}_t = (\lambda_1\mathbf{X}_{t-1} + \lambda_2\mathbf{X}_{t-2} + \cdots + \lambda_p\mathbf{X}_{t-p}) + \varepsilon_t,$$

where λ_i's are autoregression parameters.

In terms of a flat neural network architecture, the AR(p) model can be described as a functional-link network with p input nodes, q enhancement nodes, and a single output node. This will artificially increase the dimension of the input space, or the rank of the input data matrix. The network includes $p + q$ input nodes and a single output node. During the training stage, a variable number of enhancement nodes may be added as necessary. Contrary to the traditional error backpropagation models, the training of this network is fast because of the one-step learning procedure and dynamic updating algorithm mentioned above. To improve the performance in some special situations, a weighted least-square criterion may be used to optimize the weights instead of the ordinary least-squares error.

Using the stepwise updating learning, this section discusses the procedure of training the neural network for time-series forecasting. First, available data on a single time-series are split into a training set and testing set. Let the time data, $x(i + k)$, be the k-th time step after the data $x(i)$ and assume that there will be N training data points, The training stage proceeds as follows.

Step 1. *Construct Input and Output*: Build an input matrix of size $(N - p) \times p$, where p is the delay-time. The i-th row consists of $[x(i+0), x(i+1), \cdots, x(i+(p-1))]$. The target output vector $\mathbf{Y}_n$ will be produced using $[x(i + p), \cdots, x_i(N)]'$.

Step 2. *Obtain the weight matrix*: Find the pseudoinverse of $\mathbf{A}_n$ and the weight matrix $\mathbf{W}_n = \mathbf{A}_n^+\mathbf{Y}_n$. This will give the linear least-square fit with p lags, or AR(p). Predictions can be produced either single step ahead or iterated prediction. The network outputs are then compared to the actual continuation of the data using testing data. The error will be large most of the time, especially when we deal with a non-linear time-series.

Step 3. *Add a new enhancement node if the error is above the desired level*: If the error is above the desired level, a new hidden node will be added. The weights from input nodes to the enhancement node can be randomly generated, but a numerical rank check may be necessary to ensure that the added input node will increase the rank of augmented matrix by one. At this time the pseudoinverse of the new matrix can be updated by using Equation (5.4).

Step 4. *Stepwise update the weight matrix*: After entering a new input pattern to the input matrix, (i.e., adding $\mathbf{a}'$ to $\mathbf{A}_n$ and forming $\mathbf{A}_{n+1}$), the new weight matrix $\mathbf{W}_{n+1}$ can be obtained or updated, using either Equation (5.3) or the Q-R decomposition algorithm. Then testing data is applied again to check the error level.

Step 5. *Looping for further training*: Repeat by going to Step 3 until the desired error level is achieved.

It is worth noting that having more enhancement nodes does not necessarily mean better performance. Particularly, a larger than necessary number of enhancement nodes usually would make the augmented input matrix very ill-conditioned and therefore prone to computational error. Theoretically, the rank of the expanded input matrix will be increased by one, which is not the case as observed in practice. Suppose the expanded input matrix has singular value decomposition $\mathbf{A} = \mathbf{U} \sum \mathbf{V}'$, where $\mathbf{U}$, $\mathbf{V}$ are orthogonal matrices, and $\sum$ a diagonal matrix whose diagonal entries give the singular values of $\mathbf{A}$ in ascending order. Let the condition number of $\mathbf{A}$ be the ratio of the largest singular value over the the smallest one. If the small singular values are not rounded off to zeros, the conditional number would be huge. In other words, the matrix would be extremely ill-conditioned. The least-square solution resulting from the pseudoinverse would be very sensitive to small perturbations which are not desirable. A possible solution would be to cut off any small singular values (and therefore reduce the rank). If the error is not under the desired level after training, extra input nodes will be produced based on the original input nodes and the enhanced input nodes, where the weights are fixed. This is similar to the idea of 'cascade-correlation' network structure [27]. But one step learning is utilized here, which is much more efficient.

5.7   Examples and Discussion

The proposed time-series forecasting model is tested on several time-series data including an infrared laser data set, a chaotic time-series, a monthly flour price data set, and a non-linear system identification. The following examples not only show the effectiveness of the proposed method but also demonstrate a relatively fast way of forecasting time-series. The non-linear system identification of discrete-time single-input, single-output (SISO), multiple-input, multiple-output (MIMO) plants can be described by the difference equations [16]. The most common equation for system identification is

$$y_p(k+1) = f[y_p(k), \cdots, y_p(k-n+1)] + g[u(k), \cdots, u(k-m+1)],$$

where $[u(k), y_p(k)]$ represents the input-output pair of the plant at time k and f and g are differentiable functions.

The system identification model extends the input dimension, that is the addition of the state variables. The training concept is similar to the one-dimensional (i.e., time) time-series prediction. The proposed algorithm can be also applied to multi-lag, MIMO systems easily as shown in Example 4.

Example 1   This is one of the data sets used in a competition of time-series prediction held in 1992 [28]. The training data set contains 1000 points of the fluctuations in a far-infrared laser as shown in Figure 4. The goal is to predict the continuation of the time-series beyond the sample data. During the course of the competition, the physical background of the data set was withheld to avoid biasing the final prediction results. Therefore we are not going to use any information other than the time-series itself to build our network model. To determine the size of network, first we use simple linear net as a preliminary fit, i.e., AR(p), where p is the value of so-called lag. After comparing the single step error versus the value of p, it's noted that optimal choice for the lag value lies between 10 to 15. So we use an AR(15) model and add nonlinear enhancement nodes as needed. Training starts with a simple linear network with 15 inputs and 1 output node. Enhancement nodes are added one at a time and weights are updated using Equation (5.4), as described in Section 3. After about 80 enhancement nodes are added, the network can perform single step predictions exceptionally well. Since the goal is to predict multiple steps beyond the training data, iterated prediction is also produced. Figure 4 shows 60 steps iterated prediction into the future, as is compared to the actual continuation of the time-series. The whole procedure including training and producing predictions took just about less than 20 seconds on a DEC alpha machine, compared to the huge computation with over 1000 parameters to adapt and overnight training time using back-propagation training algorithm. To compare the prediction with previous work [28], the *normalized mean squared error* (NMSE) is defined as

$$\frac{1}{\hat{\sigma}_\tau^2} \frac{1}{N} \sum_{k \in \tau} (y_k - \hat{y}_k)^2,$$

where $k = 1, 2, \cdots, N$ denotes the points in the test set τ, $\hat{\sigma}_\tau^2$ denotes the sample variance of the observed value in τ, y_k and $\hat{y}_k$ are target and predicted values, respectively. A network with 25 lags and 50 enhancement nodes are used for predicting 50 steps and 100 steps ahead using 1000 data points for training. For 50 steps ahead prediction, the NMSE is about 4.15×10^{-4}, and the NMSE for 100 steps ahead prediction is about 8.1×10^{-4}. The results are better than those previously done, shown in Table 1, in both speed (such as hours, days, or weeks) and accuracy [28] (See Table 2 of reference [28], page 64).

Table 1   Previous Results for Example 1

methods	type	computer	time	NMSE(100)	-log(lik.)
conn	1-12-12-1; lag25,5,5	SPARC 2	12 hrs	0.028	3.5
loc lin	low-pass embd, 8dim 4nn	DEC3100	20 min	0.080	4.8
conn	feedforward, 200-100-1	CRAY Y-MP	3 hrs	0.77	5.5
conn	feedforward, 50-20-1	SPARC 1	3 weeks	1.0	6.1
visual	look for similar stretches	SR Iris	10 sec	1.5	6.2
visual	look for similar stretches	SR Iris	-	0.45	6.2
conn	feedforward, 50-350-50-50	386 PC	5 days	0.38	6.4
conn	recurrent, 4-4c-1	VAX 8530	1 hr	1.4	7.2
tree	k-d tree, AIC	VAX 6420	20 min	0.62	7.3
loc lin	21dim 30nn	SPARC 2	1 min	0.71	10
loc lin	3dim time delay	Sun	10 min	1.3	-
conn	feedforward	SPARC 2	20 hrs	1.5	-
conn	feedforward, weight-decay	SPARC 1	30 min	1.5	-
lin	Wiener filter, width 100	MIPS 3230	30 min	1.9	-

Example 2   Time series produced by iterating the logist map

$$f(x) = \alpha x(1-x), \ or \ x(n+1) = \alpha x(n)(1-x(n)),$$

is probably the simplest system capable of displaying deterministic chaos. This first-order difference equation, also known as the Feigenbaum equation, has been extensively studied as a model of biological populations with non-overlapping generations, where $x(n)$ represent the normalized population of n-th generation and α is a parameter that determines the dynamics of the population. The behavior of the time-series depends critically on the value of the bifurcation parameter α. If $\alpha < 1$, the map has a single fixed point and the output or population dies away to zero. For $1 < \alpha < 3$, the fixed point at zero becomes unstable and a new stable fixed point appears. So the output converges to a single nonzero value. As the value of α increases beyond 3, the output begins to oscillate first between two values, then four values, then eight values and so on, until α reaches a value of about 3.56 when the output becomes chaotic. The α is set to 4 for producing the tested time-series data from the above map. The logistic map of the time-series equation (the solid curve)

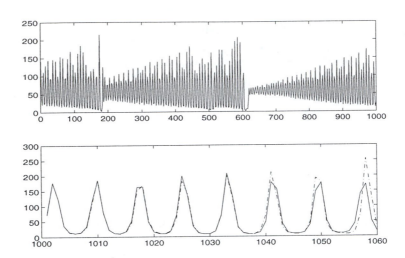

Figure 4(a)   Prediction of the time-series 60 steps of the future

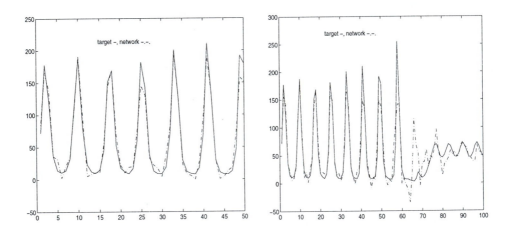

Figure 4(b)   Network prediction (first 50 points) (left), 4(c) Network prediction
(first 100 points)

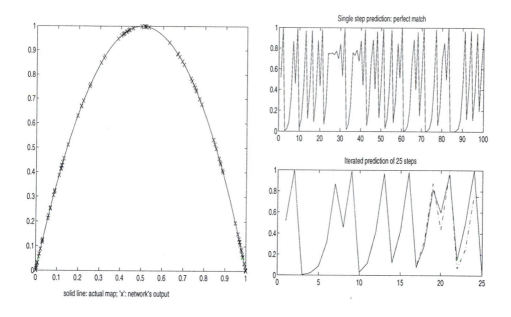

solid line: actual map; 'x': network's output

Figure 5(a)   Actual quadratic map and network's prediction (left), 5(b)   Single step prediction (right top), 5(c)   Iterated prediction of 25 steps for the future (right bottom)

and the output predicted by the neural network (the 'x' curve) is shown in Figure 5(a). A short segment of the time-series is shown in Figure 5 (b). The network is trained to predict the $(n + 1)$-th value based only on the value at n. The training set consists of 100 consecutive pairs of (x_t, x_{t+1}) time-series values. With just five enhancement nodes, the network can do a single step prediction pretty well after training. To produce multiple steps ahead prediction, ten enhancement nodes can push the iterated prediction up to 20 steps into the future with a reasonable error level (see Figure 5 (c)).

Example 3   As the third example, we tested the model on a trivariate time-series $\mathbf{X}_t = \{x_t, y_t, z_t, t = 1, 2, \cdots, T\}$, where T ranges up to 100. The data used are logarithms of the indices of monthly flour prices for Buffalo (x_t), Kansas City (y_t) and Minneapolis(z_t) over the period from 8/72 to 11/80 [29]. First we trained the network with 8 enhancement nodes using first 90 data. The next 10 data sets are then tested in one-lag prediction, starting from t = 91. To compare the prediction with previous work, the *mean squared error* (MSE) is defined as $\frac{1}{N} \sum_{k \in \tau} (y_k - \hat{y}_k)^2$, where $k = 1, 2, \cdots, N$ denotes the points in the test set τ, y_k and predicted values, respectively. Figure 6(a) shows the flour price indices. Figure 6(b) shows the networking modeling and target output. The prediction and the error are given in Figure 6(c) and Figure 6(d), respectively. The training MSEs for Minneapolis, Kansas City, and Buffalo are are 0.0039, 0.0043, and 0.0051, respectively. The

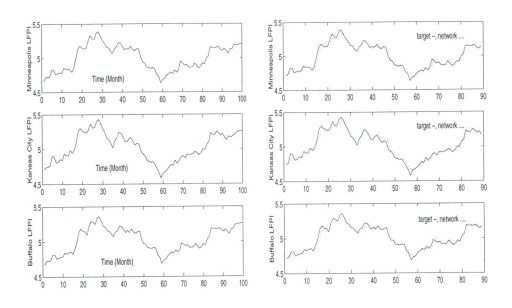

Figure 6(a)   Flour price indices (left), 6(b) Network modeling

prediction MSEs for Minneapolis, Kansas City, and Buffalo are are 0.0053, 0.0055, and 0.0054, respectively. The result is better than previous work using multi-layer network. We also trained the network with six inputs coupled (combined) with 10 enhancement nodes using the first 90 triplets from the data. The network performs well even in multi-lag prediction, or iterated prediction. This is shown in Figure 6(e). We also observe that, even though more lags or more enhancement nodes would achieve better fit during the training stage, they do not necessary improve prediction performance, especially in the case of multi-lag.

Example 4   The model is also used for a MIMO nonlinear system. The two-dimensional input-output vectors of the plant were assumed to be $u(k) = [u_1(k), u_2(k)]'$ and $y(k) = [y_{p1}(k), y_{p2}(k)]'$. The difference equation describing the plant was assumed to be of the form [16],

$$\begin{bmatrix} y_{p1}(k+1) \\ y_{p2}(k+1) \end{bmatrix} = \begin{bmatrix} f_1[y_{p1}(k), y_{p2}(k), u_1(k), u_2(k)] \\ f_2[y_{p1}(k), y_{p2}(k), u_1(k), u_2(k)], \end{bmatrix}$$

where the known functions f_1 and f_2 have the forms:

$$f_1(y_{p1}, y_{p2}, u_1, u_2) = \frac{0.8y_{p1}^3 + u_1^2 u_2}{2 + y_{p2}^2},$$

and

$$f_2(y_{p1}, y_{p2}, u_1, u_2) = \frac{y_{p1} - y_{p1}y_{p2} + (u_1 - 0.5)(u_2 + 0.8)}{1 + y_{p2}^2}.$$

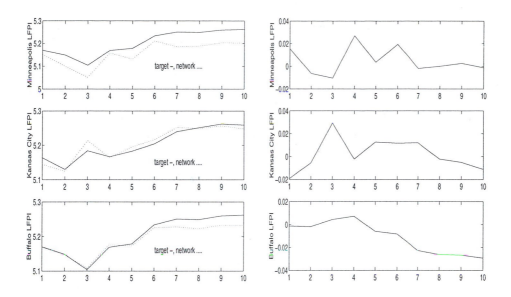

Figure 6(c)   Network prediction (one-lag) (left), 6(d) Network prediction error (one-lag)

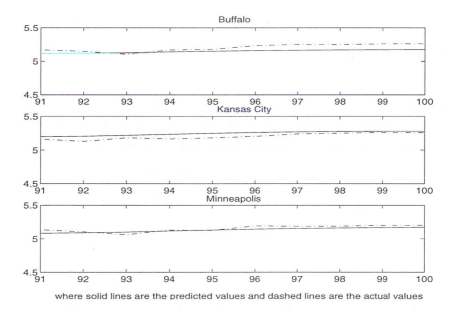

where solid lines are the predicted values and dashed lines are the actual values

Figure 6(e)   Iterated prediction (multi-lag) of flour price indices of three cities, where
solid lines are the predicted values and the dashed lines are the actual values

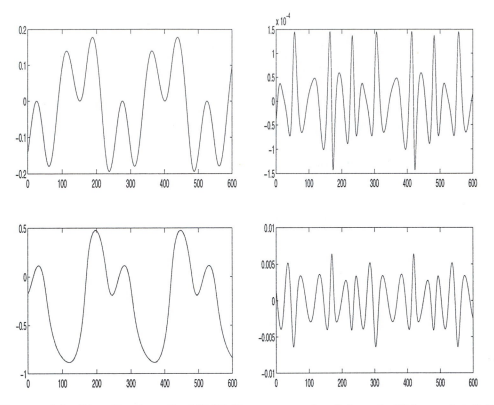

Figure 7(a)   Identification of a MIMO System, y_{p1} plot (left top), 7(b) y_{p2} plot (left bottom), 7(c) The difference for $(y_{p1} - \hat{y}_{p1})$ (right top), 7(d) The difference for $(y_{p2} - \hat{y}_{p2})$

The stepwise update algorithm with 5 enhancement nodes is used to train the above system. Using $u_1(k) = sin(2\pi k/250)$ and $u_2(k) = cos(2\pi k/250)$, the responses are shown in Figure 7. Figure 7(a) is the plot for y_{p1} and $\hat{y}_{p1}$ and Figure 7(b) is the plot for y_{p2} and $\hat{y}_{p2}$. The dashed-line and solid-line are also overlapped in this case. Figure 7(c) and (d) show the plots of $y_{p1} - \hat{y}_{p1}$ and $y_{p2} - \hat{y}_{p2}$, respectively. The training time is again very fast – about 30 seconds in a DEC workstation.

5.8   Conclusions

In summary, the algorithm described in this chapter is simple and fast and easy to update. Several examples show the promising result. There are two points that we want to emphasize: (1) The learning algorithm for functional-link network is very fast and efficient. The fast learning makes it possible for the trial-error approach to fine-tune some hard-to-determine parameters (e.g., the number of enhancement (hidden) nodes), and the dimension of the state space, or the AR parameter p. The training algorithm allows us to update the weight matrix in real-time if additional enhancement nodes are added to the system. Meanwhile, the weights can also be updated easily if new observations are added to the system. This column-wise (additional neurons) and row-wise (additional observations) update scheme is very attractive to real-time processes. (2) The easy updating of the weights in the proposed approach saves time and resources to re-train the network from scratch. This is especially beneficial when the data set is huge.

References

1. P. J. Werbos, Beyond regression: New tools for prediction and analysis in the behavioral science, *Ph.D. Dissertation*, Harvard University, Nov. 1974.

2. P. J. Werbos, Backpropagation through time: What it does and how to do it, *Proceedings of the IEEE*, Vol. 78, No. 10, pp. 1550-1560, Oct. 1990.

3. A. Cichocki and R. Unbehauen, *Neural Networks for Optimization and Signal Processing*, John Wiley and Sons, New York, 1992.

4. L. F. Wessels and E. Barnard, Avoiding false local minima by proper initialization of connections, *IEEE Transactions on Neural Networks*, Vol. 3, No. 6, pp. 899-905, 1992.

5. R. A. Jacobs, Increased rates of convergence through learning rate adaptation, *Neural Networks*, Vol. 1, pp. 295-307, 1988.

6. H. Drucker and Y. Le Cun, Improving generalization performance using double backpropagation, *IEEE Transactions on Neural Networks*, Vol. 3, No. 6, pp. 991-997, 1992.

7. S. J. Perantonis and D. A. Karras, An efficient constrained learning algorithm with momentum acceleration, *Neural Networks*, Vol. 8, No. 2, pp. 237-249, 1994.

8. D. A. Karras and S. J. Perantonis, An efficient constrained training algorithm for feedforward networks, *IEEE Transactions on Neural Networks*, Vol. 6, No. 6, pp. 1420-1434, Nov. 1995.

9. D. S. Chen and C. Jain, A robust back propagation learning algorithm for function approximation, *IEEE Transactions on Neural Networks*, Vol. 5, No. 3, pp. 467-479, May 1994.

10. Y. H. Pao and Y. Takefuji, Functional-link net computing, Theory, system architecture, and functionalities, *IEEE Computer*, Vol. 3, pp. 76-79, 1991.

11. Y. H. Pao, G. H. Park, and D. J. Sobajic, Learning and generalization characteristics of the random vector functional-link net, *Neurocomputing*, Vol. 6, pp. 163-180, 1994.

12. B. Igelnik and Y. H. Pao, Stochastic choice of basis functions in adaptive function approximation and the functional-link net," *IEEE Transactions on Neural Networks*, Vol. 6, No. 6, pp. 1320-1329, 1995.

13. S. Chen, C. F. N. Cowan, and P. M. Grant, Orthogonal least squares learning algorithm for radial basis function networks, *IEEE Transactions on Neural Networks*, Vol. 2, No. 2, pp. 302-309, March 1991.

14. D. S. Broomhead and D. Lowe, Multivariable functional interpolation and adaptive methods, *Complex Systems*, 2, pp. 321-355, 1988.

15. Y. H. Pao, G. H. Park, and D. J. Sobajic, Learning and generalization characteristics of the random vector functional-link net, *Neurocomputing*, Vol. 6, pp. 163-180, 1994.

16. K. S. Narendra and K. Parthasarathy, Identification and control of dynamical systems using neural networks, *IEEE Transactions on Neural Networks*, Vol. 1, No. 1, pp. 4-27, March 1990.

17. C. L. P. Chen, A rapid supervised learning neural network for function interpolation and approximation, *IEEE Transactions on Neural Networks*, Vol. 7, No. 5, pp. 1220-1230, Sept. 1996.

18. G. H. Golub and C. F. Van Loan, *Matrix Computations*, 3rd Edition, Johns Hopkins University Press, Baltimore, 1996.

19. M. H. Hassoun, *Fundamentals of Artificial Neural Networks*, MIT Press, Boston, 1995.

20. A. Ben-Israel and T. N. E. Greville, *Generalized Inverses: Theory and Applications*, John Wiley & Sons, New York, 1974.

21. F. H. Kishi, On line computer control techniques and their application to re-entry aerospace vehicle control, in *Advances in Control Systems Theory and Applications*, C. T. Leondes, ed., pp. 245-257, Academic Press, New York, 1964.

22. B. Widrow, Generalization and information storage in networks of adaline neuron, *Self-Organizing Systems*, pp. 435-461, M. C. Jovitz et al. eds., 1962.

23. C. R. Johnson, Jr., *Lectures on Adaptive Parameter Estimation*, Prentice-Hall, Englewood Cliffs, 1988.

24. M. J. L. Orr, Regularization in the selection of radial basis function centers, *Neural Computation*, Vol. 7, pp. 606-623, 1995.

25. V. R. Vemuri and R. D. Rogers, eds., *Artificial Neural Networks: Forecasting Time Series*, IEEE Computer Society Press, New York, 1993.

26. A. Khotanzad, R. Hwang, A. Abaye, and D. Maratukulam, An adaptive modular artificial neural network hourly load forecaster and its implementation at electric utilities, *IEEE Transactions on Power Systems*, Vol. 10, No. 3, pp. 1716-1922, 1995.

27. S. E. Fahlman and C. Lebiere, The cascade-correlation learning architecture, *Advances in Neural Information Processing Systems I*, 1989.

28. A. S. Weigend and N. A. Gershenfeld, eds., *Time Series Prediction, Forecasting the Future and Understanding the Past*, Addison-Wesley, New York, 1994.

29. K. Chakraborty, K. Mehrotra, C. Mohan, and S. Ranka, Forecasting the behavior of multivariate time series using neural networks, *Neural Networks*, Vol. 5, pp. 961-970, 1992.

Chapter 6

Basic Structure of Fuzzy Neural Networks

In this chapter we shall discuss the structure of fuzzy neural networks. We start with general definitions of multifactorial functions. And we show that a fuzzy neuron can be formulated by means of standard multifactorial function. We also give definitions of a fuzzy neural network based on fuzzy relationship and fuzzy neurons. Finally, we describe a learning algorithm for a fuzzy neural network based on $\bigvee$ and $\bigwedge$ operations.

6.1   Definition of Fuzzy Neurons

Neural networks alone have demonstrated their ability to classify, recall, and associate information [1]. In this chapter, we shall incorporate fuzziness to the networks. The objective to include the fuzziness is to extend the capability of the neural networks to handle "vague" information than "crisp" information only. Previous work has shown that fuzzy neural networks have achieved some level of success both fundamentally and practically [1-10]. As indicated in reference [1], there are several ways to classify fuzzy neural networks: (1) a fuzzy neuron with crisp signals used to evaluate fuzzy weights, (2) a fuzzy neuron with fuzzy signals which is combined with fuzzy weights, and (3) a fuzzy neuron described by fuzzy logic equations.

In this chapter, we shall discuss a fuzzy neural network where both inputs and outputs can be either a crisp value or a fuzzy set. To do this we shall first introduce multifactorial function [11, 12]. We have pointed out from Chapter 4 that one of the basic functions of neurons is that the input to a neuron is synthesized first, then activated, where the basic operators to be used as synthesizing are " $+$ " and " $\cdot$ " denoted by ($+, \cdot$) and called synthetic operators. However, there are divers styles of synthetic operators such as $(\vee, \wedge)$, $(\vee, \cdot)$, $(+, \wedge)$, etc. More general synthetic operators will be multifactorial functions, so we now briefly introduce the concept of multifactorial functions.

In $[0,1]^m$, a natural partial ordering " $\leq$ " is defined as follows:

$$\left(\forall X, Y \in [0,1]^m\right)\left(X \leq Y \Longrightarrow (x_j \leq y_j, \quad j = 1, 2, \cdots, m)\right)$$

A multifactorial function is actually a projective mapping from an m-ary space to a one-ary space, denoted by M_m. On many occasions, it is possible to transform the spaces into closed unit intervals:

$$M_m : [0,1]^m \longrightarrow [0,1], (x_1, \cdots, x_m) \longmapsto M_m(x_1, \cdots, x_m) \qquad (6.1)$$

that we call standard multifactorial functions. For what follows, we will focus our discussion on standard multifactorial functions.

The standard multifactorial functions can be classified into two groups:

Group 1. Additive Standard Multifactorial (ASM) functions

The functions in this group satisfy the condition: $\forall (x_1, x_2, \cdots, x_m) \in [0,1]^m$,

$$\bigwedge_{j=1}^{m} x_j \leq M_m(x_1, x_2, \cdots, x_m) \leq \bigvee_{j=1}^{m} x_j \qquad (6.2)$$

which means that the synthesized value should not be greater than the largest of component states and should not be less than the smallest of the component states. The following is its normal definition.

Definition 1   A mapping $M_m : [0,1]^m \longrightarrow [0,1]$ is called an m-ary Additive Standard Multifactorial function (ASM$_m$-func) if it satisfies the following axioms:

$(m.1)$ $X \leq Y \Longrightarrow M_m(X) \leq M_m(Y)$;

$(m.2)$ $\displaystyle\bigwedge_{j=1}^{m} x_j \leq M_m(X) \leq \bigvee_{j=1}^{m} x_j$;

$(m.3)$ $M_m(x_1, x_2, \cdots, x_m)$ is a continuous function of each variable x_j.

The set of all ASM$_m$-func is denoted by $\mathcal{M}_m \overset{\triangle}{=} \{M_m | \ M_m$ is a $m-$ dimensional ASM$_m-$ func$\}$. Clearly, when $m = 1$, an ASM$_m$-func is an identity mapping from $[0,1]$ to $[0,1]$. Also $(m.2)$ implies $M_m(a, \cdots, a) = a$.

Example 1 The follwing mappings are examples of ASM$_m$-funcs from $[0,1]^m$ to $[0,1]$:

$$\bigwedge : X \longmapsto \bigwedge(X) := \bigwedge_{j=1}^{m} x_j, \qquad (6.3)$$

$$\bigvee : X \longmapsto \bigvee(X) := \bigvee_{j=1}^{m} x_j, \qquad (6.4)$$

$$\sum : X \longmapsto \sum(X) := \sum_{j=1}^{m} w_j x_j, \qquad (6.5)$$

where $w_j \in [0,1]$, and $\sum\limits_{j=1}^{m} w_j = 1$.

$$M_m(X) = \bigvee_{j=1}^{m} w_j x_j, \tag{6.6}$$

where $w_j \in [0,1]$, and $\bigvee\limits_{j=1}^{m} w_j = 1$.

$$M_m(X) = \sum_{j=1}^{m} (w_j \wedge x_j), \tag{6.7}$$

where $w_j \in [0,1]$ and $\sum_{j=1}^{m} w_j = 1$.

$$M_m(X) = \bigvee_{j=1}^{m} (w_j \vee x_j), \tag{6.8}$$

where $w_j \in [0,1]$ and $\bigvee_{j=1}^{m} w_j \leq \bigvee_{j=1}^{m} x_j$.

$$M_m(X) = \bigwedge_{j=1}^{m} (w_j \wedge x_j), \tag{6.9}$$

where $w_j \in [0,1]$ and $\bigwedge_{j=1}^{m} w_j \geq \bigwedge_{j=1}^{m} x_j$.

$$M_m(X) := \bigvee_{j=1}^{m} (w_j \wedge x_j), \tag{6.10}$$

where $w_j \in [0,1]$, and $\bigvee\limits_{j=1}^{m} w_j = 1$.

$$M_m(X) := \left(\prod_{j=1}^{m} x_j \right)^{\frac{1}{m}}, \tag{6.11}$$

$$M_m(X) := \left(\sum_{j=1}^{m} w_j x_j^p \right)^{\frac{1}{p}}, \tag{6.12}$$

where $p > 0$, $w_j \in [0,1]$ and $\sum\limits_{j=1}^{m} w_j = 1$.

Group 2. Non-additive Standard Multifactorial (NASM) functions

This group of functions does not satisfy axiom $(m.2)$. That is, the synthesized value can exceed the boundaries of axiom $(m.2)$, i.e.,

$$M_m(X) \geq \bigvee_{j=1}^{m} x_j \quad or \quad M_m(X) \leq \bigwedge_{j=1}^{m} x_j. \tag{6.13}$$

For example, a department is led and managed by three people; each of them has a strong leading ability. But for some reason, they cannot work smoothly among themselves. Hence, the collective leading ability (a multifactorial score) falls below the individual's, i.e.,

$$M_3(x_1, x_2, x_3) \leq \bigvee_{j=1}^{3} x_j,$$

where x_j is the leading ability of the individual i, $i = 1, 2, 3$, and $M_3(x_1, x_2, x_3)$ is the multifactorial leading ability indictor for the group of three.

On the other hand, it is possible for the three management people to work together exceedingly well. This implies that the combined leadership score can be higher than any one of the three individual's, i.e.,

$$M_3(x_1, x_2, x_3) \geq \bigvee_{j=1}^{3} x_j.$$

It has the same meaning in the Chinese old saying: "Three cobblers with their wits combined can exceed *Chukeh Liang*, the master minded".

Definition 2 A mapping $M_m : [0,1]^m \longmapsto [0,1]$ is called an m-ary Non-Additive Standard Multifactorial function, denoted by NASM$_m$-func, if it satisfies axioms $(m.1)$, $(m.3)$, and the following axiom:

$(m.2')$ $M_m(X) \leq \bigwedge_{j=1}^{m} x_j$ or $M_m(X) \geq \bigvee_{j=1}^{m} x_j.$

The set of all NASM$_m$-funcs is denoted by $\mathcal{M}'_m$.

Example 2   The mapping $\prod : [0,1]^m \longrightarrow [0,1]$ defined as the following is a NASM$_m$-func:

$$(x_1, \cdots, x_m) \longmapsto \prod(x_1, \cdots, x_m) := \prod_{j=1}^{m} x_j. \qquad (6.14)$$

Next, we shall use the these definitions to define fuzzy neurons.

Definition 3   A fuzzy neuron is regarded as a mapping FN :

$$FN : [0,1]^m \rightarrow [0,1]$$
$$X \longmapsto FN(X) \overset{\triangle}{=} \varphi(M_m(X) - \theta), \qquad (6.15)$$

where $M_m \in \mathcal{M}_m, \theta \in [0,1]$ and φ is a mapping or an activation function, $\varphi : \Re \rightarrow [0,1]$ with $\varphi(u) = 0$ when $u \leq 0$; and $\Re$ is the field of all real numbers. And a neural network formed by fuzzy neurons is called a fuzzy neural network.

Figure 1 illustrates the working mechanism of a fuzzy neuron.

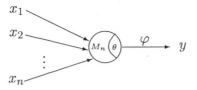

Figure 1   Illustration of a fuzzy neuron

Example 3   The following mappings from $[0,1]^n$ to $[0,1]$ are all fuzzy neurons:

$$y = \varphi\left(\sum_{i=1}^{n} w_i x_i - \theta \right), \tag{6.16}$$

where $w_i \in [0,1]$ and $\sum_{i=1}^{n} w_i = 1$.

$$y = \varphi\left(\bigvee_{i=1}^{n} (w_i x_i) - \theta \right), \tag{6.17}$$

where $w_i \in [0,1]$ and $\bigvee_{i=1}^{n} w_i = 1$.

$$y = \varphi\left(\bigvee_{i=1}^{n} (w_i \wedge x_i) - \theta \right), \tag{6.18}$$

where $w_i \in [0,1]$ and $\bigvee_{i=1}^{n} w_i = 1$.

$$y = \varphi\left(\sum_{i=1}^{n} (w_i \wedge x_i) - \theta \right), \tag{6.19}$$

where $w_i \in [0,1]$ and $\sum_{i=1}^{n} w_i = 1$.

$$y = \varphi\left(\bigvee_{i=1}^{n} (w_i \vee x_i) - \theta \right), \tag{6.20}$$

where $w_i \in [0,1]$ and $\bigvee_{i=1}^{n} w_i \leq \bigvee_{i=1}^{n} x_i$.

$$y = \varphi\left(\bigwedge_{i=1}^{n} (w_i \wedge x_i) - \theta \right), \tag{6.21}$$

where $w_i \in [0,1]$ and $\bigwedge_{i=1}^{n} w_i \leq \bigwedge_{i=1}^{n} x_i$.

Example 4   In Equation (6.17), if we let $(\forall\, i)(w_i = 1), \theta = 0$ and $\varphi = id$, where id is an identity function, i.e., $(\forall\, x)(id(x) = x)$, we have a special fuzzy neuron:

$$y = \bigvee_{i=1}^{n} x_i, \tag{6.22}$$

In the same way, from (6.21) we have another special fuzzy neuron:

$$y = \bigwedge_{i=1}^{n} x_i. \tag{6.23}$$

6.2 Fuzzy Neural Networks

In this section, we shall discuss fuzzy neural networks. We first use the concept of fuzzy relationship followed by the definition of fuzzy neurons. We also discuss a learning algorithm for a fuzzy neural network.

6.2.1 Neural Network Representation of Fuzzy Relation Equations

We consider a typical kind of fuzzy relation equation:

$$\mathbf{X} \circ \mathbf{R} = \mathbf{B} \tag{6.24}$$

where $\mathbf{X} = (x_1, x_2, \cdots, x_n)$ is the input vector, $\mathbf{R} = (r_{ij})_{n \times m}$ is the matrix of coefficients, and $\mathbf{B} = (b_1, b_2, \cdots, b_m)$ is the constant matrix. Commonly, the operator "$\circ$" can be defined as follows:

$$\bigvee_{i=1}^{n} (x_i \wedge r_{ij}) = b_j, \quad j = 1, 2, \cdots, m \tag{6.25}$$

At first, it is easy to realize that the equation can be represented by a network shown as Figure 2, where the activation functions of the neurons $f_1, f_2, \cdots, f_m$ are all taken as identity functions and the threshold values are zero.

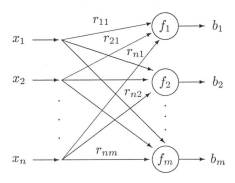

Figure 2   A network representing fuzzy relation equations

Equation (6.25) can be solved using a a fuzzy δ learning algorithm descirbed in Section 6.3. However, if the operator is not $\vee$ and $\wedge$, then Equation (6.24) is difficult to solve. In this way, we consider to re-structure the network as shown in Figure 3, where the activation function of the neuron f is also taken as an identity function and the threshold value is zero. We can interpret the network as: given a group of training samples:

$$\{((r_{1j}, r_{2j}, \cdots, r_{nj}), b_j) \mid j = 1, 2, \cdots, m\} \tag{6.26}$$

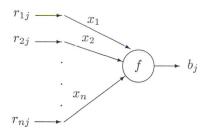

Figure 3   Another network representing fuzzy relation equations

find the weight vectors $(x_1, x_2, \cdots, x_n)$. In this way, we can solve the problem by using an adequate learning algorithm.

In Equation (6.25), if operator "$\wedge$" is replaced by operator "$\cdot$" , i.e.,

$$\bigvee_{i=1}^{n} (x_i \cdot r_{ij}) = b_j, \ j = 1, 2, \cdots, m \tag{6.27}$$

then Equation (6.24) is a generalized fuzzy relation equation.

If synthetic operator $(\vee, \cdot)$ is replaced by $(\oplus, \cdot)$, where "$\oplus$" is so-called bounded sum, i.e.,

$$\bigoplus_{i=1}^{N} (x_i \cdot r_{ij}) = b_j, \ \ j = 1, 2, \cdots, m \tag{6.28}$$

then Equation (6.24) is "almost" the same as a usual system of linear equations.

Especially, if "$\oplus$" is replaced by "$+$", i.e.,

$$\sum_{i=1}^{n} r_{ij} x_i = b_j, \ \ j = 1, 2, \cdots, m \tag{6.29}$$

then it is a system of linear equations. Of course, r_{ij} and b_j must not be in $[0,1]$ (it has already exceeded the definition of fuzzy relation equations). In other words, a system of linear equations can be also represented by a network.

6.2.2   A Fuzzy Neural Network Based on $FN(\vee, \wedge)$

Obviously, there are different types of operations for neurons existing in a fuzzy neural network. For example, from Definition 4 we have different ASM_m-func mappings, M_m, that generate different results for a neuron. A commonly used fuzzy neural network is to take $\wedge$ operator on an input and weights followed by a $\vee$ operator on all inputs. Here we consider that such a fuzzy neural network, shown in Figure 4, based on $FN(\vee, \wedge)$ from Definition 4. The network is known to have fuzzy associative memories ability.

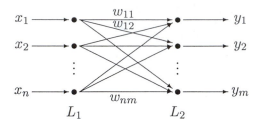

Figure 4   A fuzzy neural network base on $FN(\vee, \wedge)$

Clearly, based on the definition, the relation between input and output of this network is as follows:

$$\begin{cases} y_1 = (w_{11} \wedge x_1) \vee (w_{21} \wedge x_2) \vee \cdots \vee (w_{n1} \wedge x_n) \\ y_2 = (w_{12} \wedge x_1) \vee (w_{22} \wedge x_2) \vee \cdots \vee (w_{n2} \wedge x_n) \\ \cdots \\ y_m = (w_{1m} \wedge x_1) \vee (w_{2m} \wedge x_2) \vee \cdots \vee (w_{nm} \wedge x_n) \end{cases} \tag{6.30}$$

Rewriting Equation (6.30) as a matrix form, we have

$$\mathbf{Y} = \mathbf{X} \circ \mathbf{W}, \tag{6.31}$$

where $\mathbf{Y} = (y_1, y_2, \cdots, y_m)$, $\mathbf{X} = (x_1, x_2, \cdots, x_n)$ and

$$\mathbf{W} = \begin{pmatrix} w_{11} & w_{12} & \cdots & w_{1m} \\ w_{21} & w_{22} & \cdots & w_{2m} \\ & \cdots & & \\ w_{n1} & w_{n2} & \cdots & w_{nm} \end{pmatrix}.$$

For given a set of samples:

$$\{(\mathbf{a}_s, \mathbf{b}_s)|s = 1, 2, \cdots, p\}, \tag{6.32}$$

where $\mathbf{a}_s = (a_{s1}, a_{s2}, \cdots, a_{sn})$, $\mathbf{b}_s = (b_{s1}, b_{s2}, \cdots, b_{sm})$, $s = 1, 2, \cdots, p$, we can obtain a weight matrix $\mathbf{W}$ by means of the following system of fuzzy relation equations:

$$\begin{cases} \mathbf{a}_1 \circ \mathbf{W} = \mathbf{b}_1 \\ \mathbf{a}_2 \circ \mathbf{W} = \mathbf{b}_2 \\ \cdots \\ \mathbf{a}_p \circ \mathbf{W} = \mathbf{b}_p \quad . \end{cases} \tag{6.33}$$

If we collect $\mathbf{a}_s$ and $\mathbf{b}_s$, respectively, we have

$$\mathbf{A} = \begin{pmatrix} \mathbf{a}_1 \\ \mathbf{a}_2 \\ \vdots \\ \mathbf{a}_p \end{pmatrix} = \begin{pmatrix} a_{11} \ a_{12} \ \cdots \ a_{1n} \\ a_{21} \ a_{22} \ \cdots \ a_{2n} \\ \cdots \\ a_{p1} \ a_{p2} \ \cdots \ a_{pn} \end{pmatrix}$$

and

$$\mathbf{B} = \begin{pmatrix} \mathbf{b}_1 \\ \mathbf{b}_2 \\ \vdots \\ \mathbf{b}_p \end{pmatrix} = \begin{pmatrix} b_{11} \ b_{12} \ \cdots \ b_{1m} \\ b_{21} \ b_{22} \ \cdots \ b_{2m} \\ \cdots \\ b_{p1} \ b_{p2} \ \cdots \ b_{pm} \end{pmatrix} ;$$

then Equation (6.33) can be expressed by a single matrix equation as follows:

$$\mathbf{A} \circ \mathbf{W} = \mathbf{B}. \tag{6.34}$$

that is:

$$\begin{pmatrix} a_{11} \ a_{12} \ \cdots \ a_{1n} \\ a_{21} \ a_{22} \ \cdots \ a_{2n} \\ \cdots \\ a_{p1} \ a_{p2} \ \cdots \ a_{pn} \end{pmatrix} \circ \begin{pmatrix} w_{11} \ w_{12} \ \cdots \ w_{1m} \\ w_{21} \ w_{22} \ \cdots \ w_{2m} \\ \cdots \\ w_{n1} \ w_{n2} \ \cdots \ w_{nm} \end{pmatrix} = \begin{pmatrix} b_{11} \ b_{12} \ \cdots \ b_{1m} \\ b_{21} \ b_{22} \ \cdots \ b_{2m} \\ \cdots \\ b_{p1} \ b_{p2} \ \cdots \ b_{pm} \end{pmatrix} \tag{6.35}$$

Equation (6.21) is a fuzzy relation equation and it is not difficult to solve. We shall discuss next a fuzzy learning algorithm.

6.3   A Fuzzy δ Learning Algorithm

We now briefly describe procedures for the fuzzy δ learning algorithm [13].

Step 1   *Randomize w_{ij} initial values w_{ij}^0.*

$$w_{ij} = w_{ij}^0 \quad (i = 1, 2, \cdots, n, j = 1, 2, \cdots, m).$$

Often we can assign $(w_{ij}^0 = 1)$, $(\forall \ i, j)$.

Step 2   *Collect a pair of sample $(\mathbf{a}_s, \mathbf{b}_s)$. Let $s = 1$.*

Step 3   *Calculate the outputs incurred by $\mathbf{a}_s$. Let $k = 1$,*

$$b'_{sj} = \bigvee_{i=1}^{n} (w_{ij} \wedge a_{si}), \quad j = 1, 2, \cdots, m. \tag{6.36}$$

Step 4 *Adjust weights.* Let

$$\delta_{sj} = b'_{sj} - b_{sj}, \quad j = 1, 2, \cdots, m.$$

Update weights, i.e., calculate $(k+1)$th weights based on kth weights:

$$w_{ij}(k+1) = \begin{cases} w_{ij}(k) - \eta\delta_{sj}, & w_{ij}(t) \wedge a_{si} > b_{sj} \\ w_{ij}(k), & \text{otherwise,} \end{cases} \tag{6.37}$$

where $0 < \eta \le 1$ is the learning rate.

Step 5 *Looping.* Go to Step 3 until the following condition holds:

$$(\forall\ ij)(w_{ij}(k) - w_{ij}(k+1) < \varepsilon), \tag{6.38}$$

where $\varepsilon > 0$ is a small number for stopping the algorithm. Set $k = k + 1$.

Step 6 *Repeat a new input.* Let $s = s + 1$ and go to Step 2 until $s = p$.

We give the following example with four inputs.

Example 5 Given samples $\mathbf{a}_s, \mathbf{b}_s, \quad s = 1, 2, 3, 4$:

$$\begin{aligned}
\mathbf{a}_1 &= (0.3, 0.4, 0.5, 0.6), & \mathbf{b}_1 &= (0.6, 0.4, 0.5), \\
\mathbf{a}_2 &= (0.7, 0.2, 1.0, 0.1), & \mathbf{b}_2 &= (0.7, 0.7, 0.7), \\
\mathbf{a}_3 &= (0.4, 0.3, 0.9, 0.8), & \mathbf{b}_3 &= (0.8, 0.4, 0.5), \\
\mathbf{a}_4 &= (0.2, 0.1, 0.2, 0.3), & \mathbf{b}_4 &= (0.3, 0.3, 0.3).
\end{aligned}$$

So we have $\mathbf{A} = \begin{pmatrix} 0.3\ 0.4\ 0.5\ 0.6 \\ 0.7\ 0.2\ 1.0\ 0.1 \\ 0.4\ 0.3\ 0.9\ 0.8 \\ 0.2\ 0.1\ 0.2\ 0.3 \end{pmatrix}, \mathbf{B} = \begin{pmatrix} 0.6\ 0.4\ 0.5 \\ 0.7\ 0.7\ 0.7 \\ 0.8\ 0.4\ 0.5 \\ 0.3\ 0.3\ 0.3 \end{pmatrix}$

and

$$\begin{pmatrix} 0.3\ 0.4\ 0.5\ 0.6 \\ 0.7\ 0.2\ 1.0\ 0.1 \\ 0.4\ 0.3\ 0.9\ 0.8 \\ 0.2\ 0.1\ 0.2\ 0.3 \end{pmatrix} \circ \begin{pmatrix} w_{11}\ w_{12}\ w_{13} \\ w_{21}\ w_{22}\ w_{23} \\ w_{31}\ w_{32}\ w_{33} \\ w_{41}\ w_{42}\ w_{43} \end{pmatrix} = \begin{pmatrix} 0.6\ 0.4\ 0.5 \\ 0.7\ 0.7\ 0.7 \\ 0.8\ 0.4\ 0.5 \\ 0.3\ 0.3\ 0.3 \end{pmatrix}. \tag{6.39}$$

When $0.5 < \eta \le 1$ and $\varepsilon = 0.0001$, at $k = 80$ we have stable $\mathbf{W}$ as follows:

$$\mathbf{W} = \begin{pmatrix} 1.0\ 1.0\ 1.0 \\ 1.0\ 1.0\ 1.0 \\ 0.7\ 0.4\ 0.5 \\ 1.0\ 0.4\ 0.5 \end{pmatrix}.$$

We run several tests and find out that most values in $\mathbf{W}$ are the same, except w_{31}, w_{33}, and w_{43}. The following table details the difference.

Table 1   Results for Different Tests

η	k	w_{31}	w_{33}	w_{43}
0.5	63	0.700001	0.500001	0.400001
0.4	81	0.700001	0.500001	0.420001
0.3	110	0.700002	0.500002	0.440001
0.3	167	0.700003	0.500004	0.460003
0.1	324	0.700009	0.500009	0.480008

6.4   The Convergence of Fuzzy δ Learning Rule

In this section, we shall prove that fuzzy δ learning is convergent.

Theorem 1   Let $\{\mathbf{W}(k)|k = 1, 2, \cdots\}$ be the weight matrix sequence in fuzzy δ learning rule. Then $\mathbf{W}(k)$ must be convergent.

Proof   From Expression (6.33), we have the following two cases:
Case 1:   If $w_{ij}(k) \wedge a_{si} > b_{sj}$, then

$$b'_{sj} = \bigvee_{i=1}^{n} (w_{ij}(k) \wedge a_{si}) > b_{sj}.$$

Therefore

$$\delta_{sj} = b'_{sj} - b_{sj} > 0.$$

As $\eta > 0$, we know that

$$w_{ij}(k+1) = w_{ij}(k) - \eta \delta_{sj} < w_{ij}(k).$$

Case 2:   If $w_{ij}(k) \wedge a_{si} \leq b_{sj}$, then

$$w_{ij}(k+1) = w_{ij}(k).$$

Hence, based on the two cases, we always have

$$w_{ij}(k+1) \leq w_{ij}(k),$$

which means that the sequence $\{\mathbf{W}(k)\}$ is a monotonous decrease sequence.
    Besides, $\{\mathbf{W}(k)\}$ is bounded, because

$$\mathbf{0} \subset \mathbf{W}(k) \subset \mathbf{I},$$

where $\mathbf{0}$ is a null matrix and $\mathbf{I}$ is an unit matrix. Clearly, $\{\mathbf{W}(k)\}$ must be convergent.   **Q.E.D.**

6.5   Conclusions

In this chapter, we introduced the basic structure of a fuzzy neuron and fuzzy neural networks. First, we described what a fuzzy neuron is by means of multifactorial functions. Then the definition of fuzzy neural networks was given by using fuzzy neurons. We also described a fuzzy δ learning algorithm to solve the weights of the $FN(\bigvee, \bigwedge)$ type of fuzzy neural network. An example was also given. At last, we proved that the fuzzy δ learning algorithm must be convergent.

References

1. C. T. Lin and C. S. G. Lee, *Neural Fuzzy Systems*, Prentice-Hall, Englewood Cliffs, 1996.

2. B. Kosko, Fuzzy cognitive, *Journal of Man-machine Studies*, Vol. 24, pp. 65-75, 1986.

3. B. Kosko, *Neural Networks and Fuzzy Systems*, Prentice-Hall, Englewood Cliffs, 1990.

4. S. G. Raniuk and L. O. Hall, Fuzznet: towards a fuzzy connectionist expert system development tool, *Proceedings of IJCNN-90-WASH-DC*, pp. 483-486, 1989.

5. G. A. Carpenter, S. Grossberg, and D. B. Rosen, Fuzzy ART : Fast stable learning and categorization of analog patterns by an adaptive resonance system, *Neural Networks*, Vol. 4, pp. 759-771, 1991.

6. G. A. Carpenter, S. Grossberg, and D. B. Rosen, Fuzzy ART: An adaptive resonance algorithm for rapid stable classification of analog patterns, *International Joint Conference on Neural Networks*, IEEE Service Center, Piscataway, NJ, 1991.

7. G. A. Carpenter, S. Grossberg, and D. B. Rosen, A neural networks realization of Fuzzy ART, *Technical Report CAS/CNS-91-021*, Boston University, Boston.

8. T. Yamakawa and A. Tomoda, A fuzzy neuron and its application to pattern recognition, *3rd IFSA Congress*, pp. 330-338, 1989.

9. T. Furura, A. Kokubu, and T. Sakamoto, NFS: Neuro-fuzzy inference system, *37th meeting of IPSJ 3J-4*, pp. 1386-1387, 1988.

10. H. Takagi, Fusion technology of fuzzy theory and neural networks-survey and future directions, *Proceedings of the International Conference on Fuzzy Logic and Neural Networks*, Japan, 1990.

11. H. X. Li, Multifactorial functions in fuzzy sets theory, *Fuzzy Sets and Systems*, Vol. 35, pp. 69-84, 1990.

12. H. X. Li, Multifactorial fuzzy sets and multifactorial degree of nearness, *Fuzzy Sets and Systems*, Vol. 19, No. 3, pp. 291-298, 1986.

13. X. Li, *Fuzzy Neural Networks and Its Applications*, Guizhou Scientific and Technologic Press, Guizhou, China, 1994.

Chapter 7

Mathematical Essence and Structures of Feedback Neural Networks and Weight Matrix Design

This chapter focuses on mathematical essence and structures of neural networks and fuzzy neural networks, especially on discrete feedback neural networks. We begin with review of Hopfield networks and discuss the mathematical essence and the structures of discrete feedback neural networks. First, we discuss a general criterion on the stability of networks, and we show that the energy function commonly used can be regarded as a special case of the criterion. Second, we show that the stable points of a network can be converted as the fixed points of some function, and the weight matrix of the feedback neural networks can be solved from a group of systems of linear equations. Last, we point out the mathematical base of the outer-product learning method and give several examples of designing weight matrices based on multifactorial functions.

7.1 Introduction

In previous chapters, we have discussed in detail the mathematical essence and structures of feedforward neural networks. Here, we study the mathematical essence and structures of feedback neural networks, namely, the Hopfield networks [1]. Figure 1 illustrates a single-layer Hopfield net with n neurons, where $u_1, u_2, \cdots, u_n$ represent the n neurons. As for the feedback networks, $x_1, x_2, \cdots, x_n$ stand for n input variables as well as n output variables that feedbacks to the inputs. $b_1, b_2, \cdots, b_n$ are outer input variables, which usually are treated as "the first impetus", then they are removed and the network will continue to evolve itself. w_{ij} $(i, j = 1, 2, \cdots, n)$ are connection weights, $w_{ij} = w_{ji}$ and $w_{ii}{=}0$. The activation functions of the neurons are denoted by φ_i, where the threshold values are θ_i.

For a time series $t_0, t_1, t_2, \cdots, t_k, \cdots$, the state equations of the network are:

$$x_i(t_{k+1}) = \varphi_i\Big(\sum_{j=1}^{n} w_{ij}(t_k)x_j(t_k) - \theta_i \Big), \quad i = 1, 2, \cdots, n, \ k = 0, 1, 2, \cdots.$$

the following:

$$x_i(k+1) = \varphi_i\Big(\sum_{j=1}^{n} w_{ij}(k)x_j(k) - \theta_i\Big), \quad i = 1, 2, \cdots, n, \ k = 0, 1, 2, \cdots \tag{7.1}$$

where $w_{ij}(0)$ and $x_j(0) \overset{\triangle}{=} b_j$ are the initial values of $w_{ij}(k)$ and $x_j(k)$, respectively.

For the sake of convenience, the threshold values θ_i can be merged into φ_i (see Chapter 3). So expression (7.1) becomes the following:

$$x_i(k+1) = \varphi_i\Big(\sum_{j=1}^{n} w_{ij}(k)x_j(k)\Big), \quad i = 1, 2, \cdots, n, \ k = 0, 1, 2, \cdots \tag{7.2}$$

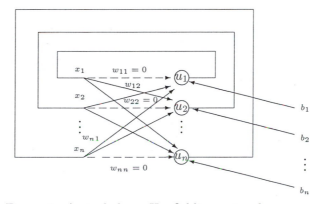

Figure 1   A single-layer Hopfieldian network

When the activation functions φ_i $(i = 1, 2, \cdots, n)$ are all invertible, Expression (7.2) has the following form:

$$\varphi_i^{-1}\big(x_i(k+1)\big) = \sum_{j=1}^{n} w_{ij}(k)x_j(k), \quad i = 1, 2, \cdots, n, \ k = 0, 1, 2, \cdots \tag{7.3}$$

Particularly, when $\varphi_i(i = 1, 2, \cdots, n)$ are all identical functions, Equation (7.3) becomes more simple form:

$$x_i(k+1) = \sum_{j=1}^{n} w_{ij}(k)x_j(k), \quad i = 1, 2, \cdots, n, \ k = 0, 1, 2, \cdots \tag{7.4}$$

Now if we write $\mathbf{W}(k) = (w_{ij}(k))_{n \times n}$, $\mathbf{X}(k) = (x_1(k), x_2(k), \cdots, x_n(k))^T$ and $\mathbf{\Phi} = (\varphi_1, \varphi_2, \cdots, \varphi_n)^T$, Equation (7.2) can be written as the following matrix form:

$$\mathbf{X}(k+1) = \mathbf{\Phi}(\mathbf{W}(k)\mathbf{X}(k)) \tag{7.5}$$

where for Φ acting on a vector $\mathbf{A}=(a_1, a_2, \cdots, a_n)^T$ means that $\Phi(\mathbf{A}) = (\varphi_1(a_1), \varphi_2(a_2), \cdots, \varphi_n(a_n))^T$. From (7.5) we have:

$$
\begin{aligned}
\mathbf{X}(k+1) &= \Phi\Big(\mathbf{W}(k)\Phi(\mathbf{W}(k-1)\mathbf{X}(k-1))\Big) \\
&= \Phi\Big(\mathbf{W}(k)\Phi(\mathbf{W}(k-1)\Phi(\mathbf{W}(k-2)\mathbf{X}(k-1)))\Big) \\
&= \cdots = \Phi\Big(\mathbf{W}(k)\Phi(\mathbf{W}(k-1)\Phi(\cdots \mathbf{W}(0)\mathbf{X}(0)\cdots))\Big) \ .
\end{aligned}
\tag{7.6}
$$

Actually, Equation (7.5) is an iteration procedure and from (7.6) we know that the iteration procedure is just a modifying procedure of the weight matrix:

$$
\mathbf{W}(0) \longrightarrow \mathbf{W}(1) \cdots \longrightarrow \mathbf{W}(k) \longrightarrow \cdots
\tag{7.7}
$$

Especially when the activation functions φ_i are all linear functions, $\varphi_i(x) = a_i x + b_i$, i.e., $\Phi(\mathbf{X}) = \mathbf{A}\mathbf{X} + \mathbf{B}$, where $\mathbf{A}=(a_1, a_2, \cdots, a_n)^T$, $\mathbf{B}=(b_1, b_2, \cdots, b_n)^T$ and the product of two vectors means Hardmard's product, for example, $\mathbf{A}\mathbf{X} \triangleq (a_1 x_1, a_2 x_2, \cdots, a_n x_n)$. In this case, Equation (7.6) changes to the following form:

$$
\begin{aligned}
\mathbf{X}(k+1) &= \mathbf{W}(k)\mathbf{W}(k-1)\cdots\mathbf{W}(1)\mathbf{W}(0)\mathbf{A}^{k+1}\mathbf{X}(0) + (\mathbf{A}^k + \mathbf{A}^{k-1} + \cdots + \mathbf{A} + \mathbf{I})\mathbf{B} \\
&= \Big(\prod_{j=0}^{k} \mathbf{W}(j)\Big)\mathbf{A}^{k+1}\mathbf{X}(0) + \mathbf{B}\sum_{j=0}^{k} \mathbf{A}^j
\end{aligned}
\tag{7.8}
$$

where $\mathbf{I}=(1, 1, \cdots, 1)^T$ and $\mathbf{A}^c = \mathbf{I}$.

Furthermore, when $\mathbf{W}(k) = \mathbf{W}(0)$ $(k = 1, 2, \cdots)$, expression (7.8) becomes a simple form:

$$
\mathbf{X}(k+1) = \Big(\mathbf{W}(0)\mathbf{A}\Big)^{k+1}\mathbf{X}(0) + \mathbf{B}\sum_{j=0}^{k} \mathbf{A}^j \ .
\tag{7.9}
$$

Note 1. In (7.5), denote $f(\mathbf{X}(k)) \triangleq \Phi\Big(\mathbf{W}(k)\mathbf{X}(k)\Big)$, and equation (7.5) becomes an equation of function:

$$
\mathbf{X}(k+1) = f(\mathbf{X}(k))
\tag{7.10}
$$

which is just an interaction function. If (7.10) is convergent, i.e., there exists a number k_0 such that $\mathbf{X}(k_0 +1)=\mathbf{X}(k_0)$, then $\mathbf{X}(k)=f(\mathbf{X}(k))$ for all $k \geq k_0$, which means the stable points of the net are the fixed points of function $f(\mathbf{X})$. In other words, the stable points of the net are the solutions of the following function equation:

$$
\mathbf{X} = f(\mathbf{X}) \ .
\tag{7.11}
$$

This inspires us to study the stable points of feedback neural networks by means of the fixed point theory in functional analysis.

Note 2. In expression (7.9), when the activation functions φ_i are all identical functions, we have

$$
\mathbf{X}(k) = \Big(\mathbf{W}(0)\Big)^{k}\mathbf{X}(0) \ .
\tag{7.12}
$$

In particular, when the dimension $n = 1$, and we denote $x_k \triangleq \mathbf{X}(k)$ and $a = \mathbf{W}(0)$, then (7.12) is written as

$$x_k = x_0 a^k . \tag{7.13}$$

Clearly, $x_0 = 0$ is a fixed point of the mapping and its locus is $x_k = x_0$ that does not relate to k. When $a > 1$, x_k increases according to index law and the motion locus is divergent; when $0 \leq a < 1$, x_k decreases according to index law, and $x_k \to 0$, when $k \to \infty$, which means that the locus is attracted to the fixed point $x_0 = 0$. If $a = -1$, then it expresses an oscillatory system and the oscillatory period is 2. For general n dimension case, the weight matrix $\mathbf{W}(k)$ and the initial value $\mathbf{X}(0)$ decisively influence the system.

Note 3. When all the activation functions φ_i are proportional functions, $\varphi_i(x) = a_i x$, i.e., $\mathbf{\Phi}(\mathbf{X}) = \mathbf{AX}$, (7.8) has

$$\mathbf{X}(k+1) = \Big(\prod_{j=0}^{k} \mathbf{W}(j) \Big) \mathbf{A}^{k+1} \mathbf{X}(0) . \tag{7.14}$$

It is important that (7.2) becomes the following:

$$x_i(k+1) = \sum_{j=1}^{n} w_{ij}(k)\varphi_i(x_j(k)), \quad i = 1, 2, \cdots, n, \ k = 0, 1, 2, \cdots \tag{7.15}$$

In the next section, our discussion will follow this expression.

7.2 A General Criterion on the Stability of Networks

We now consider the stability of discrete Hopfield network. In this case, as referred to in Equation (7.15), we define the limit the ranges of φ_i ($i = 1, 2, \cdots, n$) to [-1, 1].

In Equation (7.15), the right side of the equation is regarded as the output of neuron u_i, at time k, then we denote $y_i(k)$ as $x_i(k+1)$, i.e.,

$$y_i(k) = \sum_{j=1}^{n} w_{ij}(k)\varphi_i(x_j(k)) . \tag{7.16}$$

Regard $y_1(k), y_2(k), \cdots, y_n(k)$ as the n-dimension truth-value of the network output, $(y_1(k), y_2(k), \cdots, y_n(k))$. If $(y_1(k), y_2(k), \cdots, y_n(k))$ is projected on a one-dimension space linearly, we have an aggregated truth-value output of the system at time k, or called the whole quantity of the truth-value of the system output at time k, denoted by

$$E(k) = \sum_{i=1}^{n} a_i y_i(k) \tag{7.17}$$

where a_i $(i = 1, 2, \cdots, n)$ are weights. Because $\varphi_j(x_i(k))$ may be viewed as the activation degree (or called as excitation degree) of $x_i(k)$ with respect to φ_j, we can naturally consider $\varphi_j(x_i(k))$ as i-th weight, i.e., $a_i = \varphi_j(x_i(k))$,

So (7.17) becomes the following:

$$
\begin{aligned}
E(k) &= \sum_{i=1}^{n} \varphi_j(x_i(k)) y_i(k) = \sum_{i=1}^{n} \varphi_j(x_i(k)) \sum_{j=1}^{n} w_{ij}(k) \varphi_i(x_j(k)) \\
&= \sum_{i=1}^{n} \sum_{j=1}^{n} w_{ij}(k) \varphi_i(x_j(k)) \varphi_j(x_i(k))
\end{aligned}
\tag{7.18}
$$

It is easy to think that the stable points of the network should make $E(k)$ reach its maximal values. Therefore, we may reasonably consider Equation (7.18) as a criterion of the network stability.

Note 4. If $E'(k) = -E(k)$, that is

$$
E'(k) = -\sum_{i=1}^{n} \sum_{j=1}^{n} w_{ij}(k) \varphi_i(x_j(k)) \varphi_j(x_i(k))
\tag{7.19}
$$

then $E'(k)$ can be considered as the whole quantity of the false-value of the system output at time k. Of course, the stable points of the network should make $E'(k)$ reach its minimal values. So Equation (7.19) is also a criterion of the network stability.

Note 5. In Equation (7.18), the terms "$\varphi_i(x_j(k)) \cdot \varphi_j(x_i(k))$" means a kind of mutual action or influence between $\varphi_i(x_j(k))$ and $\varphi_j(x_i(k))$. If we add an independent action of $\varphi_i(x_j(k))$ to Equation (7.18), then (7.18) becomes the following:

$$
E(k) = \sum_{i=1}^{n} \sum_{j=1}^{n} w_{ij}(k) \varphi_i(x_j(k)) \varphi_j(x_i(k)) + \sum_{i=1}^{n} \sum_{j=1}^{n} \theta_{ij}(k) \varphi_i(x_j(k)) \ .
\tag{7.20}
$$

In the same way, (7.19) has the following form:

$$
E'(k) = -\sum_{i=1}^{n} \sum_{j=1}^{n} w_{ij}(k) \varphi_i(x_j(k)) \varphi_j(x_i(k)) - \sum_{i=1}^{n} \sum_{j=1}^{n} \theta_{ij}(k) \varphi_i(x_j(k)) \ ,
\tag{7.21}
$$

where the constants, $\theta_{ij}(k)$, can be viewed as weights that join $\varphi_i(x_j(k))$ together.

Note 6. If we ignore the action of $\varphi_i(x_j(k))$ when $i \neq j$, Equation (7.21) is simplified as follows:

$$
E''(k) = -\sum_{i=1}^{n} \sum_{j=1}^{n} w_{ij}(k) \varphi_i(x_i(k)) \varphi_j(x_j(k)) - \sum_{i=1}^{n} \theta_i(k) \varphi_i(x_i(k)) \ ,
\tag{7.22}
$$

where $\theta_{ij}(k)$ is also simplified as θ_i. By adding a constant $\frac{1}{2}$ to the above equation, we have

$$E^*(k) = -\frac{1}{2}\sum_{i=1}^{n}\sum_{j=1}^{n} w_{ij}(k)\varphi_i(x_i(k))\varphi_j(x_j(k)) - \sum_{i=1}^{n}\theta_i(k)\varphi_i(x_i(k)) \ . \qquad (7.23)$$

If we define $v_i(k) \overset{\triangle}{=} \varphi_i(x_i(k))$, then Equation (7.23) becomes

$$E^*(k) = -\frac{1}{2}\sum_{i=1}^{n}\sum_{j=1}^{n} w_{ij}(k)v_i(k)v_j(k) - \sum_{i=1}^{n}\theta_i(k)v_i(k) \ . \qquad (7.24)$$

If we omit the discrete time variable k, then (7.24) can be written as the following:

$$E^* = -\frac{1}{2}\sum_{i=1}^{n}\sum_{j=1}^{n} w_{ij}(k)v_iv_j - \sum_{i=1}^{n}\theta_i v_i \ . \qquad (7.25)$$

Obviously, Equation (7.25) is just the energy function proposed by Hopfield in 1984 [1-3]. This means that the energy function is a special case of Equation (7.21), and Equation (7.20) can be regarded as a general criterion on the stability of networks.

7.3 Generalized Energy Function

First of all, let us recall the physics background of the energy function in the Hopfield network. In fact, in a lump of ferromagnet, the rotation of a ferromagnetic molecule has two directions, denoted by 1 and -1, respectively. Let $p_1, p_2, \cdots, p_n$ be n ferromagnetic molecules. Then $p_i \in \{1, -1\}$ $(i = 1, 2, \cdots, n)$, where a ferromagnetic molecule and its rotation are denoted by the same sign p_i. We use J_{ij} to represent the action between p_i and p_j. Obviously, $J_{ij} = J_{ji}$. The Hamiltonian function of this kind of material should have the following form:

$$H = -\frac{1}{2}\sum_{i=1}^{n}\sum_{j=1}^{n} J_{ij}p_ip_j - \sum_{i=1}^{n} H_ip_i \ , \qquad (7.26)$$

where H_i is the disturbance of outer random field being to p_i. The action result of whole molecules in the ferromagnet should make H minimal. So Hopfield defines an energy function of the network as follows, being similar to Equation (7.26):

$$E = -\frac{1}{2}\sum_{i=1}^{n}\sum_{j=1}^{n} w_{ij}(k)v_iv_j - \sum_{i=1}^{n}\theta_i v_i \ ,$$

which is Equation (7.25), where v_i is the output of i-th neuron and θ_i is also outer random action.

It is easy to comprehend the fact that, for a piece of ferromagnet, the outer random action, $\sum_{i=1}^{n}\theta_i v_i$ is important, but for a neural network, it is not so important that it can be omitted. Besides, the "energy" of an object means the whole of its molecules action is mutually attractive. So for a single-layer Hopfield network, a simplified enery function can be considered as follows.

$$E_1(k) \overset{\triangle}{=} \sum_{i=1}^{n} \sum_{j=1}^{n} w_{ij}(k) x_i(k) x_j(k) \quad . \tag{7.27}$$

And for the case of continuous time variable, we have

$$E_1(t) \overset{\triangle}{=} \sum_{i=1}^{n} \sum_{j=1}^{n} w_{ij}(t) x_i(t) x_j(t) \quad . \tag{7.28}$$

Extending this concept to fuzzy neurons, we assume that the activation functions φ_i are nonnegative, especially $\varphi_i(R) \subset [0, 1]$, i.e., the ranges of φ_i are $[0, 1]$. Let $\top$ be a triangular norm (T-norm) and $\bot$ be its co-T-norm. We suggest a following generalized energy function:

$$E_2(k) \overset{\triangle}{=} \bot_{i=1}^{n} \bot_{j=1}^{n} \Big(w_{ij}(k) \top x_i(k) \top x_j(k) \Big) \quad . \tag{7.29}$$

Example 1. Take $\bot = \vee$ and $\top = \wedge$. We have

$$E_3(k) = \bigvee_{i=1}^{n} \bigvee_{j=1}^{n} \Big(w_{ij}(k) \wedge x_i(k) \wedge x_j(k) \Big) \quad . \tag{7.30}$$

Example 2. The following equations are also the special cases of Equation (7.29):

$$E_4(k) = \bigvee_{i=1}^{n} \bigvee_{j=1}^{n} \Big[w_{ij}(k)(x_i(k) \wedge x_j(k)) \Big], \tag{7.31}$$

$$E_5(k) = \bigvee_{i=1}^{n} \bigvee_{j=1}^{n} \Big[w_{ij}(k) \wedge (x_i(k) x_j(k)) \Big], \tag{7.32}$$

$$E_6(k) = \bigvee_{i=1}^{n} \bigvee_{j=1}^{n} \Big[w_{ij}(k) x_i(k) x_j(k) \Big], \tag{7.33}$$

$$E_7(k) = \bigvee_{i=1}^{n} \bigvee_{j=1}^{n} \Big[w_{ij}(k) \wedge x_i(k) \wedge x_j(k) \Big], \tag{7.34}$$

$$E_8(k) = \sum_{i=1}^{n} \sum_{j=1}^{n} \Big[w_{ij}(k)(x_i(k) \wedge x_j(k)) \Big], \tag{7.35}$$

and

$$E_9(k) = \sum_{i=1}^{n} \sum_{j=1}^{n} \Big[w_{ij}(k) \wedge (x_i(k) x_j(k)) \Big]. \tag{7.36}$$

7.4   Learning Algorithm of Discrete Feedback Neural Networks

In this section, we describe the learning of the Hopfield network. Let $\mathbf{V}_l = (v_{l1}, v_{l2}, \cdots, v_{ln})^T$, $(l = 1, 2, \cdots, p)$ be p sample points memorized by the network, which should be the fixed points of the function $f(\mathbf{X})$ (see Equation (7.11)), then we have $f(\mathbf{V}_l) = \mathbf{V}_l$ $(l = 1, 2, \cdots, p)$, that is (from Equation (7.5)), $\mathbf{\Phi}(\mathbf{WV}_l) = \mathbf{V}_l$ $(l = 1, 2, \cdots, p)$. If there exists the inverse function of $\mathbf{\Phi}$, $\mathbf{\Phi}^{-1} = (\varphi_1^{-1}, \varphi_2^{-1}, \cdots, \varphi_n^{-1})^T$, then

$$\mathbf{WV}_l = \mathbf{\Phi}^{-1}(\mathbf{V}_l), \quad l = 1, 2, \cdots, p \ . \tag{7.37}$$

The above equation means the following:

$$
\begin{aligned}
w_{11}v_{11} + w_{12}v_{12} + \cdots + w_{1n}v_{1n} &= \varphi_1^{-1}(v_{11}), \\
w_{21}v_{11} + w_{22}v_{12} + \cdots + w_{2n}v_{1n} &= \varphi_2^{-1}(v_{12}), \\
&\cdots\cdots \\
w_{n1}v_{11} + w_{n2}v_{12} + \cdots + w_{nn}v_{1n} &= \varphi_n^{-1}(v_{1n}), \\
w_{11}v_{21} + w_{12}v_{22} + \cdots + w_{1n}v_{2n} &= \varphi_1^{-1}(v_{21}), \\
w_{21}v_{21} + w_{22}v_{22} + \cdots + w_{2n}v_{2n} &= \varphi_2^{-1}(v_{22}), \\
&\cdots\cdots \\
w_{n1}v_{21} + w_{n2}v_{22} + \cdots + w_{nn}v_{2n} &= \varphi_n^{-1}(v_{22}), \\
&\cdots\cdots \\
w_{11}v_{p1} + w_{12}v_{p2} + \cdots + w_{1n}v_{pn} &= \varphi_1^{-1}(v_{p1}), \\
w_{21}v_{p1} + w_{22}v_{p2} + \cdots + w_{2n}v_{pn} &= \varphi_2^{-1}(v_{p2}), \\
&\cdots\cdots \\
w_{n1}v_{p1} + w_{n2}v_{p2} + \cdots + w_{nn}v_{pn} &= \varphi_2^{-1}(v_{pn}).
\end{aligned}
$$

Let us rearrange the above equations as follows:

$$
\left.
\begin{aligned}
v_{11}w_{11} + v_{12}w_{12} + \cdots + v_{1n}w_{1n} &= \varphi_1^{-1}(v_{11}) \\
v_{21}w_{11} + v_{22}w_{12} + \cdots + v_{2n}w_{1n} &= \varphi_1^{-1}(v_{21}) \\
&\cdots\cdots \\
v_{p1}w_{11} + v_{p2}w_{12} + \cdots + v_{pn}w_{1n} &= \varphi_1^{-1}(v_{p1})
\end{aligned}
\right\} \tag{7.38.1}
$$

$$
\left.
\begin{aligned}
v_{11}w_{21} + v_{12}w_{22} + \cdots + v_{1n}w_{2n} &= \varphi_2^{-1}(v_{12}) \\
v_{21}w_{21} + v_{22}w_{22} + \cdots + v_{2n}w_{2n} &= \varphi_2^{-1}(v_{22}) \\
&\cdots\cdots \\
v_{p1}w_{21} + v_{p2}w_{22} + \cdots + v_{pn}w_{2n} &= \varphi_2^{-1}(v_{p2})
\end{aligned}
\right\} \tag{7.38.2}
$$

$$\cdots\cdots$$

$$
\left.
\begin{aligned}
v_{11}w_{n1} + v_{12}w_{n2} + \cdots + v_{1n}w_{nn} &= \varphi_1^{-1}(v_{1n}) \\
v_{21}w_{n1} + v_{22}w_{n2} + \cdots + v_{2n}w_{nn} &= \varphi_1^{-1}(v_{2n}) \\
&\cdots\cdots \\
v_{p1}w_{n1} + v_{p2}w_{n2} + \cdots + v_{pn}w_{nn} &= \varphi_1^{-1}(v_{pn})
\end{aligned}
\right\} \ . \tag{7.38.n}
$$

Obviously, (7.38.1)–(7.38.n) are n independent systems of linear equations. In order to solve the weight, let $\mathbf{V} \overset{\triangle}{=} (v_{ij})_{p\times n}$, $\mathbf{W}_i \overset{\triangle}{=} (w_{i1}, w_{i2}, \cdots, w_{in})^T$ and

$$\mathbf{B}_i \overset{\triangle}{=} (\varphi_i^{-1}(v_{1i}), \varphi_i^{-1}(v_{2i}), \cdots, \varphi_i^{-1}(v_{pi}))^T, \ i = 1, 2, \cdots, n.$$

then $(7.38.1)$-$(7.38.n)$ can be written as matrix forms:

$$\mathbf{V}\mathbf{W}_i = \mathbf{B}_i, \quad i = 1, 2, \cdots, n \tag{7.39}$$

Generally speaking, the systems of linear equations shown as (7.39) are contradictory systems. So we should view them as a problem of optimization to solve these systems of linear equations, based on the previous chapter. In fact, let

$$E(\mathbf{W}_l) \stackrel{\triangle}{=} \frac{1}{2} \sum_{i=1}^{n} \left(\sum_{j=1}^{n} v_{ij} w_{ij} - \varphi_l^{-1}(v_{il}) \right)^2, \tag{7.40}$$

and

$$\nabla E(\mathbf{W}_l) = \left(\frac{\partial E(\mathbf{W}_l)}{\partial w_{l1}}, \frac{\partial E(\mathbf{W}_l)}{\partial w_{l2}}, \cdots, \frac{\partial E(\mathbf{W}_l)}{\partial w_{ln}}, \right)^T, \tag{7.41}$$

where $l = 1, 2, \cdots, n$. According to the previous chapter, we have the following interaction equation of gradient descent algorithm:

$$\mathbf{W}_l^{(k+1)} = \mathbf{W}_l^{(k)} - \mu_l^{(k)} \nabla E(\mathbf{W}_l^{(k)}), \quad l = 1, 2, \cdots, n \tag{7.42}$$

where the step factors $\mu_l^{(k)}$ hold the condition:

$$D(\mathbf{W}_l^{(k)} - \mu_l^{(k)} \nabla E(\mathbf{W}_l^{(k)})) = \min_\mu E(\mathbf{W}_l^{(k)} - \mu \nabla E(\mathbf{W}_l^{(k)})). \tag{7.43}$$

Also similar to the previous chapter, we can get the explicit scheme of $\mu_l^{(k)}$ as follows:

$$\mu_l^{(k)} = \frac{(\nabla E(\mathbf{W}_l^{(k)}))^T \nabla E(\mathbf{W}_l^{(k)})}{(\nabla E(\mathbf{W}_l^{(k)}))^T \mathbf{Q} \nabla E(W_l^{(k)})}, \tag{7.44}$$

where $\mathbf{Q} = (q_{ij})_{n \times n}$ and $q_{ij} = \sum_{k=1}^{p} v_{ki} v_{kj}$.

Note 7. When $\mathbf{Q}$ is a positive definite matrix, the unique standing point can be found, for each l, as the following:

$$\mathbf{W}_l^* = -\mathbf{Q}^{-1} \mathbf{D}_l \tag{7.45}$$

where $\mathbf{D}_l = (d_{l1}, d_{l2}, \cdots, d_{ln})^T$ and $d_{li} = -\sum_{k=1}^{p} v_{ki} \varphi_i^{-1}(v_{ki})$.

Obviously, after $\mathbf{W}_l (l = 1, 2, \cdots, n)$ are determined, we obtain the weight matrix $\mathbf{W}$:

$$\mathbf{W} = [\mathbf{W}_1 \, \mathbf{W}_2 \, \cdots \, \mathbf{W}_n]^T, \tag{7.46}$$

that is, $\mathbf{W}$ is represented as a block-wised matrix consisted of $\mathbf{W}_l$ $(l = 1, 2, \cdots, n)$.

7.5 Design Method of Weight Matrices Based on Multifactorial Functions

In this section, we use multifactorial functions discussed in Chapter 6 to design weight matrices. We give the following simple and commonly used multifactorial functions as examples.

Example 3. The following mappings are examples of ASM$_m$-func (m-ary Additive Standard Multifactorial Function) from $[0,1]^m$ to $[0,1]$:

$$\wedge : (x_1, \cdots, x_m) \longmapsto \wedge(x_1, \cdots, x_m) \stackrel{\triangle}{=} \bigwedge_{j=1}^{m} x_j; \tag{7.47}$$

$$\vee : (x_1, \cdots, x_m) \longmapsto \vee(x_1, \cdots, x_m) \stackrel{\triangle}{=} \bigvee_{j=1}^{m} x_j; \tag{7.48}$$

$$\sum : (x_1, \cdots, x_m) \longmapsto \sum(x_1, \cdots, x_m) \stackrel{\triangle}{=} \sum_{j=1}^{m} a_j x_j, \tag{7.49}$$

where $a_j \in [0,1]$ and $\sum_{j=1}^{m} a_j = 1$;

$$M_m(x_1, \cdots, x_m) \stackrel{\triangle}{=} \bigvee_{j=1}^{m} (a_j \wedge x_j), \tag{7.50}$$

where $a_j \in [0,1]$ and $\bigvee_{j=1}^{m} a_j = 1$;

$$M_m(x_1, \cdots, x_m) \stackrel{\triangle}{=} \bigvee_{j=1}^{m} (a_j x_j), \tag{7.51}$$

where $a_j \in [0,1]$ and $\bigvee_{j=1}^{m} a_j = 1$;

$$M_m(x_1, \cdots, x_m) \stackrel{\triangle}{=} \left(\prod_{j=1}^{m} x_j \right)^{\frac{1}{m}}; \tag{7.52}$$

$$M_m(x_1, \cdots, x_m) \stackrel{\triangle}{=} \left(\sum_{j=1}^{m} a_j x_j^p \right)^{\frac{1}{p}}, \tag{7.53}$$

where $p > 0$, $a_j \in [0,1]$ and $\sum_{j=1}^{m} a_j = 1$.

Example 4. The mapping $\prod : [0,1]^m \to [0,1]$ defined as the following is a NASM$_m$-func (m-ary Nonadditive Standard Multifactorial Function):

$$(x_1, \cdots, x_m) \longmapsto \prod(x_1, \cdots, x_m) \stackrel{\triangle}{=} \prod_{j=1}^{m} x_j \quad . \tag{7.54}$$

With these examples, we now consider the design of weight matrixes. Suppose there exist p sample points memorized by the network, $V_l = (v_{l1}, v_{l2}, \cdots, v_{ln})^T$, $l = 1, 2, \cdots, n$. Take a p-ary multifactorial function M_p (additive or nonadditive) and a 2-ary multifactorial function M_2 (additive or non-additive). We can form the connection weights w_{ij} according to the following expression:

$$w_{ij} = \begin{cases} M_P(M_2(v_{1i}, v_{1j}), M_2(v_{2i}, v_{2j}), \cdots, M_2(v_{pi}, v_{pj})) & , \quad i \neq j \\ 0 & , \quad i = j \end{cases} \qquad (7.55)$$

where M_2 is required to hold symmetry: $M_2(x, y) = M_2(y, x)$; for instance, Equations (7.47), (7.48), (7.52) and (7.54) are of symmetry, which can ensure that the w_{ij} formed by (7.55) are of symmetry: $w_{ij} = w_{ji}$.

Example 5. M_2 has taken the form shown in Equation (7.54) from Example 4 and M_p for the one of (7.49). We have

$$w_{ij} = \begin{cases} \displaystyle\sum_{l=1}^{p} a_l v_{li} v_{lj} & , \quad i \neq j \\ 0 & , \quad i = j \end{cases} \qquad (7.56)$$

In particular, when $a_l = \frac{1}{p}$ $(l = 1, 2, \cdots, p)$, (7.56) becomes the following:

$$w_{ij} = \begin{cases} \displaystyle\frac{1}{p}\sum_{l=1}^{p} v_{li} v_{lj} & , \quad i \neq j \\ 0 & , \quad i = j \end{cases} \qquad (7.57)$$

Furthermore, if $\frac{1}{p}$ is replaced by a general parameter α, then Equation (7.57) is generalized as follows:

$$w_{ij} = \begin{cases} \displaystyle\alpha\sum_{l=1}^{p} v_{li} v_{lj} & , \quad i \neq j \\ 0 & , \quad i = j \end{cases} \qquad (7.58)$$

which is just the well-known expression, called the outer-produce weight matrix design method.

Example 6. M_2 is taken for $\wedge$ shown as Expression (7.47), i.e., $M_2(x, y) = x \wedge y$, and M_p still for Expression (7.49). Then

$$w_{ij} = \begin{cases} \displaystyle\sum_{l=1}^{p} a_l(v_{li} \wedge v_{lj}) & , \quad i \neq j \\ 0 & , \quad i = j \end{cases} \qquad (7.59)$$

Similar to (7.57) and (7.58), we also have

$$w_{ij} = \begin{cases} \displaystyle\frac{1}{p}\sum_{l=1}^{p} (v_{li} \wedge v_{lj}) & , \quad i \neq j \\ 0 & , \quad i = j \end{cases} \qquad (7.60)$$

and

$$w_{ij} = \begin{cases} \alpha \sum_{l=1}^{p} (v_{li} \wedge v_{lj}) & , \quad i \neq j \\ 0 & , \quad i = j \end{cases} \quad (7.61)$$

so that Expression (7.61) can be viewed as a kind of generalized or semi-fuzzy outer-product weight matrix design method.

Example 7. If taking $M_2 = \wedge$ and $M_p = \vee$ that is $M_p(x_1, x_2, \cdots, x_p) = \bigvee_{l=1}^{p} x_l$, then we have

$$w_{ij} = \begin{cases} \bigvee_{l=1}^{p} (v_{li} \wedge v_{lj}) & , \quad i \neq j \\ 0 & , \quad i = j \end{cases} \quad (7.62)$$

which should be regarded as a kind of fuzzy outer-product weight matrix design method.

Example 8. Let $M_2 = \prod$, i.e., $M_2(x, y) = xy$ and $M_p = \vee$. We have

$$w_{ij} = \begin{cases} \bigvee_{l=1}^{p} (v_{li} v_{lj}) & , \quad i \neq j \\ 0 & , \quad i = j \end{cases} \quad (7.63)$$

which can be regarded as a kind of generalized or semi-fuzzy outer-product weight matrix design method.

Example 9. M_2 is taken for the form of Expression (7.52), i.e., $M_2(x, y) = \sqrt{xy}$ and $M_P = \sum$, where $a_l = \frac{1}{p} (l = 1, 2, \cdots, p)$. We have

$$w_{ij} = \begin{cases} \dfrac{1}{p} \sum_{l=1}^{p} \sqrt{v_{li} v_{lj}} & , \quad i \neq j \\ 0 & , \quad i = j \end{cases} \quad (7.64)$$

which is another kind of generalized or semi-fuzzy outer-product weight matrix design method.

7.6   Conclusions

In this chapter, we concentrated on discrete feedback neural networks. We summarize the result as follows.

(a) It is pointed out that, if the relation between the input and output of a feedback neural network is considered to be a function, the stable points of the network are just the fixed points of the function. This means that feedback neural networks can be studied by means of fixed point theory in mathematics.

(b)   A general criterion on the stability of feedback neural networks is given. The Hopfield energy function can be regarded as a special case of the general criterion. We introduce the generalized energy function and give examples.

(c)   The mathematical essence on learning algorithm of discrete feedback neural networks is revealed and a speed learning algorithm is given.

(d)   We introduce the design method of weight matrix based on multifactorial functions. The well-known outer-product weight matrix design method is only a special case of our method. Also, we study the generalized or semi-fuzzy or fuzzy outer-product weight matrix design methods.

References

1. J. J. Hopfield, Neurons with graded response have collective computational properties like those of two-state neurons, *Proceedings of the National Academy of Sciences,* USA, 81, pp. 3088-3092, 1984.

2. B. Kosko, *Neural Networks and Fuzzy Systems,* Prentice-Hall, Englewood Cliffs, 1992.

3. Y. H. Pao, *Adaptive Pattern Recognition and Neural Networks,* Addison-Wesley, New York, 1989.

Chapter 8

Generalized Additive Weighted Multifactorial Function and its Applications to Fuzzy Inference and Neural Networks

In this chapter, a new family of multifactorial function, called generalized additive weighted multifactorial function, is proposed and discussed in detail. First, its properties in n-dimensional space are discussed and then our results are extended to the infinite dimensional space. Second, the implication of its constant coefficients is explained by fuzzy integral. Finally, its application in fuzzy inference is discussed and we show that it is a usual kind of composition operator in fuzzy neural networks.

8.1 Introduction

Chapter 6 has detailed definitions and properties of multifactorial functions. Mulitfactorial functions, which can be used to compose the "states", are very effective methods in multi-criteria fuzzy decision-making [1]. In addition, the multifactorial function is used to define fuzzy perturbation function [2]. In [3, 4], by means of multifactorial functions, multifactorial fuzzy sets and the multifactorial degree of nearness are given and they are used to deal with multifactorial pattern recognition and clustering analysis with fuzzy characteristics.

The simple additive weighted aggregation operator (SAW) is a usual ASM-func and is used widely in many aspects. In the following, we generalize SAW into a very general class of standard multifactorial functions called generalized additive weighted multifactorial functions $(M(\bot, \top))$. We provide the conditions under which the generalized additive weighted multifactorial function is ASM-func, considering the continuous t-conorms restricted to Archimedean ones or the Maximum operator $\vee$. We then extend the results to the infinite-dimensional multifactorial function. The implication of its constant coefficients is discussed by using fuzzy integral as a tool, especially, the condition under which the constant coefficient implies the weight value. Finally, we show an application in fuzzy inference.

8.2   On Multifactorial Functions

In this section, we shall first briefly review multifactorial functions discussed in Chapter 6. Let $f_1, f_2, \cdots, f_n$ be mutually independent factors with their own state space, X_i; and $A_1, A_2, \cdots, A_n$ be fuzzy sets of $X_1, X_2, \cdots, X_n$. As usual, they are described by their membership functions, $\varphi_1, \varphi_2, \cdots, \varphi_n$, where $\varphi_i : X_i \to [0, 1]$ $(i = 1, \cdots, n)$.

A mapping $M \colon X_1 \times X_2 \times \cdots \times X_n \longrightarrow [0, 1]$ can be viewed as a n-dimensional multifactorial function. And a mapping $M \colon [0, 1]^n \longrightarrow [0, 1]$ is regarded as a n-dimensional standard multifactorial function.

Let $X = X_1 \times X_2 \times \cdots \times X_n$, $\varphi = (\varphi_1, \varphi_2, \cdots, \varphi_n)$, and M be a n-dimensional standard multifactorial function. Denote $M' = M \circ \varphi$. M' is then a n-dimensional multifactorial function.

In $[0, 1]^n$ a partial "$\geq$" is defined by

$$X \geq Y \quad \text{iff} \quad (x_j \geq y_j,\ j = 1, \cdots, n),$$

where $X = (x_1, \cdots, x_n)$, $Y = (y_1, \cdots, y_n) \in [0, 1]^n$.

Definition 1   A mapping $M \colon [0, 1]^n \longrightarrow [0, 1]$ is called a n-dimensional Additive Standard Multifactorial (ASM) function if it satisfies:
$(m.1)$ $X \geq Y$ implies $M(X) \geq M(Y)$;
$(m.2)$ $\min_j(x_j) \leq M(X) \leq \max_j(x_j)$; and
$(m.3)$ $M(X)$ is continuous in all its arguments.
We can simply denote it by ASM-func.

Define
$$\mathcal{C}_n = \{M | M \text{ is a } n - \text{dimensional ASM-func}\}.$$

Clearly, if M satisfies the conditions:
(i) $X \geq Y$ implies $M(X) \geq M(Y)$,
(ii) $M(a, \cdots, a) = a$, and
(iii) $M(X)$ is continuous in all its arguments,
then $M \in \mathcal{C}_n$. Moreover, if

$$\bigwedge_{j=1}^{n} x_j \leq M(x_1, x_2, \cdots, x_n) \leq \bigvee_{j=1}^{n} x_j,$$

then $M(a, a, \cdots, a) = a$.

8.3   Generalized Additive Weighted Multifactorial Functions

Definition 2   A mapping $M : [0, 1]^n \longrightarrow [0, 1]$ is called a n-dimensional generalized additive weighted multifactorial function, if

$$M(x_1, \cdots, x_n) = \perp_{j=1}^{n}(a_j \top x_j),$$

where $a_j \in [0,1]$ and $j = 1, \cdots, n$. Denote it by $M(\perp, \top)$, where $\perp(\top)$ is t-conorm (t-norm).

It is easy to know that the following two laws are true:

Distributive Law:
$$a\top(b_1 \perp b_2 \perp \cdots \perp b_n) = (a\top b_1)\perp(a\top b_2)\perp \cdots \perp(a\top b_n), \qquad D(1)$$
$$a\perp(b_1 \top b_2 \top \cdots \top b_n) = (a\perp b_1)\top(a\perp b_2)\top \cdots \top(a\perp b_n). \qquad D(2)$$

Idempotent Law:
$$a\perp a\perp \cdots \perp a = a, \qquad\qquad id(1)$$
$$a\top a\top \cdots \top a = a. \qquad\qquad id(2)$$

Theorem 1
   (a) $D(1) \Rightarrow id(1) \Rightarrow \perp = \vee,$
   (b) $D(2) \Rightarrow id(2) \Rightarrow \top = \wedge.$

Proof   (a)   i)   Take $b_1 = b_2 = \cdots = b_n = 1$. By $D(1)$ we have

$$a\top 1 = (a\top 1)\perp(a\top 1)\perp \cdots \perp(a\top 1),$$

and therefore

$$a\perp a\perp \cdots \perp a = a.$$

ii) $\forall X = (x_1, x_2, \cdots, x_n)$, if $x_n \leq x_{n-1} \leq \cdots \leq x_1$, then

$$x_1 = x_1 \perp 0 \perp 0 \perp \cdots \perp 0 \leq x_1 \perp x_2 \perp \cdots \perp x_n$$
$$\leq x_1 \perp x_1 \perp \cdots \perp x_1 = x_1;$$

hence

$$\perp(x_1, x_2, \cdots, x_n) = x_1 = \bigvee_{j=1}^{n} x_j,$$

and thus $\perp = \vee$.

   (b)   It follows in the similar manner.   **Q.E.D.**

Theorem 2   Let $\top$ be any continuous t-norm. For any $a \in [0,1]$, we have

$$\bigvee_{j=1}^{n} (a\top x_j) = \top(a, \bigvee_{j=1}^{n} x_j).$$

Proof   Since $\bigvee_{j=1}^{n} \top(a, x_j) \geq \top(a, x_j)$, $\forall j \in \{1, \cdots, n\}$, we have

$$\bigvee_{j=1}^{n} \top(a, x_j) \geq \top(a, \bigvee_{j=1}^{n} x_j). \qquad (8.1)$$

For all $j \in \{1, 2, \cdots, n\}$, we have

$$\top(a, x_j) \leq \top(a, \bigvee_{j=1}^{n} x_j).$$

therefore

$$\bigvee_{j=1}^{n} \top(a, x_j) \le \top(a, \bigvee_{j=1}^{n} x_j), \tag{8.2}$$

By using (1) and (2), the result follows.   **Q.E.D.**

From Theorem 1 and Theorem 2, we have the following theorem. However, we will not discuss the proof in detail.

Theorem 3
(a) $D(1) \Leftrightarrow id(1) \Leftrightarrow \bot = \vee.$
(b) $D(2) \Leftrightarrow id(2) \Leftrightarrow \top = \wedge.$

Theorem 4   Let $\top$ be any continuous t-norm. If $\bigvee_{j=1}^{n} a_j = 1$, then $M(\vee, \top) \in \mathcal{C}_n$.

Proof   Clearly $M(\vee, \top)$ satisfies (m.1) and (m.3). By Theorem 3, we have

$$\bigvee_{j=1}^{n} (a_j \top x) = \top(x, \bigvee_{j=1}^{n} a_j) = \top(x, 1) = x, \quad \forall x \in [0, 1].$$

The result is true obviously.   **Q.E.D.**

If $\bot$ is Archimedean and continuous, we have the following conclusion.

Theorem 5   If $\bot_{j=1}^{n} a_j = 1$, then $M(\bot, \wedge) \notin \mathcal{C}_n$.

Let $M(\bot, \top) \triangleq M_\bot$, and $\top(\bot)$ be continuous.

Theorem 6 $X \ge Y$ implies $M_\bot(X) \ge M_\bot(Y)$.

Theorem 7   If $\bot_{j=1}^{n} a_j = 1$, for any t-norm $\top$, then $M_\bot \in \mathcal{C}_n$ iff $\bot = \vee$.

Proof   Necessity: if $\bot_{j=1}^{n} a_j = 1$ and $M_\bot \in \mathcal{C}_n$, $\forall a \in [0, 1]$ $(j = 1, 2, \cdots, n)$, only if $\bot_{j=1}^{n} a_j = 1$, then

$$\bot_{j=1}^{n}(a \top a_1, a \top a_2, \cdots, a \top a_n) = a.$$

Take $a_1 = a_2 = \cdots = a_n = 1$. Then $\bot(1, \cdots, 1) = 1$, and therefore

$$\bot(a, a, \cdots, a) = a.$$

By Theorem 3, we have $\bot = \vee$.
Sufficiency: it is clearly by Theorem 4.   **Q.E.D.**

The following discussion assumes that $\top(\bot)$ is Archimedean continuous.

Theorem 8   Let $\bot$ be a strict t-conorm with normal additive generator g and $\top$ be a strict t-norm with a multiplicative generator h. For all $a \in [0, 1]$, if $\bot_{j=1}^{n} a_j = 1$, and $g \triangleq h$, then $M(a, \cdots, a) = a$, where $g \triangleq h$ implies g is also a multiplication generator of $\top$.

Proof If $g \stackrel{\triangle}{=} h$, then $f = h^\lambda$ with $\lambda \geq 0$. For all $x, y \in [0, 1]$, we have

$$x \top y = h^{-1}(h(x)h(y)) = g^{-1}(g(x)g(y)).$$

Since $\perp_{j=1}^n a_j = 1$, $\sum_{j=1}^n g(a_j) = g(1)$. However

$$\perp_{j=1}^n (a_j \top x_j) = g^{-1}(\sum_{i=1}^n g(a_j)g(x_j)).$$

Thus

$$\begin{aligned} M_\perp(a, a, \cdots, a) &= g^{-1}(\sum_{j=1}^n g(a_j)g(a)) \\ &= g^{-1}(g(a)\sum_{j=1}^n g(a_j)) \\ &= a \top 1 \\ &= a. \end{aligned}$$

Q.E.D.

Theorem 9 Let $\perp$ be a strict t-conorm with normal additive generator g and $\top$ be a strict t-norm with a multiplicative generator h. If $(\perp, \top)$ satisfies $D(1)$, then $g \stackrel{\triangle}{=} h$.

The proof is shown in [5].

Corollary 1 Let (g, h) be the generator group of $(\perp, \top)$. If $\sum_{j=1}^n g(a_j) = g(1)$ and $g \stackrel{\triangle}{=} h$, then $M_\perp \in \mathcal{C}_n$.

Corollary 2 If $(\perp, \top)$ is strict, $\perp$ and $\top$ satisfy $D(1)$, and $\perp_{j=1}^n a_j = 1$, then $M_\perp \in \mathcal{C}_n$.

By using Corollary 1, we can generate ASM-func.

Example 1 Assume $\sum_{j=1}^n a_j = 1$, then $M(\oplus, \cdot) \in \mathcal{C}_n$.
 g is the additive generator of $\oplus$ with $g(x) = x$, and

$$\begin{aligned} x \oplus y &= g^{(-1)}(g(x) + g(y)) = g^{(-1)}(x + y) \\ &= \min\{x + y, 1\}. \end{aligned}$$

h is the multiplication generator of "$\cdot$" with $h(x) = x$, and

$$x \cdot y = h^{-1}(h(x)h(y)) = x \cdot y.$$

By Corollary 1, clearly $g \stackrel{\triangle}{=} h$, thus the result follows obviously.

8.4   Infinite Dimensional Multifactorial Functions

Let $X_n \in [0,1]^n$. Write

$$M_\infty(X) = \lim_{n \to \infty} M(X_n),$$

M_∞ is called infinite-dimensional multifactorial function, and we denote

$$\mathcal{C}_\infty = \{M_\infty \,|\, M_\infty \text{ is ASM-func}\}.$$

Theorem 10 $\forall X \in [0,1]^\infty$, if

$$M(x_1, x_2) \leq M(x_1, x_2, x_3) \leq \cdots \leq M(x_1, x_2, \cdots, x_n) \leq \cdots \leq 1,$$

or

$$M(x_1, x_2) \geq M(x_1, x_2, x_3) \geq \cdots \geq M(x_1, x_2, \cdots, x_n) \geq \cdots \geq 0,$$

then $\{M(X_n)\}_{n=2}^\infty$ is convergent.

Theorem 11
  (i)   For all continuous $\top$, $\top_\infty$ exists.
  (ii)   For all continuous $\bot$, $\bot_\infty$ exists.
  (iii)   If $M(X_n) = \bot_{j=1}^n(a_j \top x_j)$, then M_∞ exists.

Proof   (i)   For any $X = (x_1, x_2, \cdots, x_n)$,

$$\begin{aligned}
0 \leq \top(x_1, x_2, \cdots, x_n) \\
= \top(\top(x_1, x_2, \cdots, x_{n-1}), x_n) \\
\leq \min(\top(x_1, x_2, \cdots, x_{n-1}), x_n) \\
\leq \top(x_1, x_2, \cdots, x_{n-1}) \\
\leq 1.
\end{aligned}$$

So the conclusion (i) is true.
  (ii)

$$\begin{aligned}
0 \leq \bot(x_1, x_2, \cdots, x_{n-1}) \\
\leq \max(\bot(x_1, x_2, \cdots, x_{n-1}, x_n) \\
\leq \bot(x_1, x_2, \cdots, x_n) \\
\leq 1.
\end{aligned}$$

This means the conclusion (ii) is true.
  (iii)

$$\begin{aligned}
0 \leq M(X_{n-1}) = \bot_{j=1}^n(a_j \top x_j) \\
\leq (\bot_{j=1}^{n-1}(a_j \top x_j)) \bot 0 \\
\leq \bot_{j=1}^{n-1}(a_j \top x_j) \bot (a_n \top x_n) \\
\leq 1,
\end{aligned}$$

which means (iii) is also true.   **Q.E.D.**

Without going into details, we also present the following theorems.

Theorem 12   If $M_n \in \mathcal{C}_n$, then $M_\infty \in \mathcal{C}_\infty$.

Theorem 13   If $\perp_{j=1}^{\infty} a_j = 1$, for all continuous $\top$, then $M_\infty(\perp, \top) \in \mathcal{C}_\infty$ iff $\perp = \vee$.

Theorem 14   Let (g, h) be the generator group of $(\perp, \top)$. If $\perp_{j=1}^{\infty} a_j = 1$ and $g \overset{\triangle}{=} h$, for any Archimedean continuous $\perp(\top)$, then $M_\infty(\perp, \top) \in \mathcal{C}_\infty$.

8.5   M($\perp, \top$) and Fuzzy Integral

Let $\perp(\top)$ be Archimedean continuous.

Theorem 15   Let $F = (\perp, \perp, \perp, \top)$, (g, h) be the generator group of $(\perp, \top)$ with strict t-norm, and μ be $\perp$-decomposable measure. If $g \overset{\triangle}{=} h, \perp_{i=1}^{n} a_i = 1$, and $\mu(\{x_i\}) = a_i$, then $M(\perp, \top) = \mathcal{F}_\mu$.

Proof   Assume that $x_{(1)}, x_{(2)}, \cdots, x_{(n)}$ is a permutation of the elements in $(x_1, x_2, \cdots, x$ such that

$$x_{(1)} \leq x_{(2)} \leq \cdots \leq x_{(n)},$$

then

$$M(\perp, \top) = g^{(-1)}(\sum_{i=1}^{n} g(a_j \top x_i))$$

$$= g^{(-1)}(\sum_{i=1}^{n} gg^{-1}(g(a_j)g(x_i)))$$

$$= g^{(-1)}(\sum_{i=1}^{n} g(x_i)g(\mu(\{x_i\}))).$$

Taking $A_{(i)} = \bigcup_{k=i}^{n}\{x_{(k)}\}$, we have

$$M(\perp, \top) = g^{(-1)}(\sum_{i=1}^{n} g(x_{(i)})(g \cdot \mu(A_{(i)}) - g \cdot \mu(A_{(i+1)})))$$

$$= g^{(-1)}[\sum_{i=1}^{n} (g(x_{(i)}) - g(x_{(i-1)}))g \cdot \mu(A_{(i)})]$$

$$= g^{(-1)}[\sum_{i=1}^{n} g(x_{(i)} -_{\perp} x_{(i-1)})g \cdot \mu(A_{(i)})]$$

$$= \perp_{i=1}^{n}(x_{(i)} -_{\perp} x_{(i-1)})\top \mu(A_{(i)}).$$

Let

$$f = \perp_{i=1}^{n}(x_{(i)} \top 1_{D_{(i)}}),$$

where $D_{(i)} = \{x_{(i)}\}$, then

$$M(\bot, \top) = (\bot) \int f \top \, d\mu. \qquad \textbf{Q.E.D.}$$

By Theorem 5, it is shown that the constant coefficient of $M(\bot, \top)$ implies the weight value when $\bot(\top)$ is Archimedean continuous and $g(x) = x$.

If $\bot = \vee$ or $\top = \wedge$, we have the following conclusions.

Property 1 If $\mu(A_{(i)}) = a_i$ and $\bigvee_{i=1}^{n} a_i = 1$, then

$$M(\vee, \wedge) = (S) \int .$$

Property 2 Let $F = (\vee, \vee, \vee, \cdot), \mu(A_{(i)}) = a_i$, and $\bigvee_{i=1}^{n} a_i = 1$, then

$$M(\vee, \cdot) = \mathcal{F}_\mu .$$

Property 3 Let $F = (\oplus, \oplus, \oplus, \wedge), \mu(\{x_i\}) = a_i, \mu$ be additive measure, and $\sum_{i=1}^{n} a_i = 1$, then

$$M(\oplus, \wedge) = \mathcal{F}_\mu .$$

Property 4 Let $F = (\oplus, \oplus, \oplus, \cdot), \mu(\{x_i\}) = a_i, \mu$ be additive measure, and $\sum_{i=1}^{n} a_i = 1$, then

$$M(\oplus, \cdot) = \mathcal{F}_\mu .$$

Properties 1–4 give the description of the different implication of the constant coefficients of $M(\vee, \wedge)$, $M(\vee, \cdot)$, $M(\oplus, \cdot)$, and $M(\oplus, \wedge)$, where the constant coefficient of $M(\oplus, \cdot)$ and $M(\oplus, \wedge)$ implies the weight value.

8.6 Application in Fuzzy Inference

Let X and Y be the universes of input variables x and output variables y, respectively. Denoting $\mathcal{A} = \{A_i\}_{1 \leq i \leq n}$ and $\mathcal{B} = \{B_i\}_{1 \leq i \leq n}$, where $A_i \in \mathcal{F}(X)$ and $B_i \in \mathcal{F}(Y)$, we regard $\mathcal{A}$ and $\mathcal{B}$ as linguistic variables so that the general form of fuzzy inference can be formed as follows:

premise :	if x is A',		
rules :	if x is A_1, then y is B_1,	$\cdots$	R_1
	if x is A_2, then y is B_2,	$\cdots$	R_2
	$\vdots$		
	if x is A_n, then y is B_n,	$\cdots$	R_n

$$(8.3)$$

conclusion :   then y is B'.

The problem is to find the output value B' for the output variable y associated with a particular value A' for the input variable x according to the inference rules $R = \bigcup_{i=1}^{n} R_i$. Our purpose here is to investigate the relationship between the multifactorial function and the reasoning process.

First, consider an individual rule:

$$
\begin{aligned}
\text{premise}: \quad &\text{if } x \text{ is } A', \\
\text{rule}: \quad &\text{if } x \text{ is } A_i, \text{ then } y \text{ is } B_i,
\end{aligned}
\tag{8.4}
$$

$$\text{conclusion}: \qquad\qquad\qquad \text{then } y \text{ is } B'_i.$$

According to the idea in [8], we use τ_i to indicate the measure of the degree to which the input to the system and the antecedent of the individual rule are compatible, and is called the firing level of the rule, denoted as $\tau_i = A' \odot A_i$. Having obtained this firing level we then use it to obtain the individual rule output value B'_i. The individual rule output B'_i is seen to be completely determined by τ_i and the rule consequent B_i. We denote this as $B'_i = \tau_i \circ B_i$. So we have

$$(\forall y \in Y)(\mu_{B'_i}(y) = \tau_i \circ \mu_{B_i}(y)),$$

where "$\circ$" is a pointwise and likewise operation. Thus, there exists a function denoted as h such that

$$\mu_{B'_i}(y) = h(\tau_i, \mu_{B_i}(y)).$$

More generally, h is a binary operator which shall satisfy the following conditions:
(i). $h(1, b) = b, \quad \forall b \in [0, 1]$.
(ii). $h(0, b) = g, \quad \forall b \in [0, 1]$, where g is the identity under the aggregation.
(iii). If $b > b'$, then $h(a, b) \geq h(a, b')$.
(iv). If $a > a'$, then

$$
\begin{cases}
h(a, b) \geq h(a', b), & b \geq g, \\
h(a, b) \leq h(a', b), & b \leq g.
\end{cases}
$$

Yager has proved that for $g = 0$ any t-norm, T, satisfies the four conditions. So (iv) can be represented as follows :

$$\mu_{B'_i}(y) = \mathsf{T}(\tau_i, \mu_{B_i}(y)). \tag{8.5}$$

We now look at (iii). The overall system output B' can be seen as the aggregation of these individual rules output $B'_1, B'_2, \cdots, B'_n$. We denote this process

$$\mu_{B'}(y) = M(\mu_{B'_1}(y), \mu_{B'_2}(y), \cdots, \mu_{B'_n}(y)), \tag{8.6}$$

where M shall be the standard multifactorial function which satisfies the following conditions:
(i). Monotonicity : $(\forall Y_1, Y_2)(Y_1 \geq Y_2 \Rightarrow M(Y_1) \geq M(Y_2))$.
(ii). Commutativity: $M(X) = M(\sigma(X))$, where σ is any permutation of X.

(iii). If there exists i such that $\tau_i = 1$ and $\tau_j = 0\, (j \neq i)$, then

$$M(g, \cdots, g, b, g, \cdots, g) = b,$$

where g is a fixed identity.

(iv). If $\tau_i = 0$, then

$$M(b_1, \cdots, b_{i-1}, g, b_{i+1}, \cdots, b_n) = M(b_1, \cdots, b_{i-1}, b_{i+1}, \cdots, b_n).$$

Clearly, for $g = 0$, t-conorm, $\perp$, satisfies the four condition, and (iii) can be represented as follows.

$$\mu_{B'}(y) = \perp(\mu_{B'_1}(y), \cdots, \mu_{B'_n}(y)).$$

Combining with (5), we have

$$\mu_{B'}(y) = \perp_{i=1}^{n}(\tau_i \top \mu_{B_i}(y)),$$

and therefore $M(\perp, \top)$ is useful to implement the reasoning process.

If the input is a singleton, x is x_0, and the antecedent of the form x is A_i, then the usual case is to use $\tau_i = \mu_{A_i}(x_0)$. So we have

$$\mu_{B'}(y) = \perp_{i=1}^{n}(\mu_{A_i}(x_0) \top \mu_{B_i}(y)). \tag{8.7}$$

Let $\perp = \vee$, $\top = \wedge$. We have

$$\mu_{B'}(y) = \bigvee_{i=1}^{n}(\mu_{A_i}(x_0) \wedge \mu_{B_i}(y)),$$

which is the formulation used by Mamdani [7] in his original work on fuzzy control.

Let $\perp = \oplus$, $\top = \cdot$. We have

$$\mu_{B'}(y) = \oplus_{i=1}^{n}(\mu_{A_i}(x_0) \cdot \mu_{B_i}(y)),$$

which is the $(+, \cdot)$-centroid algorithm in [8] on fuzzy control.

According to the form of $M(\perp, \top)$, it seems to be similar to the fuzzy neural type models [8]. In these models, the activation function is the identical function, $\varphi(x) = x$; $\mu_{B_i}(y)$ are the weights, τ_i are the input, and $\top$ and $\perp$ are the general operations of union and intersection, respectively (see Figure 1).

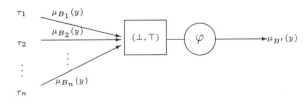

Figure 1 Fuzzy neural model for the fuzzy inference process.

However, the overall fuzzy inference process can be described in the following diagram:

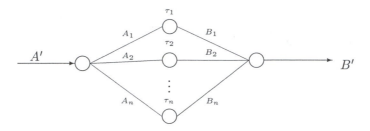

Figure 2. Fuzzy inference process.

Thus, the inference process can be represented in the form of fuzzy neural networks as follows.

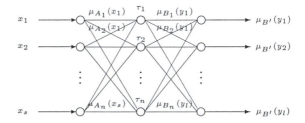

Figure 3 Fuzzy neural networks representation of the fuzzy inference.

where $X = \{x_1, x_2, \cdots, x_s\}$, $Y = \{y_1, y_2, \cdots, y_l\}$.

8.7   Conclusions

In this chapter, we introduced a new class of standard multifactorial functions, called generalized additive weighted multifactorial functions. We investigated some of their properties and the implication of their constant coefficients. In fact, this kind of multifactorial function is of importance for neural fuzzy intelligence systems. It can play an important role in fuzzy decision-making, fuzzy control, fuzzy inference, and fuzzy neural networks.

References

1. H. X. Li, P. Z. Wang, and V. C. Yen, Factor spaces theory and its applications to fuzzy information processing. (I). The basics of factor spaces, *Fuzzy Sets and Systems*, Vol. 95, pp. 147–160, 1998.

2. H. X. Li, Fuzzy perturbation analysis, Part 2, *Fuzzy Sets and Systems*, Vol. 19, pp. 165–175, 1986.

3. H. X. Li, Multifactorial fuzzy sets and multifactorial degree of nearness, *Fuzzy Sets and Systems*, Vol. 19, 291–297, 1986.

4. H. X. Li, Multifactorial functions in fuzzy sets theory, *Fuzzy Sets and Systems*, Vol. 35, pp. 69–84, 1990.

5. S. Weber, Two integrals and some modified versions–critical remarks, *Fuzzy Sets and Systems*, Vol. 20, pp. 97–125, 1986.

6. R. R. Yager, Aggregation operators and fuzzy systems modeling, *Fuzzy Sets and Systems*, Vol. 67, pp. 129–145, 1994.

7. E. H. Mamdani and S. Assilian, An experiment in linguistic synthesis with a fuzzy logic controller, *International Journal of Man-Machine Studies*, Vol. 7, pp. 1–13, 1975.

8. M. Mizumoto, The improvement of fuzzy control algorithm, Part 4 : $(+, \cdot)$-centroid algorithm, *Procceeding of Fuzzy Systems Theory* (In Japanese), 1990.

Chapter 9

The Interpolation Mechanism of Fuzzy Control

This chapter demonstrates that the commonly used fuzzy control algorithms can be regarded as interpolation functions. This means that fuzzy control method is similar to finite element method in mathematical physics, which is a kind of direct manner or numerical method in control system. We start with an introduction of Mamdanian fuzzy control algorithm first. And we prove several theorems to indicate that the Mamdanian algorithm is essential an interpolator.

9.1 Preliminary

During the past decades, we have witnessed a rapid growth of interest and experienced a variety of applications of fuzzy logic systems [1-8]. Especially, fuzzy control has been a center of focus [4,7]. In this chapter, we take a mathematical analysis of interpolation property for fuzzy control. First, we review, briefly, the Mamdanian fuzzy control algorithm. Without loss of generality, we consider controllers with two-input and one-output as an example.

Let X and Y be the universes of input variables and Z be the universe of output. Denote $\mathcal{A} = \{A_i\}_{(1 \leqslant i \leqslant n)}$, $\mathcal{B} = \{B_i\}_{(1 \leqslant i \leqslant n)}$, and $\mathcal{C} = \{C_i\}_{(1 \leqslant i \leqslant n)}$. We regard $\mathcal{A}$, $\mathcal{B}$, and $\mathcal{C}$ as linguistic variables. The n fuzzy inference rules are formed as follows:

$$\text{If } x \text{ is } A_i \text{ and } y \text{ is } B_i \text{ then } z \text{ is } C_i \ , \tag{9.1}$$

where $x \in X$, $y \in Y$, and $z \in Z$, $i = 1, 2, \cdots, n$ are base variables. According to the Mamdanian algorithm, the inference relation of i-th inference rule is a fuzzy relation from $X \times Y$ to Z, $R_i \triangleq (A_i \times B_i) \times C_i$, where

$$\mu_{R_i}(x, y, z) \triangleq (\mu_{A_i}(x) \wedge \mu_{B_i}(y)) \wedge \mu_{C_i}(z).$$

As n inference rules should be joined by "or" (corresponding to set theoretical operator "$\bigcup$"), the whole inference relation is $R = \bigcup\limits_{n=1}^{n} R_i$, i.e.,

$$\mu_R(x, y, z) = \bigvee_{i=1}^{n} \mu_{R_i}(x, y, z) = \bigvee_{i=1}^{n} [(\mu_{A_i}(x) \wedge \mu_{B_i}(y)) \wedge \mu_{C_i}(z)] \ . \tag{9.2}$$

Given $A' \in \mathcal{F}(X)$ and $B' \in \mathcal{F}(Y)$, the conclusion of inference $C' \in \mathcal{F}(Z)$ can be determined as $C' \triangleq (A' \times B') \circ R$, by use of the Composition of Relations Inference (CRI) method, where $\circ$ is the composition of fuzzy relation, and where

$$\mu_{C'}(z) = \bigvee_{(x,y)\in X \times Y} [(\mu_{A'}(x) \wedge \mu_{B'}(y)) \wedge \mu_{R_i}(x,y,z)] \ . \tag{9.3}$$

To change the crisp inputs to fuzzy sets, we use fuzzification described as follows.

$$\mu_{A'}(x) = \begin{cases} 1, & x = x' \\ 0, & x \neq x', \end{cases} \qquad \mu_{B'}(y) = \begin{cases} 1, & y = y' \\ 0, & y \neq y' \ . \end{cases}$$

Using the above expression and (9.2), we can obtain the inference result C'

$$\mu_{C'}(z) = \mu_R(x', y', z) = \bigvee_{i=1}^{n} [(\mu_{A_i}(x') \wedge \mu_{B_i}(y')) \wedge \mu_{C_i}(z)] \ . \tag{9.4}$$

The commonly used "centroid" defuzzification method that turns a fuzzy set, C', to a crisp number is:

$$z' = \int_{z \in Z} z\mu_{C'}(z)dz \Big/ \int_{z \in Z} \mu_{C'}(z)dz \ . \tag{9.5}$$

Note: In Expression (9.4), $\mu_{A_i'}(x') \wedge \mu_{B_i}(y')$ is a number that is independent to z, denoted by λ_i'. So Expression (9.4) can be rewritten as the following simple form:

$$\mu_{C'}(z) = \bigvee_{i=1}^{n} (\lambda_i' \wedge \mu_{C_i}(z)) \ . \tag{9.6}$$

By definition, for a fuzzy set to be multiplied by a number, we have $(\lambda_i' \wedge \mu_{C_i}(z)) = (\lambda_i' \cdot \mu_{C_i}(z))(z)$. So if we define $C_i' \triangleq \lambda_i' C_i$, it is easy to know that,

$$C' = \bigcup_{i=1}^{n} C_i' = \bigcup_{i=1}^{n} \lambda_i' C_i \ , \tag{9.7}$$

where λ_i' is regarded as the agreement degree of ith rule with respect to (x', y'). These C_i can also be considered as expansion coefficients so that Expression (9.7) is just an expansion on C_i $(i = 1, 2, \cdots, n)$.

For the sake of convenience, we introduce some concepts. Given a universe X, let $\mathcal{A} = \{A_i\}_{(1 \leqslant i \leqslant n)}$ be a family of normal fuzzy sets on X, (i.e., $(\forall i)(\exists x_i \in X)(\mu_{A_i}(x_i) = 1))$ and x_i be the peak point of A_i.

$\mathcal{A}$ is a fuzzy partition of X, if it satisfies the condition:

$$(\forall i, j)(i \neq j \Rightarrow x_i \neq x_j) \text{ and } (\forall x \in X)(\sum_{i=1}^{n} \mu_{A_i}(x) = 1) \ , \tag{9.8}$$

where A_i is called a base element of $\mathcal{A}$. Then we also call $\mathcal{A}$ a group of base elements of X. In the paper, when $X = [a, b] \subset R$ (R is the set of real numbers), we always assume that $a < x_1 < x_2 \cdots < x_n < b$.

9.2 The Interpolation Mechanism of Mamdanian Algorithm with One Input and One Output

Let X and Y be the universe of input and output variables, respectively, and $\mathcal{A} = \{A_i\}_{(1 \leqslant i \leqslant n)}$ and $\mathcal{B} = \{B_i\}_{(1 \leqslant i \leqslant n)}$ be a fuzzy partition of X and Y, respectively. Define X and Y as real number intervals, i.e., $X = [a, b]$ and $Y = [c, d]$ holding $a < x_1 < x_2 < \cdots < x_n < b$. Let x_i and y_i be the peak points of A_i and B_i, respectively. When X and Y are general measurable sets, the following result is also true. Moreover, we always regard A_i and B_i as integrable functions.

Theorem 1   Under the above conditions, there exists a group of base functions $\mathcal{A}' = \{A_i\}_{(1 \leqslant i \leqslant n)}$ such that the Mamdanian algorithm with one input and one output is approximately an univariate piecewise interpolation function taking A_i' as its base functions, as follows:

$$F(x) = \sum_{i=1}^{n} \mu_{A_i'}(x) y_i \qquad (9.9)$$

and $\mathcal{A}'$ is just a fuzzy partition of X. Especially, when $\{y_i\}_{(1 \leqslant i \leqslant n)}$ is an equidistant partition, $\mathcal{A}'$ degenerates into $\mathcal{A}$, i.e.,

$$F(x) = \sum_{i=1}^{n} \mu_{A_i}(x) y_i \quad . \qquad (9.10)$$

Proof.   According to the Mamdanian algorithm, $\mathcal{A}$ and $\mathcal{B}$ form the inference rules:

$$\text{if } x \text{ is } A_i \text{ then } y \text{ is } B_i \qquad (9.11)$$

and the inference relation of ith rule is $R_i = A_i \times B_i$, and then the whole inference relation is $R = \bigcup_{i=1}^{n} R_i$, i.e.,

$$\mu_R(x, y) = \bigvee_{i=1}^{n} \mu_{R_i}(x, y) = \bigvee_{i=1}^{n} (\mu_{A_i}(x) \wedge \mu_{B_i}(y)) \quad . \qquad (9.12)$$

For a given input $x' \in X$, being similar to Expression (9.4), we have that

$$\mu_{B'}(y) = \bigvee_{i=1}^{n} (\mu_{A_i}(x') \wedge \mu_{B_i}(y)).$$

And by following Expression (9.5), we can obtain a crisp response value y'

$$y' = \int_{c}^{d} y \mu_{B'}(y) dy \bigg/ \int_{c}^{d} \mu_{B'}(y) dy \quad . \qquad (9.13)$$

Let $h_1 = y_1 - c$, $h_i = y_i - y_{i-1}$ $(i = 2, 3, \cdots, n)$ and $h = \max\{h_i | 1 \leqslant i \leqslant n\}$. Because $\mathcal{A}$ and $\mathcal{B}$ are fuzzy partitions, they have the Kronecker's property, i.e.,

$$\mu_{A_i}(x_j) = \delta_{ij} = \mu_{B_i}(y_j) = \begin{cases} 1, & i = j \\ 0, & i \neq j \end{cases}.$$

Based on the definition of definite integral, we have that

$$y' = \frac{\int_c^d y\mu_{B'}(y)dy}{\int_c^d \mu_{B'}(y)dy} \approx \frac{\sum\limits_{i=1}^{n} y_i \mu_{B'}(y_i)h_i}{\sum\limits_{i=1}^{n} \mu_{B'}(y_i)h_i}$$

$$= \frac{\sum\limits_{i=1}^{n} h_i[\vee_{k=1}^{n}(\mu_{A_k}(x') \wedge \mu_{B_k}(y_i))]y_i}{\sum\limits_{i=1}^{n} h_i[\vee_{k=1}^{n}(\mu_{A_k}(x') \wedge \mu_{B_k}(y_i))]}$$

$$= \frac{\sum\limits_{i=1}^{n} h_i \mu_{A_i}(x')y_i}{\sum\limits_{i=1}^{n} h_i \mu_{A_i}(x')}.$$

Denoting $\alpha_i(x') \triangleq h_i / \sum\limits_{j=1}^{n} h_j \mu_{A_j}(x')$, and $\mu_{A'_i}(x') \triangleq \alpha_i(x')\mu_{A_i}(x')$ then

$$y' \approx \sum_{i=1}^{n} \alpha_i(x')\mu_{A_i}(x')y_i = \sum_{i=1}^{n} \mu_{A'_i}(x')y_i \qquad (9.14)$$

which means that y' is approximately represented as an univariate piecewise interpolation function taking A'_i as its base functions. Note that we denote $\mathcal{A}' = \{A'_i\}_{(1 \leqslant i \leqslant n)}$. If we denote $F(x) = \sum\limits_{i=1}^{n} \mu_{A'_i}(x)y_i$, then we have Expression (9.9).

Then it is easy to prove that $\mathcal{A}'$ satisfies the conditions of fuzzy partition. At last, when $\{y_i\}_{(1 \leqslant i \leqslant n)}$ is an equidistant partition, as $(\forall i)(h_i = h')$ by Expression (9.8), we have

$$\alpha_i(x) = h_i / \sum_{j=1}^{n} h_j \mu_{A_j}(x) = 1 / \sum_{j=1}^{n} \mu_{A_j}(x) = 1$$

So $A'_i = A_i$, i.e., $\mathcal{A}' = \mathcal{A}$. Thus, $F(x) = \sum\limits_{i=1}^{n} \mu_{A_i}(x)y_i$, which is just Expression (9.10) **Q.E.D.**

Note: From $A'_i(x_i) = \alpha_i(x_i)\mu_{A_i}(x_i)$ we know that $\mu_{A'_i}(x)$ is a kind of weighted form of $\mu_{A_i}(x)$. For $\alpha_i(x)$ is only related to h_i and $\mu_{A_i}(x)$, $\mu_{A'_i}(x)$ is identical to $\mu_{A_i}(x)$. Besides, the antecedents of the inferences, A_i, are just the base functions of interpolation and only the peak points, y_i, (the consequents of inferences, B_i) appear in the interpolation equation. Under the Kronecker's property, the shape of membership functions of the consequents has nothing to do with the interpolation. This provides us with a very important guideline for designing a fuzzy controller.

9.3   The Interpolation Mechanism of Mamdanian Algorithm with Two Inputs and One Output

Similar to Section 9.2, let X, Y, and Z be real number intervals : $X = [a, b]$, $Y = [c, d]$, and $Z = [e, f]$, where Z_i is the peak point of C_i. And suppose that A_i, B_i, and C_i are integrable functions.

Theorem 2   Under the above conditions, there exists a group of base functions $\Phi = \{\phi_i\}_{(1 \leqslant i \leqslant n)}$ such that the Mamdanian algorithm with two inputs and one output is, approximately, a binary piecewise interpolation function taking ϕ_i as its base functions, as follows:

$$F(x, y) = \sum_{i=1}^{n} \phi_i(x, y) z_i \quad . \tag{9.15}$$

Proof.   Let $h_1 = z_1 - e$, $h_i = z_i - z_{i-1} (i = 2, 3, \cdots, n-1)$ and $h = max\{h_i|_{1 \leqslant i \leqslant n}\}$. From Expressions (9.4) and (9.5) , we have that

$$z' = \frac{\int_e^f z \mu_{C'}(z) dz}{\int_e^f \mu_{C'}(z) dz} \approx \frac{\sum_{i=1}^{n} z_i \mu_{C'}(z_i) h_i}{\sum_{i=1}^{n} \mu_{C'}(z_i) h_i}$$

$$= \frac{\sum_{i=1}^{n} h_i(\mu_{A_i}(x') \wedge \mu_{B_i}(y')) z_i}{\sum_{i=1}^{n} h_i(\mu_{A_i}(x') \wedge \mu_{B_i}(y'))}$$

$$= \sum_{i=1}^{n} \beta_i(x', y')(\mu_{A_i}(x') \wedge \mu_{B_i}(y')) z_i$$

$$= \sum_{i=1}^{n} \phi_i(x', y') z_i \quad , \tag{9.16}$$

where

$$\beta_i(x', y') \triangleq h_i / \sum_{i=1}^{n} h_i(\mu_{A_i}(x') \wedge \mu_{B_i}(y'))$$

and

$$\phi_i(x', y') \triangleq \beta_i(x', y')(\mu_{A_i}(x') \wedge \mu_{B_i}(y')).$$

If we let $F(x, y) = \sum_{i=1}^{n} \phi_i(x, y) z_i$, then we have Expression (9.15).   **Q.E.D.**

9.4 A Note on Completeness of Inference Rules

Equation (9.1) is a conventional way of expressing inference rules. However, it does not include "cross relationship" in consequence part. In this case, we define the following mapping that includes the cross relationship. And we show the interpolation function of this kind of rule thereafter. Let us define the following mapping (taking the case with two-input and one-output as an example):

$$f^* : \ A^* \times B^* \longrightarrow C^*,$$
$$(A_i^*, B_j^*) \longmapsto f^*(A_i^*, B_j^*) = C_{ij}^*, \tag{9.17}$$

where $\mathcal{A}^* = \{A_i^*\}_{(1 \leqslant i \leqslant p)}$, $\mathcal{B}^* = \{B_i^*\}_{(1 \leqslant i \leqslant q)}$, and $\mathcal{C}^* = \{C_i^*\}_{(1 \leqslant i \leqslant p, 1 \leqslant i \leqslant q)}$, which is a linguistic variable on X, Y, and Z, respectively. Taking account of the base variables of these linguistic variables, Expression (9.17) can also be regarded as the following inference rules:

$$\text{if } x \text{ is } A_i^* \text{ and } y \text{ is } B_j^* \text{ then } z \text{ is } C_{ij}^* \tag{9.18}$$

which is clearly a complete group of inference rules. According to Mamdanian algorithm, the inference relation of (i, j)th reference rule is $R_{ij}^* = A_i^* \times B_j^* \times C_{ij}^*$. So the whole inference relation of these $p \times q$ inference rules is $R^* = \bigcup\limits_{i=1}^{p} \bigcup\limits_{j=1}^{q} R_{ij}^*$, i.e.,

$$\mu_{R^*}(x, y, z) = \bigvee_{i=1}^{p} \bigvee_{j=1}^{q} (\mu_{A_i^*}(x) \wedge \mu_{B_j^*}(y) \wedge \mu_{C_{ij}^*}(z)) . \tag{9.19}$$

For a given input (x, y) being similar to the previous discussion, we have

$$\mu_{C'}(z) = \bigvee_{i=1}^{p} \bigvee_{j=1}^{q} [\mu_{A_i^*}(x') \wedge \mu_{B_j^*}(y') \wedge \mu_{C_{ij}^*}(z)] . \tag{9.20}$$

If take $k \triangleq (i-1)q + j$, $A_k \triangleq A_i^*$, $B_k \triangleq B_j^*$, $C_k \triangleq C_{ij}^*$, $z_k \triangleq z_{ij}$ (where z_{ij} is the peak point of C_{ij}^*) and $n = pq$, then Expression (9.20) changes back into Expression (9.4). By using the definition, if we write $h_{ij} \triangleq h_k$, and

$$\beta_{ij}(x, y) \triangleq h_{ij} \Big/ \sum_{i=1}^{p} \sum_{j=1}^{q} h_{ij}(\mu_{A_i^*}(x) \wedge \mu_{B_j^*}(y))$$

and

$$\phi_{ij}(x, y) \triangleq \beta_{ij}(x, y)(\mu_{A_i^*}(x) \wedge \mu_{B_j^*}(y)),$$

then $\beta_{ij}(x, y) = \beta_k(x, y)$ and $\phi_{ij}(x, y) = \phi_k(x, y)$, where $k = (i-1)q + j$. Then Expression (9.15) can be written as follows.

$$F(x, y) = \sum_{i=1}^{p} \sum_{j=1}^{q} \beta_{ij}(x, y)(\mu_{A_i^*}(x) \wedge \mu_{B_j^*}(y)) = \sum_{i=1}^{p} \sum_{j=1}^{q} \phi_{ij}(x, y) z_{ij} \tag{9.21}$$

which is a typical binary piecewise interpolation function.

9.5　The Interpolation Mechanism of $(+, \cdot)$-Centroid Algorithm

In this section, we show the interpolation property of $(+, \cdot)$-centroid algorithm proposed in [4].

Theorem 3　Under the conditions in Theorem 1, $(+, \cdot)$-centroid algorithm with one-input and one-output has the same conclusion as in Theorem 1.

Proof.　In Expression (9.12) , after $(\vee, \wedge)$ is replaced by $(+, \cdot)$, we have

$$\mu_R(x, y) = \sum_{i=1}^{n} \mu_{R_i}(x, y) = \sum_{i=1}^{n} \mu_{A_i}(x)\mu_{B_i}(y) \tag{9.22}$$

For a given input x',

$$\mu_{B'}(y) = \sum_{i=1}^{n} \mu_{A_i}(x')\mu_{B_i}(y).$$

From Expression (9.13) , we know that

$$
\begin{aligned}
y' &\approx \frac{\sum\limits_{i=1}^{n} y_i \mu_{B'}(y_i) h_i}{\sum\limits_{i=1}^{n} \mu_{B'}(y_i) h_i} = \frac{\sum\limits_{i=1}^{n} h_i \left(\sum\limits_{k=1}^{n} \mu_{A_k}(x')\mu_{B_k}(y) \right) y_i}{\sum\limits_{i=1}^{n} h_i \left(\sum\limits_{k=1}^{n} \mu_{A_k}(x')\mu_{B_k}(y) \right)} \\
&= \frac{\sum\limits_{i=1}^{n} h_i \mu_{A_i}(x) y_i}{\sum\limits_{i=1}^{n} h_i \mu_{A_i}(x')} = \sum_{i=1}^{n} \alpha_i(x')\mu_{A_i}(x') y_i \\
&= \sum_{i=1}^{n} \mu_{A_i'}(x') y_i,
\end{aligned}
$$

which is just Expression (9.14). The other results in the theorem are clear.　**Q.E.D.**

The theorem means that $(+, \cdot)$-centroid algorithm is in essence the same as Mamdanian algorithm in the case with one-input and one-output.

Theorem 4　Under the conditions in Theorem 2, there exists a group of base functions $\Psi = \{\psi_i\}_{(1 \leqslant i \leqslant n)}$ such that $(+, \cdot)$-centroid algorithm with two-input and one-output is approximately a binary piecewise interpolation function taking ψ_i as its base functions, as follows:

$$F(x, y) = \sum_{i=1}^{n} \psi_i(x, y) z_i \quad . \tag{9.23}$$

Furthermore, when $\mathcal{A}^*$ and $\mathcal{B}^*$ are fuzzy partitions and $\{z_i\}_{(1 \leqslant i \leqslant n)}$ is an equidistant partition, then,

$$F(x, y) = \sum_{i=1}^{n} \mu_{A_i}(x)\mu_{B_i}(y) z_i = \sum_{i=1}^{p} \sum_{j=1}^{q} \mu_{A_i^*}(x)\mu_{B_j^*}(y) z_{ij} \quad . \tag{9.24}$$

Proof.   In Expression (9.2) , after $(\vee, \wedge)$ is replaced by $(+, \cdot)$,

$$\mu_{R_i}(x, y, z) = \sum_{i=1}^{n} \mu_{R_i}(x, y, z) \ .$$

For a given input (x, y) ,

$$\mu_{C'}(z) = \sum_{i=1}^{n} \mu_{A_i}(x')\mu_{B_i}(y')\mu_{C_i}(z).$$

From Expression (9.5), we know that

$$z' \approx \frac{\sum\limits_{i=1}^{n} z_i \mu_{C'}(z_i) h_i}{\sum\limits_{i=1}^{n} \mu_{C'}(z_i) h_i} = \frac{\sum\limits_{i=1}^{n} h_i (\sum\limits_{k=1}^{n} h_i \mu_{A_i}(x')\mu_{B_i}(y)) z_i}{\sum\limits_{i=1}^{n} h_i \mu_{A_i}(x')\mu_{B_i}(y)}$$

$$= \sum_{i=1}^{n} \gamma_i(x', y')\mu_{A_i}(x')\mu_{B_i}(y') z_i = \sum_{i=1}^{n} \psi_i(x', y') z_i \ ,$$

where

$$\gamma_i(x', y') \triangleq h_i / \sum_{i=1}^{n} h_i \mu_{A_i}(x')\mu_{B_i}(y')$$

and

$$\psi_i(x', y') \triangleq \gamma_i(x', y')\mu_{A_i}(x')\mu_{B_i}(y') \ .$$

Let $F(x, y) = \sum\limits_{i=1}^{n} \psi_i(x, y) z_i$. Then we get Expression (9.23). If we write $\gamma_{ij}(x, y) = \gamma_k(x, y)$ and $\psi_{ij}(x, y) = \psi_k(x, y)$, where $k = (i-1)q + j$, then Expression (9.23) can be written as follows:

$$F(x, y) = \sum_{i=1}^{p}\sum_{j=1}^{q} \psi_{ij}(x, y) z_{ij} = \sum_{i=1}^{p}\sum_{j=1}^{q} \gamma_{ij}(x, y)\mu_{A_i^*}(x)\mu_{B_j^*}(y) \ , \qquad (9.25)$$

especially where $\mathcal{A}^*$ and $\mathcal{B}^*$ are fuzzy partitions and $\{z_i\}_{(1 \leq i \leq n)}$ is an equidistant partition. It is easy to prove that $\gamma_{ij}(x, y) \equiv 1$. So Expression (9.25) is true. **Q.E.D.**

9.6   The Interpolation Mechanism of Simple Inference Algorithm

The simple inference algorithm proposed in [6,7] is considered to be a simple algorithm with respect to fuzzy inference. Without loss of generality, the inference rules with two-input and one-output are defined as follows:

$$\text{If } x \text{ is } A_i \text{ and } y \text{ is } B_i \text{ then } z \text{ is } z_i \ . \qquad (9.26)$$

Then for a given input (x', y') the corresponding value z' is calculated by the following steps:

Step 1:

$$\lambda_i = \mu_{A_i}(x')\mu_{B_i}(y') \quad (\text{or} \quad \lambda_i = \mu_{A_i}(x') \wedge \mu_{B_i}(y')) \ . \tag{9.27}$$

Step 2:

$$z' = \sum_{i=1}^{n} \lambda_i z_i / \sum_{i=1}^{n} \lambda_i \ . \tag{9.28}$$

Theorem 5 Let $\mathcal{A} = \{A_i\}_{(1 \leqslant i \leqslant n)}$ be a fuzzy partition of X. The simple inference algorithm with one-input and one-output is just an unary piecewise interpolation function taking A_i as its base functions, as Expression (9.10).

Proof. Here the inference rules are: if x is A_i then y is y_i. For a given x', clearly $\lambda_i = \mu_{A_i}(x')$. From Expression (9.28) we have

$$y' = \frac{\sum_{i=1}^{n} \lambda_i y_i}{\sum_{i=1}^{n} \lambda_i} = \frac{\sum_{i=1}^{n} h_i(\sum_{i=1}^{n} \mu_{A_i}(x'))y_i}{\sum_{i=1}^{n} \mu_{A_i}(x')} = \sum_{i=1}^{n} \mu_{A_i}(x')y_i \ . \tag{9.29}$$

Taking $F(x) = \sum_{i=1}^{n} \mu_{A_i}(x)y_i$, then we get Expression (9.10).   **Q.E.D.**

Theorem 6 The simple inference algorithm with two-input and one-output is just a binary piecewise interpolation function. When $\lambda_i = \mu_{A_i}(x) \wedge \mu_{B_i}(y)$, we have

$$F(x, y) = \sum_{i=1}^{n} \theta_i(x, y)z_i \ ,$$

where

$$\theta_i(x, y) = (\mu_{A_i}(x) \wedge \mu_{B_i}(y)) / \sum_{i=1}^{n}(\mu_{A_i}(x) \wedge \mu_{B_i}(y)) \ .$$

When $\lambda_i = \mu_{A_i}(x)\mu_{B_i}(y)$ and $\mathcal{A}^*$ and $\mathcal{B}^*$ are fuzzy partitions, we have

$$F(x, y) = \mu_{A_i}(x)\mu_{B_i}(y)z_i \ .$$

Proof. When $\lambda_i = \mu_{A_i}(x) \wedge \mu_{B_i}(y)$, from Expression (9.28), we have

$$F(x, y) = z = \frac{\sum_{i=1}^{n}(\mu_{A_i}(x) \wedge \mu_{B_i}(y))z_i}{\sum_{i=1}^{n}(\mu_{A_i}(x) \wedge \mu_{B_i}(y))} = \sum_{i=1}^{n} \theta_i(x, y)z_i \ .$$

When $\lambda_i = \mu_{A_i}(x)\mu_{B_i}(y)$, and $\mathcal{A}^*$ and $\mathcal{B}^*$ are fuzzy partitions, we have

$$\sum_{k=1}^{n} \mu_{A_k}(x)\mu_{B_k}(y) = \sum_{i=1}^{p}\sum_{j=1}^{q} \mu_{A_i^*}(x)\mu_{B_j^*}(y)$$

$$= (\sum_{i=1}^{p} \mu_{A_i^*}(x))(\sum_{j=1}^{q} \mu_{B_j^*}(y)) = 1 \ .$$

Hence

$$F(x,y) = z = \frac{\sum_{i=1}^{n} \mu_{A_i}(x)\mu_{B_i}(y)z_i}{\sum_{i=1}^{n} \mu_{A_i}(x)\mu_{B_i}(y)} = \sum_{i=1}^{n} \mu_{A_i}(x)\mu_{B_i}(y)z_i \ .$$

Q.E.D.

9.7   The Interpolation Mechanism of Function Inference Algorithm

Takagi and Sugeno proposed the function inference algorithm [8], which is the generalization of the simple inference algorithm. We consider the case with two-input and one-output as an example, where the inference rules are defined as follows.

$$\text{If } x \text{ is } A_i \text{ and } y \text{ is } B_i \text{ , then } z \text{ is } z_i(x,y) \ . \tag{9.30}$$

Clearly, in Expression (9.26), if constant points z_i are replaced by variable points $z_i(x,y)$ then the expression becomes Expression (9.30).

For a given input (x',y') , $\lambda \triangleq g_i(\mu_{A_i}(x), \mu_{B_i}(y'))$, where g_i is a general operator, for example, $g_i = \wedge$ (or $\cdot$). Then being similar to Expression (9.28),

$$z' = \sum_{i=1}^{n} \lambda_i z_i(x',y') / \sum_{i=1}^{n} \lambda_i \ .$$

Theorem 7   There exists a group of base functions $\xi_i(x,y)(i = 1, 2, \cdots, n)$ such that function inference algorithm is just a binary piecewise interpolation function with variable nodal points defined as follows.

$$F(x,y) = \sum_{i=1}^{n} \xi_i(x,y)z_i(x,y) \ , \tag{9.31}$$

where variable nodal points mean that the nodal points $z_i(x,y)$ are functions of (x,y).

Proof.   If we take

$$\xi_i(x,y) = \frac{g_i(\mu_{A_i}(x), \mu_{B_i}(y'))}{\sum\limits_{i=1}^{n} g_i(\mu_{A_i}(x), \mu_{B_i}(y'))} \ ,$$

then

$$F(x,y) = z = \frac{\sum_{i=1}^{n} \lambda_i z_i(x,y)}{\sum_{i=1}^{n} \lambda_i} = \sum_{i=1}^{n} \xi_i z_i(x,y)z_i(x,y) \ .$$

When $\lambda_i = \cdot$ and $\xi_i(x,y) = \mu_{A_i}(x)\mu_{B_i}(y)$, Expression (9.31) becomes the following simple form:

$$F(x,y) = \sum_{i=1}^{n} \mu_{A_i}(x)\mu_{B_i}(y)z_i(x,y) \ . \tag{9.32}$$

Q.E.D.

Note   Usually $z_i(x, y)$ are taken as linear functions:

$$z_i(x, y) = a_i x + b_i y + c_i \quad ,$$

where coefficients a_i, b_i, and c_i should be determined by means of some methods.

9.8   A General Fuzzy Control Algorithm

From the above results, we summarize a general fuzzy control algorithm. Without loss of generality, we consider the case with two-input and one-output as an example.

Theorem 8   Let $\mathcal{C}$ be a fuzzy partition and A_i, B_i, and C_i be integrable functions. A fuzzy controller with two-input and one-output is a binary piecewise interpolation function with variable nodal points as follows.

$$F(x, y) = \sum_{i=1}^{n} \eta_i(x, y) z_i(x, y) \quad , \tag{9.33}$$

where

$$\eta_i = h_i'(\mu_{A_i}(x) \mathrm{T} \mu_{B_i}(y)) / \sum_{i=1}^{n} h_i'(\mu_{A_i}(x) \mathrm{T} \mu_{B_i}(y))$$

and "T" means a t-norm.

Clearly, the results from Theorem 1 to Theorem 7 can be regarded as corollaries of Theorem 8.

Because the operations between $\mu_{A_i}(x)$ and $\mu_{B_i}(y)$ are derived from Cartesian product between fuzzy sets and t-norm, T, this theorem is a generalized form of intersection operations of fuzzy sets.

If $T = \cdot$ and $\{h_i\}_{(1 \leqslant i \leqslant n)}$ is an equidistant partition and $z_i(x, y) \equiv z_i$, then Expression (9.33) turns into Expression (9.24). In terms of performance of interpolation performance, Expression (9.24) has a better result. Expression (9.24) can be easily generalized as the case with multi-input and multi-output.

Consider the inference rules of three inputs and two outputs: if x is A_i and y is B_i and z is C_i then u is D_i and v is E_i. It is not difficult to prove that the fuzzy control algorithm is a piecewise interpolation function of many vector variables.

$$
\begin{aligned}
F(x, y, z) &= (u, v) \\
&= \left(\sum_{i=1}^{n} \mu_{A_i}(x) \mu_{B_i}(y) \mu_{C_i}(z) u_i, \sum_{i=1}^{n} \mu_{A_i}(x) \mu_{B_i}(y) \mu_{C_i}(z) v_i \right).
\end{aligned}
\tag{9.34}
$$

9.9   Conclusions

We prove, mathematically, that the Mamdanian control algorithms are in essence some interpolation functions and a fuzzy controller is an interpolator. Extending this idea, we have proved that a fuzzy logic system is essentially equivalent to a feedforward neural network [9]. Referring to inference rules expressed as (9.1), the fuzzy sets A_i, B_i, and C_i can be determined from their peak points denoted by x_i, y_i, and z_i , which is equivalent to acquiring a group of input-output sample pairs given as follows:

$$\{(x_i, y_i), z_i | i = 1, 2, \cdots, n\} \ . \tag{9.35}$$

If A_i and B_i are regarded as a group of base functions of interpolation, an interpolation function, for example, being expressed as (9.24), can be obtained. In other words, it is the same to acquire a group of data in Expression (9.35) and the rules.

References

1. H. X. Li, The mathematical essence of fuzzy controls and fine fuzzy controllers, *Advances in Machine Intelligence and Soft-Computing*, Paul P. Wang, ed., Vol. IV, Bookwright Press, Durham, pp. 55-74, 1997.

2. W. A. Farag, V. H. Quintana, and G. Lambert-Torres, A genetic-based neuro-fuzzy approach for modeling and control of dynamical systems, *IEEE Transactions on Neural Networks,* Vol. 9, No. 5, pp. 756-767, 1998.

3. C. T. Lin and C. S. G. Lee, *Neural Fuzzy Systems: A Neural-Fuzzy Synergism to Intelligent Systems,* Prentice-Hall, Englewood Cliffs, 1996.

4. M. Mizumoto, The improvement of fuzzy control algorithm, part 4: $(+, \bullet)$-centroid algorithm, *Proceedings of Fuzzy Systems Theory* (in Japanese), pp. 6-9, 1990.

5. P. Z. Wang and H. X. Li, *Fuzzy Information Processing and Fuzzy Computers,* Science Press, Beijing, 1997.

6. T. Terano, K. Asai, and M. Sugeno, *Fuzzy Systems Theory and Its Applications,* Academic Press, Tokyo, 1992.

7. M. Sugeno, *Fuzzy Control* (in Japanese), Japan Industry Press, Tokyo, 1988.

8. T. Takagi and M. Sugeno, Fuzzy identification of systems and its applications to modeling and control, *IEEE Transactions on Systems, Man and Cybernetics,* SMC-15, pp. 1-116, 1985.

9. H. X. Li and C. L. P. Chen, The equivalence between fuzzy logic systems and feedforward neural networks, *IEEE Transactions on Neural Networks,* Vol. 11, No. 2, pp. 356-365, 2000.

Chapter 10

The Relationship between Fuzzy Controllers and PID Controllers

In this chapter, we shall discuss internal relations between fuzzy controllers and PID controllers. Based on the interpolation property of fuzzy controllers, we first point out that a fuzzy controller with one input and one output is a piecewise P controller. Then we prove that a fuzzy controller with two inputs and one output is a piecewise PD (or I) controller with mutual affection between P and D (or PI). At last, we prove that a fuzzy controller with three inputs and one output is a piecewise PID controller with mutual affection among P, I, and D. Moreover, we use a discrete difference approximation scheme to design fuzzy controllers.

10.1 Introduction

Since fuzzy logic was first introduced by Lotfi Zadeh in 1965 [1], there are tremendous studies and successes in fuzzy control applications [2-16]. Fuzzy control paradigm takes heuristical human expertise (in terms of fuzzy If-Then rules), and with the aid of nonlinear control theory, designs a controller that synthesizes these rules. Fuzzy control has been applied successfully to PID control, robust control, sliding control, and supervisory control [6-8,12], to name a few.

Because of its simplicity and robust performance, proportional-integral-derivative (PID) controllers are the most commonly used controllers in industrial process control. The transfer function of a PID controller has the following form:

$$G(s) = K_P + \frac{K_I}{s} + K_D s \ ,$$

where K_P, K_I, and K_D are the propositional, integral, and derivative gains, respectively. Or another equivalent PID controller form is:

$$u(t) = K_P[e(t) + \frac{1}{T_I} \int_0^t e(\tau)d\tau + T_D \frac{de(t)}{dt}] \ ,$$

where $T_I = \frac{K_P}{K_I}$ and $T_D = \frac{K_D}{K_P}$ are known as integral and derivative time constants, respectively.

In practice, the PID gains are usually tuned by expertises. However, some studies have shown that the gains can be tuned systematically [12-14]. In this chapter, we take a different study. Rather than focusing on the methodology of finding or designing a fuzzy PID controller, we analyze and compare the performance of PID controllers and fuzzy controllers. Detailed design of a fuzzy PID controller can be found from literature [12-14].

10.2   The Relationship of Fuzzy Controllers with One Input One Output and P Controllers

Let $X = [-E, E]$ and $V = [-U, U]$ be, respectively, the universe of error (i.e., the universe of input variable) and the universe of control quantities (i.e., the universe of output variable), and $\mathcal{A} = \{A_i\}_{(1 \leq i \leq n)}$ and $\mathcal{B} = \{B_i\}_{(1 \leq i \leq n)}$ be, respectively, a fuzzy partition (a group of base element) on X and V, where $A_i \in \mathcal{F}(X)$ and $B_i \in \mathcal{F}(V)$ are called, respectively, base elements on X and V. Let x_i and u_i be, respectively, the peak point of A_i and B_i satisfying the condition: $-E \leq x_1 < x_2 < \cdots < x_n \leq E$ and $-U \leq u_1 < u_2 < \cdots < u_n \leq U$. Regarding $\mathcal{A}$ and $\mathcal{B}$ as linguistic variables, a fuzzy inference base R can be formed as follows:

$$\text{if } x \text{ is } A_i \text{ , then } u \text{ is } B_i, \ i = 1, 2, \cdots, n, \tag{10.1}$$

where $x \in X$ and $u \in Y$, called base variables. Referring to Chapter 8, we know that a fuzzy controller is in essence a interpolator. The fuzzy controller based on the inference rules shown as Expression (10.1) can be written as an interpolation function: $u \overset{\triangle}{=} F(x) = \sum_{i=1}^{n} \mu_{A_i}(x)u_i$. Considering time variable t, we can rewrite the equations as the following:

$$u(t) \overset{\triangle}{=} F(x(t)) = \sum_{i=1}^{n} \mu_{A_i}(x(t))u_i \ . \tag{10.2}$$

Usually the group of base elements can be taken as the following linear base elements (i.e., the membership functions are of the shapes with "triangular waves" as Figure 1):

$$\mu_{A_1}(x(t)) = \begin{cases} (x(t) - x_2)/(x_1 - x_2), & x_1 \leq x(t) \leq x_2; \\ 0, & \text{otherwise}, \end{cases} \tag{10.3}$$

$$\mu_{A_i}(x(t)) = \begin{cases} (x(t) - x_{i-1})/(x_i - x_{i-1}), & x_{i-1} \leq x(t) \leq x_i; \\ (x(t) - x_{i+1})/(x_i - x_{i+1}), & x_i < x(t) \leq x_{i+1}; \\ 0, & \text{otherwise}, \end{cases} \tag{10.4}$$
$$i = 2, 3, \cdots, n - 1,$$

$$\mu_{A_n}(x(t)) = \begin{cases} 0, & \text{otherwise}; \\ (x(t) - x_{n-1})/(x_n - x_{n-1}), & x_{n-1} \leq x(t) \leq x_n. \end{cases} \tag{10.5}$$

Similarly, we also define the groups of base elements, B, in a similar way. Clearly these groups of base elements are two-phased. It is pointed out that the common inference rules of fuzzy controllers have the monotonicity. This implies the monotonicity of the control functions. Thus, we have $(\forall x \neq 0)(sign(F(x)) = -sign(F(-x)))$. So $F(0) = 0$.

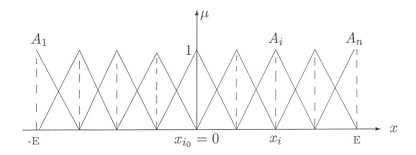

Figure 1   A group of linear bade elements

Denote $e(t)$ as control error, i.e., the error between reference input and system output, and assume that the membership functions are as the linear base elements shown in Figure 1.

Theorem 1   Let $x(t) = e(t)$, then a fuzzy controller with one input and one output is just a piecewise P controller with translation coefficients, as the following:

$$u(t) = F(x(t))$$
$$= u^{(1)}(t) + \cdots + u^{(i_0-1)}(t) + u^{(i_0)}(t) + \cdots + u^{(n-1)}(t), \qquad (10.6)$$

where

$$u^{(i)}(t) = \begin{cases} K_P^{(i)}(x(t) + T_C^{(i)}), & x_i \leq x(t) \leq x_{i+1}; \\ 0, & \text{otherwise}, \end{cases} \qquad (10.7)$$

and $i = 1, 2, \cdots, n-1$; $K_P^{(i)}$ is the proportional gain in error range $[x_i, x_{i+1}]$ and $T_C^{(i)}$ is the translation coefficient in $[x_i, x_{i+1}]$, especially $T_C^{(i_0-1)} = 0 = T_C^{(i_0)}$. Hence,

$$u^{(i_0-1)}(t) = \begin{cases} K_P^{(i_0-1)}x(t), & x_{i_0-1} \leq x(t) \leq x_{i_0}; \\ 0, & \text{otherwise}, \end{cases} \qquad (10.8)$$

and

$$u^{(i_0)}(t) = \begin{cases} K_P^{(i_0)}x(t), & x_{i_0} \leq x(t) \leq x_{i_0-1}; \\ 0, & \text{otherwise}. \end{cases} \qquad (10.9)$$

Proof When $x(t) \in [x_i, x_{i+1}]$, from Expressions (10.2)–(10.5), we have

$$
\begin{aligned}
u(t) = F(x(t)) &= \sum_{i=1}^{n} \mu_{A_i}(x(t))u_i \\
&= \mu_{A_i}(x(t))u_i + \mu_{A_{i+1}}(x(t))u_{i+1} \\
&= \frac{x(t) - x_{i+1}}{x_i - x_{i+1}}u_i + \frac{x(t) - x_i}{x_{i+1} - x_i}u_{i+1} \\
&= \frac{u_i - u_{i+1}}{x_i - x_{i+1}}x(t) + \frac{x_i u_{i+1} - x_{i+1}u_i}{x_i - x_{i+1}} \\
&= \frac{u_i - u_{i+1}}{x_i - x_{i+1}}\left(x(t) + \frac{x_i u_{i+1} - x_{i+1}u_i}{u_i - u_{i+1}}\right).
\end{aligned}
$$

Taking

$$
K_P^{(i)} \triangleq \frac{u_i - u_{i+1}}{x_i - x_{i+1}}, \quad T_C^{(i)} \triangleq \frac{x_i u_{i+1} - x_{i+1}u_i}{u_i - u_{i+1}}, \tag{10.10}
$$

and

$$
u^{(i)}(t) \triangleq \begin{cases} K_P^{(i)}(x(t) + T_C^{(i)}), & x_i \le x(t) \le x_{i+1}; \\ 0, & \text{otherwise}, \end{cases}
$$

then $u(t) = \sum_{i=1}^{n-1} u^{(i)}(t)$, that is just Expression (10.6). Moreover, from $x_{i_0} = 0$, we know that $u_{i_0} = F(x_{i_0}) = F(0) = 0$. It is easy to prove that $T_C^{(i_0-1)} = 0 = T_C^{(i_0)}$. So (10.8) and (10.9) are true. **Q.E.D.**

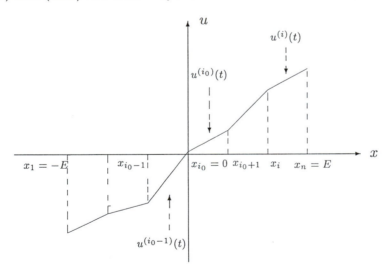

Figure 2 Piecewise P control function

Note 1 From Figure 2 we can understand that, when $x(t)$ is in the neighborhood of the origin of $x - u$ space (i.e., $x(t) \in [x_{i_0-1}, x_{i_0+1}]$), a fuzzy controller degenerates into a P controller: $u^{(i_0-1)}(t) = K_P^{(i_0-1)}x(t)$ or $u^{(i_0)} = K_P^{(i_0)}x(t)$. But when it moves further away from the origin, the controller is represented as a certain P controller with translation coefficient, which is so important that it can effectively inhibit overregulation. However, in the neighborhood of the origin, the fuzzy controller retains the weakness with more static state error as being a P controller.

Note 2   In the $x - u$ plane, a P controller, $u(t) = K_P x(t)$, is a straight line passing through the origin, which means that it is of linear regulation law. But a fuzzy controller is a polygonal line passing through the origin, which has nonlinear regulation law as it can approximate a curve. So a fuzzy controller is much better than a P controller.

10.3   The Relationship of Fuzzy Controllers with Two Inputs One Output and PD (or PI) Controllers

Let $X = [-E, E]$ and $Y = [-EC, EC]$ be the universes of input variables and $V = [-U, U]$ be the universe of output variables. Denote $\mathcal{A} = \{A_i\}_{(1 \leq i \leq p)}, \mathcal{B} = \{B_j\}_{(1 \leq j \leq q)}$ and $\mathcal{D} = \{D_{ij}\}_{(1 \leq i \leq p, 1 \leq j \leq q)}$, respectively, a fuzzy partition on X, Y, and V, where x_i, y_j, and u_{ij} is, respectively, the peak point of A_i, B_j, and D_{ij} satisfying that $-E \leq x_1 < x_2 < \cdots < x_p \leq E, -EC \leq y_1 < y_2 < \cdots < y_q \leq EC$ and $-U \leq u_{11} < u_{12} < \cdots < u_{pq} \leq U$. Here A_i are also defined by Expressions (10.3), (10.4), and (10.5) but n is replaced by p, and B_j are defined similar to them. $\mathcal{A}$, $\mathcal{B}$, and $\mathcal{D}$ form a reference rule base R:

$$\text{If } x \text{ is } A_i \text{ and } y \text{ is } B_j \text{ , then } u \text{ is } D_{ij} \text{ ,} \tag{10.11}$$

where $i = 1, 2, \cdots, p, j = 1, 2, \cdots, q$. With reference to Chapter 8, the fuzzy controller with two inputs and one output based on Expression (10.11) can be shown as the following interpolation function:

$$u(t) \overset{\triangle}{=} F(x(t), y(t)) = \sum_{i=1}^{p}\sum_{j=1}^{q} \mu_{A_i}(x(t))\mu_{B_j}(y(t))u_{ij} \ . \tag{10.12}$$

Theorem 2   Let $x(t) = e(t)$ and $y(t) = \frac{de(t)}{dt}$, then a fuzzy controller with two inputs and one output is a piecewise PD controller with mutual affection between $x(t)$ and $y(t)$ and with translation coefficients as the following:

$$u(t) \overset{\triangle}{=} F(x(t), y(t)) = \sum_{i=1}^{p-1}\sum_{j=1}^{q-1} u_{i,j}(t) \ , \tag{10.13}$$

where

$$u^{(i,j)}(t) = \begin{cases} K_P^{(i,j)}(x(t) + T_D^{(i,j)}y(t) + T_{PD}^{(i,j)}x(t)y(t) + T_C^{(i,j)}), & \begin{aligned} &x_i \leq x(t) \leq x_{i+1}, \\ &y_j \leq y(t) \leq y_{j+1}; \end{aligned} \\ 0, & \text{otherwise} \end{cases}$$
$$\tag{10.14}$$

and $i = 1, 2, \cdots, p - 1, j = 1, 2, \cdots, q - 1; K_P^{(i,j)}, T_D^{(i,j)}, T_{PD}^{(i,j)}$, and $T_C^{(i,j)}$ are, respectively, the proportional gain, the differential time constant, the time constant of mutual affection between $x(t)$ and $y(t)$, and the translation coefficient, in $[x_i, x_{i+1}] \times [y_j, y_{j+1}]$. Especially let (x_{i_0}, y_{j_0}) be the origin of $x - y$ plane, i.e., $x_{i_0} = 0 = y_{j_0}$, then

$$T_C^{(i_0,j_0)} = T_C^{(i_0,j_0-1)} = T_C^{(i_0-1,j_0-1)} = T_C^{(i_0-1,j_0)} = 0;$$

thus, when $i = i_0 - 1, i_0$ and $j = j_0 - 1, j_0$, Expression (10.14) can be simplified as follows:

$$u^{(i,j)}(t) = \begin{cases} K_P^{(i,j)}(x(t) + T_D^{(i,j)}y(t) + T_{PD}^{(i,j)}x(t)y(t)), & x_i \leq x(t) \leq x_{i+1}, \\ & y_j \leq y(t) \leq y_{j+1}; \quad (10.15) \\ 0, & \text{otherwise .} \end{cases}$$

Proof   When $(x(t), y(t)) \in [x_i, x_{i+1}] \times [y_j, y_{j+1}]$, by means of the structures of the group of linear base elements, we have

$$u(t) = F(x(t), y(t)) = \sum_{i=1}^{p} \sum_{j=1}^{q} \mu_{A_i}(x(t))\mu_{B_j}(y(t))u_{ij}$$

$$= \mu_{A_i}(x(t))\mu_{B_j}(y(t))u_{ij} + \mu_{A_i}(x(t))\mu_{B_{j+1}}(y(t))u_{ij+1}$$
$$+ \mu_{A_{i+1}}(x(t))\mu_{B_j}(y(t))u_{i+1j} + \mu_{A_{i+1}}(x(t))\mu_{B_{j+1}}(y(t))u_{i+1j+1}$$
$$= \frac{x(t) - x_{i+1}}{x_i - x_{i+1}} \cdot \frac{y(t) - y_{j+1}}{y_j - y_{j+1}} u_{ij} + \frac{x(t) - x_{i+1}}{x_i - x_{i+1}} \cdot \frac{y(t) - y_j}{y_{j+1} - y_j} u_{ij+1}$$
$$+ \frac{x(t) - x_i}{x_{i+1} - x_i} \cdot \frac{y(t) - y_{j+1}}{y_j - y_{j+1}} u_{i+1j} + \frac{x(t) - x_i}{x_{i+1} - x_i} \cdot \frac{y(t) - y_j}{y_{j+1} - y_j} u_{i+1j+1}$$
$$= \frac{y_j(u_{ij+1} - u_{i+1j+1}) + y_{j+1}(u_{ij+1} - u_{ij})}{(x_i - x_{i+1})(y_j - y_{j+1})} [x(t)$$
$$+ \frac{x_i(u_{i+1j} - u_{i+1j+1}) + x_{i+1}(u_{ij+1} - u_{ij})}{y_j(u_{ij+1} - u_{i+1j+1}) + y_{j+1}(u_{i+1j} - u_{ij})} y(t)$$
$$+ \frac{u_{ij} - u_{ij+1} - u_{i+1j} + u_{i+1j+1}}{y_j(u_{ij+1} - u_{i+1j+1}) + y_{j+1}(u_{i+1j} - u_{ij})} x(t)y(t)$$
$$+ \frac{u_{ij}x_{i+1}y_{j+1} - u_{ij+1}x_{i+1}y_j - u_{i+1j}x_iy_{j+1} + u_{i+1j+1}x_iy_j}{y_j(u_{ij+1} - u_{i+1j+1}) + y_{j+1}(u_{i+1j} - u_{ij})}].$$

Taking

$$K_P^{(i,j)} \triangleq \frac{y_j(u_{ij+1} - u_{i+1j+1}) + y_{j+1}(u_{i+1j} - u_{ij})}{(x_i - x_{i+1})(y_j - y_{j+1})}, \tag{10.16}$$

$$T_D^{(i,j)} \triangleq \frac{x_i(u_{i+1j} - u_{i+1j+1}) + x_{i+1}(u_{ij+1} - u_{ij})}{y_j(u_{ij+1} - u_{i+1j+1}) + y_{j+1}(u_{i+1j} - u_{ij})}, \tag{10.17}$$

$$T_{PD}^{(i,j)} \triangleq \frac{u_{ij} - u_{ij+1} - u_{i+1j} + u_{i+1j+1}}{y_j(u_{ij+1} - u_{i+1j+1}) + y_{j+1}(u_{i+1j} - u_{ij})} \tag{10.18}$$

and

$$T_C^{(i,j)} \triangleq \frac{u_{ij}x_{i+1}y_{j+1} - u_{ij+1}x_{i+1}y_j - u_{i+1j}x_i y_{j+1} + u_{i+1j+1}x_i y_j}{y_j(u_{ij+1} - u_{i+1j+1}) + y_{j+1}(u_{i+1j} - u_{ij})}, \qquad (10.19)$$

we have

$$u(t) = K_P^{(i,j)}(x(t) + T_D^{(i,j)}y(t) + T_{PD}^{(i,j)}x(t)y(t) + T_C^{(i,j)})$$
$$= u^{(i,j)}(t) .$$

Moreover, $F(0,0) = 0$ by the monotonicity of control function. So $u_{i_0 j_0} = F(x_{i_0}, y_{j_0}) = F(0,0) = 0$. Hence, it is easy to know that

$$T_C^{(i_0,j_0)} = T_C^{(i_0,j_0-1)} = T_C^{(i_0-1,j_0-1)} = T_C^{(i_0-1,j_0)} = 0,$$

based on expression (10.19).   **Q.E.D.**

Note 3   It is not difficult to learn that a PD controller,

$$u(t) = K_P(x(t) + T_D y(t)),$$

is a plane passing through the origin in xyu space, which means that it is of linear regulation law. However, a fuzzy controller is a piecewise quadratic surface passing through the origin in the space, in which the whole surface can approximate a nonlinear regulation law. So its whole merit is much better than the PD controller. But in the neighborhood of the origin, as the translation coefficients are zero, there exists (i,j) such that

$$u(t) = u^{(i,j)}(t) = K_P^{(i,j)}(x(t) + T_D^{(i,j)}y(t) + T_{PD}^{(i,j)}x(t)y(t)). \qquad (10.20)$$

At this region, $|x(t)|$ and $|y(t)|$ are so small that the higher order terms $T_{PD}^{(i,j)}x(t)y(t)$ can be omitted, so we have

$$u(t) = u^{(i,j)}(t) \approx K_P^{(i,j)}(x(t) + T_D^{(i,j)}y(t)). \qquad (10.21)$$

This means that the fuzzy controller is approximately a PD controller in the neighborhood of the origin. Hence, the weakness of PD controllers is inherited by the fuzzy controller in the neighborhood.

Note 4   This gives us a conclusion about the effectiveness of fuzzy controllers. In the earlier stage of control processing, the effect of a fuzzy controller is better than the effect of a PID controller, especially on inhibiting overregulation; but when $(x(t), y(t))$ is in the neighborhood of the origin, from Expression (10.21), we know that a fuzzy controller is approximately a PD controller. And of course the PD controller cannot be better than a PID controller.

If we write $Y = [-EI, EI]$, similar to Theorem 2, we have the following result.

Theorem 3   Let $x(t) = e(t)$ and $y(t) = \int_0^t e(\tau)d\tau$, then a fuzzy controller with two inputs and one output is just a piecewise PI controller with mutual affection between $x(t)$ and $y(t)$ and with translation coefficients:

$$u(t) = F(x(t), y(t)) = \sum_{i=1}^{p-1}\sum_{j=1}^{q-1} u^{(i,j)}(t),$$

where

$$u^{(i,j)}(t) = \begin{cases} K_P^{(i,j)}\left(x(t) + \dfrac{y(t)}{T_I^{(i,j)}} + T_{PI}^{(i,j)}x(t)y(t) + T_C^{(i,j)}\right), & x_i \le x(t) \le x_{i+1}, \\ & \\ & y_j \le y(t) \le y_{j+1}; \\ 0, & \text{otherwise,} \end{cases}$$

and $i = 1, 2, \cdots, p-1, j = 1, 2, \cdots, q-1; T_I^{(i,j)}$ and $T_{PI}^{(i,j)}$ are, respectively, the integral time constant and the time constant of mutual affection between $x(t)$ and $y(t)$ in $[x_i, x_{i+1}] \times [y_j, y_{j+1}]$ where

$$T_I^{(i,j)} \triangleq \frac{y_j(u_{ij+1} - u_{i+1j+1}) + y_{j+1}(u_{ij+1} - u_{ij})}{u_{ij} - u_{ij+1} - u_{i+1j} + u_{i+1j+1}}.$$

Moreover, there is the result similar to Expression (10.15).

Note 5   This fuzzy controller is seldom used because it is hard to acquire the reference rules about I. Clearly, its merit is better than a PI controller. In the neighborhood of the origin , although $|x(t)|$ is very small as the translation coefficients are zero, $|y(t)| = |\int_0^t e(\tau)d\tau|$ may not be very small. But it can be regarded as $y(t) \approx I = \text{const}$. So there exists (i, j) such that

$$\begin{aligned} u(t) = u^{(i,j)}(t) &= K_P^{(i,j)}(x(t) + y(t)/T_I^{(i,j)} + T_{PI}^{(i,j)}x(t)y(t)) \\ &\approx K_P^{(i,j)}(x(t) + y(t)/T_I^{(i,j)} + IT_{PI}^{(i,j)}x(t)) \\ &= K_P^{(i,j)}[(1 + IT_{PI}^{i,j})x(t) + y(t)/T_I^{(i,j)}] \\ &= K_P^{(i,j)}(1 + IT_{PI}^{i,j})x(t) + y(t)/[T_I^{(i,j)}(1 + IT_{PI}^{(i,j)})] \\ &= \overset{*}{K}_p^{(i,j)}(x(t) + y(t)/\overset{*}{T}_I^{(i,j)}), \end{aligned}$$

where we define

$$\overset{*}{K}_p^{(i,j)} \triangleq K_P^{(i,j)}(1 + IT_{PI}^{(i,j)})$$

and

$$\overset{*}{T}_I^{(i,j)} \triangleq T_I^{(i,j)}(1 + IT_{PI}^{(i,j)}).$$

This means that, in the neighborhood of the origin, the controller is also approximately a PI controller, and the fuzzy controllers inherit the weakness of PI controllers.

10.4   The Relationship of Fuzzy Controllers with Three Inputs One Output and PID Controllers

Let $X = [-E, E], Y = [-EI, EI]$ and $Z = [-EC, EC]$ be the universes of input variables and $V = [-U, U]$ be the universe of output variables. $\mathcal{A} = \{A_i\}_{(1 \leq i \leq p)}, \mathcal{B} = \{B_i\}_{(1 \leq j \leq q)}, \mathcal{C} = \{C_s\}_{(1 \leq s \leq r)}$, and $\mathcal{D} = \{D_{ijs}\}_{(1 \leq i \leq p, 1 \leq j \leq q, 1 \leq s \leq r)}$ are, respectively, a partition on X, Y, Z, and V, where x_i, y_j, z_s, and u_{ijs} are, respectively, the peak point of A_i, B_j, C_s, and D_{ijs} satisfying that $-E \leq x_1 < x_2 < \cdots < x_p \leq E, -EI \leq y_1 < y_2 < \cdots < y_q \leq EI, -EC \leq z_1 < z_2 < \cdots < z_r \leq EC$, and $-U \leq u_{111} < u_{112} < \cdots < u_{pqr} \leq U$. The membership functions about A_i, B_j, and C_s are defined like Expressions (10.3), (10.4), and (10.5). $\mathcal{A}$, $\mathcal{B}$, $\mathcal{C}$, and $\mathcal{D}$ can form a reference rule base R:

$$\text{If } x \text{ is } A_i \text{ and } y \text{ is } B_j \text{ and } z \text{ is } C_s, \text{then } u \text{ is } D_{ijs} . \tag{10.22}$$

Again, with reference to Chapter 8, a fuzzy controller with three inputs and one output based on (10.22) can be shown as the following interpolation function:

$$u(t) \triangleq F(x(t), y(t), z(t))$$
$$= \sum_{i=1}^{p} \sum_{j=1}^{q} \sum_{s=1}^{r} \mu_{A_i}(x(t)) \mu_{B_j}(y(t)) \mu_{C_s}(z(t)) u_{ijs}. \tag{10.23}$$

Theorem 4   Let $x(t) = e(t)$, and

$$y(t) = \int_0^t e(\tau) d\tau, \quad \text{and} \quad z(t) = \frac{de(t)}{dt},$$

then a fuzzy controller with three inputs and one output is just a piecewise PID controller with mutual affection among $x(t)$, $y(t)$, and $z(t)$ and with translation coefficients, as the following:

$$u(t) = F(x(t), y(t), z(t)) = \sum_{i=1}^{p-1} \sum_{j=1}^{q-1} \sum_{s=1}^{r-1} u^{(i,j,s)}(t), \tag{10.24}$$

where, when $(x(t), y(t), z(t)) \in [x_i, x_{i+1}] \times [y_j, y_{j+1}] \times [z_s, z_{s+1}]$,

$$u^{(i,j,s)}(t) = K_P^{(i,j,s)} \Big(x(t) + y(t)/T_I^{(i,j,s)} + T_D^{(i,j,s)} z(t)$$
$$+ T_{PI}^{(i,j,s)} x(t) y(t) + T_{PD}^{(i,j,s)} x(t) z(t) + T_{ID}^{(i,j,s)} y(t) z(t)$$
$$+ T_{PID}^{(i,j,s)} x(t) y(t) z(t) + T_C^{(i,j,s)} \Big), \tag{10.25}$$

otherwise $u^{(i,j,s)}(t) = 0$. Let $x_{i_0} = y_{j_0} = z_{s_0} = 0$, then

$$
\begin{aligned}
T_C^{(i_0,j_0,s_0)} &= T_C^{(i_0,j_0,s_0-1)} = T_C^{(i_0,j_0,s_0-1)} \\
&= T_C^{(i_0,j_0-1,s_0)} = T_C^{(i_0-1,j_0,s_0)} \\
&= T_C^{(i_0,j_0-1,s_0-1)} = T_C^{(i_0-1,j_0-1,s_0)} \\
&= T_C^{(i_0-1,j_0,s_0-1)} = T_C^{(i_0-1,j_0-1,s_0-1)} \\
&= 0.
\end{aligned}
$$

Proof   When $(x(t), y(t), z(t)) \in [x_i, x_{i+1}] \times [y_j, y_{j+1}] \times [z_s, z_{s+1}]$ taking note of the linear structures of A_i, B_j, and C_s, we have

$$
\begin{aligned}
u(t) &= F(x(t), y(t), z(t)) \\
&= \sum_{i=1}^{p} \sum_{j=1}^{q} \sum_{s=1}^{r} \mu_{A_i}(x(t))\mu_{B_j}(y(t))\mu_{C_s}(z(t))u_{ijs} \\
&= \mu_{A_i}(x(t))\mu_{B_j}(y(t))\mu_{C_s}(z(t))u_{ijs} \\
&\quad + \mu_{A_i}(x(t))\mu_{B_j}(y(t))\mu_{C_{s+1}}(z(t))u_{ijs+1} \\
&\quad + \mu_{A_i}(x(t))\mu_{B_{j+1}}(y(t))\mu_{C_s}(z(t))u_{ij+1s} \\
&\quad + \mu_{A_i}(x(t))\mu_{B_{j+1}}(y(t))\mu_{C_{s+1}}(z(t))u_{ij+1s+1} \\
&\quad + \mu_{A_{i+1}}(x(t))\mu_{B_j}(y(t))\mu_{C_s}(z(t))u_{i+1js} \\
&\quad + \mu_{A_{i+1}}(x(t))\mu_{B_j}(y(t))\mu_{C_{s+1}}(z(t))u_{i+1js+1} \\
&\quad + \mu_{A_{i+1}}(x(t))\mu_{B_j}(y(t))\mu_{C_s}(z(t))u_{i+1js} \\
&\quad + \mu_{A_{i+1}}(x(t))\mu_{B_j}(y(t))\mu_{C_{s+1}}(z(t))u_{i+1js+1} \\
&\quad + \mu_{A_{i+1}}(x(t))\mu_{B_{j+1}}(y(t))\mu_{C_s}(z(t))u_{i+1j+1s} \\
&\quad + \mu_{A_{i+1}}(x(t))\mu_{B_{j+1}}(y(t))\mu_{C_{s+1}}(z(t))u_{i+1j+1s+1} \\
&= \frac{x(t) - x_{i+1}}{x_i - x_{i+1}} \cdot \frac{y(t) - y_{j+1}}{y_j - y_{j+1}} \cdot \frac{z(t) - z_{s+1}}{z_s - z_{s+1}} u_{ijs} \\
&\quad + \frac{x(t) - x_{i+1}}{x_i - x_{i+1}} \cdot \frac{y(t) - y_{i+1}}{y_j - y_{j+1}} \cdot \frac{z(t) - z_s}{z_{s+1} - z_s} u_{ijs+1} \\
&\quad + \frac{x(t) - x_{i+1}}{x_i - x_{i+1}} \cdot \frac{y(t) - y_j}{y_{j+1} - y_j} \cdot \frac{z(t) - z_{s+1}}{z_s - z_{s+1}} u_{ij+1s} \\
&\quad + \frac{x(t) - x_{i+1}}{x_i - x_{i+1}} \cdot \frac{y(t) - y_j}{y_{j+1} - y_j} \cdot \frac{z(t) - z_s}{z_{s+1} - z_s} u_{ij+1s+1} \\
&\quad + \frac{x(t) - x_i}{x_{i+1} - x_i} \cdot \frac{y(t) - y_{j+1}}{y_j - y_{j+1}} \cdot \frac{z(t) - z_{s+1}}{z_s - z_{s+1}} u_{i+1js} \\
&\quad + \frac{x(t) - x_i}{x_{i+1} - x_i} \cdot \frac{y(t) - y_j}{y_j - y_{j+1}} \cdot \frac{z(t) - z_s}{z_{s+1} - z_s} u_{i+1js+1} \\
&\quad + \frac{x(t) - x_i}{x_{i+1} - x_i} \cdot \frac{y(t) - y_j}{y_{j+1} - y_j} \cdot \frac{z(t) - z_{s+1}}{z_s - z_{s+1}} u_{i+1j+1s}
\end{aligned}
$$

$$+ \frac{x(t) - x_i}{x_{i+1} - x_i} \cdot \frac{y(t) - y_j}{y_{j+1} - y_j} \cdot \frac{z(t) - z_s}{z_{s+1} - z_s} u_{i+1j+1s+1}$$

$$= \frac{1}{(x_i - x_{i+1})(y_j - y_{j+1})(z_s - z_{s+1})} \Big[(y_{j+1}z_{s+1}u_{ijs} - y_{j+1}z_s u_{ijs+1}$$

$$- y_j z_{s+1} u_{ij+1s} + y_j z_s u_{ij+1s+1} - y_{j+1}z_{s+1}u_{i+1js} + y_{j+1}z_s u_{i+1js+1}$$

$$+ y_j z_{s+1} u_{i+1j+1s} - y_j z_s u_{i+1j+1s+1})x(t) + (x_{i+1}z_{s+1}u_{ijs} - x_{i+1}z_s u_{ijs+1}$$

$$- x_{i+1}z_{s+1}u_{ij+1s} + x_{i+1}z_s u_{ij+1s+1} - x_i z_{s+1}u_{i+1js} + x_i z_s u_{i+1js+1}$$

$$+ x_i z_{s+1} u_{i+1j+1s} - x_i z_s u_{i+1j+1s+1})y(t) + (x_{i+1}y_{j+1}u_{ijs} - x_{i+1}y_{j+1}u_{ijs+1}$$

$$- x_{i+1}y_j u_{ij+1s} + x_{i+1}y_j u_{ij+1s+1} - x_i y_{j+1}u_{i+1js} + x_i y_{j+1}u_{i+1js+1}$$

$$+ x_i y_j u_{i+1j+1s} - x_i y_j u_{i+1j+1s+1})z(t) + (-z_{s+1}u_{ijs} + z_s u_{ijs+1}$$

$$+ z_{s+1}u_{ij+1s} - z_s u_{ij+1s+1} + z_{s+1}u_{i+1js} - z_s u_{i+1js+1}$$

$$- z_{s+1}u_{i+1j+1s} + z_s u_{i+1j+1s+1})x(t)y(t) + (-y_{j+1}u_{ijs} + y_{j+1}u_{ijs+1}$$

$$+ y_j u_{ij+1s} - y_j u_{ij+1s+1} + y_j u_{i+1js} - y_{j+1}u_{i+1js+1}$$

$$- y_j u_{i+1j+1s} + y_j u_{i+1j+1s+1})x(t)z(t) + (-x_{i+1}u_{ijs} + x_{i+1}u_{ijs+1}$$

$$+ x_{i+1}u_{ij+1s} - x_{i+1}u_{ij+1s+1} + x_i u_{i+1js} - x_i u_{i+1js+1}$$

$$- x_i u_{i+1j+1s} + x_i u_{i+1j+1s+1})y(t)z(t) + (u_{ijs} - u_{ijs+1} - u_{ij+1s} + u_{ij+1s+1}$$

$$- u_{i+1js} + u_{i+1js+1} + u_{i+1j+1s} - u_{i+1j+1s+1})x(t)y(t)z(t)$$

$$+ (-x_{i+1}y_{j+1}z_{s+1}u_{ijs} + x_{i+1}y_{j+1}z_s u_{ijs+1} + x_{i+1}y_j z_{s+1}u_{ij+1s}$$

$$- x_{i+1}y_j z_s u_{ij+1s+1} + x_i y_{j+1}z_{s+1}u_{i+1js} - x_i y_{j+1}z_s u_{i+1js+1}$$

$$- x_i y_j z_{s+1}u_{i+1j+1s} + x_i y_j z_s u_{i+1j+1s+1}) \Big].$$

Write

$$M(i, j, s) \triangleq y_{j+1}z_{s+1}u_{ijs} - y_{j+1}z_s u_{ijs+1} - y_j z_{s+1}u_{ij+1s}$$

$$+ y_j z_s u_{ij+1s+1} - y_{j+1}z_{s+1}u_{i+1js} + y_{j+1}z_s u_{i+1js+1}$$

$$+ y_j z_{s+1}u_{i+1j+1s} - y_j z_s u_{i+1j+1s+1},$$

and take

$$K_P^{(i,j,s)} \triangleq M(i, j, s) \Big/ (x_i - x_{i+1})(y_j - y_{j+1})(z_s - z_{s+1}), \tag{10.26}$$

$$T_I^{(i,j,s)} \triangleq M(i, j, s) \Big/ \Big(x_{i+1}z_{s+1}u_{ijs} - x_{i+1}z_s u_{ijs+1}$$

$$- x_{i+1}z_{s+1}u_{ij+1s} + x_{i+1}z_s u_{ij+1s+1} - x_i z_{s+1}u_{i+1js}$$

$$+ x_i z_s u_{i+1js+1} + x_i z_{s+1}u_{i+1j+1s}$$

$$- x_i z_s u_{i+1j+1s+1}\Big), \tag{10.27}$$

$$T_D^{(i,j,s)} \triangleq \Big(x_{i+1}y_{j+1}u_{ijs} - x_{i+1}y_{j+1}u_{ijs+1} - x_{i+1}y_j u_{ij+1s}$$

$$+ x_{i+1}y_j u_{ij+1s+1} - x_i y_{j+1}u_{i+1js} + x_i y_{j+1}u_{i+1js+1}$$

$$+ x_i y_j u_{i+1j+1s} - x_i y_j u_{i+1j+1s+1}\Big) \Big/ M(i, j, s), \tag{10.28}$$

$$T_{PI}^{(i,j,s)} \triangleq \Big(-z_{s+1}u_{ijs} + z_s u_{ijs+1} + z_{s+1}u_{ij+1s}$$
$$- z_s u_{ij+1s+1} + z_{s+1}u_{i+1js} - z_s u_{i+1js+1}$$
$$- z_{s+1}u_{i+1j+1s} + z_s u_{i+1j+1s+1} \Big) \Big/ M(i,j,s), \qquad (10.29)$$

$$T_{PD}^{(i,j,s)} \triangleq \Big(-y_{j+1}u_{ijs} + y_{j+1}u_{ijs+1} + y_j u_{ij+1s} - y_j u_{ij+1s+1} + y_j u_{i+1js}$$
$$- y_{j+1}u_{i+1js+1} - y_j u_{i+1j+1s} + y_j u_{i+1j+1s+1} \Big) \Big/ M(i,j,s), \qquad (10.30)$$

$$T_{ID}^{(i,j,s)} \triangleq \Big(-x_{i+1}u_{ijs} + x_{i+1}u_{ijs+1} + x_{i+1}u_{ij+1s}$$
$$- x_{i+1}u_{ij+1s+1} + x_i u_{i+1js} - x_i u_{i+1js+1}$$
$$- x_i u_{i+1j+1s} + x_i u_{i+1j+1s+1} \Big) \Big/ M(i,j,s), \qquad (10.31)$$

$$T_{PID}^{(i,j,s)} \triangleq \Big(u_{ijs} - u_{ijs+1} - u_{ij+1s} + u_{ij+1s+1} - u_{i+1js} + u_{i+1js+1}$$
$$+ u_{i+1j+1s} - u_{i+1j+1s+1} \Big) \Big/ M(i,j,s) , \qquad (10.32)$$

and

$$T_{C}^{(i,j,s)} \triangleq \Big(-x_{i+1}y_{j+1}z_{s+1}u_{ijs} + x_{i+1}y_{j+1}z_s u_{ijs+1} + x_{i+1}y_j z_{s+1}u_{ij+1s}$$
$$- x_{i+1}y_j z_s u_{ij+1s+1} + x_i y_{j+1}z_{s+1}u_{i+1js} - x_i y_{j+1}z_s u_{i+1js+1}$$
$$- x_i y_j z_{s+1}u_{i+1j+1s} + x_i y_j z_s u_{i+1j+1s+1} \Big) \Big/ M(i,j,s) \qquad (10.33)$$

we have $u(t) = u^{(i,j,s)}(t)$. So (10.24) is true. Moreover, as

$$u_{i_0 j_0 s_0} = F(x_{i_0}, y_{j_0}, z_{s_0}) = F(0,0,0) = 0,$$

we have

$$T_C^{(i_0,j_0,s_0)} = T_C^{(i_0,j_0,s_0-1)} = \cdots = T_C^{(i_0-1,j_0-1,s_0-1)} = 0.$$

Q.E.D.

Note 6 In $xyzu$ space, a PID controller is a hyperplane passing through the origin, which is of linear regulation law; while a fuzzy controller is a piecewise cubic surface, where the whole surface has nonlinear regulation law. So the effect of fuzzy controller is much better than the effect of a PID, in the whole control processing. However, in the neighborhood of the origin, the fuzzy controller is approximately a PID controller.

10.5 The Difference Schemes of Fuzzy Controllers with Three Inputs and One Output

In this section, we rewrite the fuzzy controller equations to difference equations in discrete time domain. Without loss of generality, let us use a fuzzy controller with three inputs and one output as an example.

Let kT $(k = 0, 1, 2, \cdots)$ be sampling time points where T is sampling period, and

$$y(t) = \int_0^t e(\tau)d\tau \approx T \sum_{i=0}^k e(k) \triangleq \sum_{i=0}^k e(kT)T, \tag{10.34}$$

and

$$z(t) = \frac{de(t)}{dt} \approx \frac{e(k) - e(k-1)}{T} \triangleq \frac{e(kT) - e[(k-1)T]}{T}, \tag{10.35}$$

where $e(kT)$ is simpling denoted by $e(k)$.

10.5.1 Positional Difference Scheme

Theorem 5   The fuzzy controller with three inputs and one output as expression (10.23) (or (10.24)) has the following positional algorithm:

$$u(k) = F(x(k), y(k), z(k)) = \sum_{i=1}^{p-1}\sum_{j=1}^{q-1}\sum_{s=1}^{r-1} u^{(i,j,s)}(k), \tag{10.36}$$

where, when $(x(t), y(t), z(t)) \in [x_i, x_{i+1}] \times [y_j, y_{j+1}] \times [z_s, z_{s+1}]$, we have

$$u^{(i,j,s)}(k) = K_P^{(i,j,s)}e(k) + K_I^{(i,j,s)}\sum_{l=0}^k e(l)$$

$$+ K_D^{(i,j,s)}[e(k) - e(k-1)] + K_{PI}^{(i,j,s)}e(k)\sum_{l=0}^k$$

$$e(l) + K_{PD}^{(i,j,s)}e(k)[e(k) - e(k-1)]$$

$$+ K_{ID}^{(i,j,s)}[e(k) - e(k-1)]\sum_{l=0}^k e(l)$$

$$+ K_{PID}^{(i,j,s)}e(k)[e(k) - e(k-1)]\sum_{l=0}^k e(l) + K_C^{(i,j,s)}, \tag{10.37}$$

and when $(x(t), y(t), z(t)) \notin [x_i, x_{i+1}] \times [y_j, y_{j+1}] \times [z_s, z_{s+1}]$, $u^{(i,j,s)}(k) = 0$, in which

$$K_I^{(i,j,s)} \triangleq \frac{TK_P^{(i,j,s)}}{T_I^{(i,j,s)}}, \quad K_D^{(i,j,s)} \triangleq \frac{K_P^{(i,j,s)}T_D^{(i,j,s)}}{T}, \tag{10.38}$$

$$K_{PI}^{(i,j,s)} \triangleq TK_P^{(i,j,s)}T_{PI}^{(i,j,s)}, \quad K_{PD}^{(i,j,s)} \triangleq \frac{K_P^{(i,j,s)}T_{PD}^{(i,j,s)}}{T}, \tag{10.39}$$

$$K_{ID}^{(i,j,s)} \triangleq K_P^{(i,j,s)}T_{ID}^{(i,j,s)}, \quad K_{PID}^{(i,j,s)} \triangleq \frac{K_P^{(i,j,s)}T_{PID}^{(i,j,s)}}{T}, \tag{10.40}$$

and

$$K_C^{(i,j,s)} \triangleq K_P^{(i,j,s)} T_C^{(i,j,s)} , \qquad (10.41)$$

we have

$$\begin{aligned}
K_C^{(i_0,j_0,s_0)} &= K_C^{(i_0,j_0,s_0-1)} = K_C^{(i_0,j_0-1,s_0)} \\
&= K_C^{(i_0-1,j_0,s_0)} = K_C^{(i_0,j_0-1,s_0-1)} \\
&= K_C^{(i_0-1,j_0-1,s_0)} = K_C^{(i_0-1,j_0,s_0-1)} \\
&= K_C^{(i_0-1,j_0-1,s_0-1)} \\
&= 0.
\end{aligned}$$

Proof   Substituting (10.34) and (10.35) into (10.25), it is not difficult to see that the theorem is true.   **Q.E.D.**

10.5.2   Incremental Difference Scheme

Let

$$\Delta u(k) \triangleq u(k) - u(k-1) = F(x(k), y(k), z(k)) - F(x(k-1), y(k-1), z(k-1)),$$

$$\Delta e(k) = e(k) - e(k-1)$$

and

$$\Delta^2 e(k) = \Delta e(k) - \Delta e(k-1).$$

From Theorem 5 we can obtain the following result:

Theorem 6   The fuzzy controller with three inputs and one output as expression (10.23) (or (10.24)) has the following incremental equations:

$$\Delta u(k) = \Delta F(x(k), y(k), z(k)) = \sum_{i=1}^{p-1} \sum_{j=1}^{q-1} \sum_{s=1}^{r-1} \Delta u^{(i,j,s)}(k), \qquad (10.42)$$

where, when $(x(t), y(t), z(t)) \in [x_i, x_{i+1}] \times [y_j, y_{j+1}] \times [z_s, z_{s+1}]$,

$$\begin{aligned}
\Delta u^{(i,j,s)}(k) =& K_P^{(i,j,s)} \Delta e(k) + K_I^{(i,j,s)} e(k) + K_D^{(i,j,s)} \Delta^2 e(k) \\
&+ K_{PI}^{(i,j,s)} [e^2(k) + \Delta e(k)] \sum_{l=0}^{k-1} e(l) + K_{PD}^{(i,j,s)} [e(k)\Delta e(k) \\
&- e(k-1)\Delta e(k-1)] + K_{ID}^{(i,j,s)} [e(k)\Delta e(k) - \Delta e(k-1)] \\
&+ K_{PID}^{(i,j,s)} [e^2(k)\Delta e(k) - e(k-1)\Delta e(k-1)] \sum_{l=0}^{k} e(l), \qquad (10.43)
\end{aligned}$$

and when $(x(t), y(t), z(t)) \notin [x_i, x_{i+1}] \times [y_j, y_{j+1}] \times [z_s, z_{s+1}]$, $\Delta u^{(i,j,s)}(k) = 0$.

Proof   This proof is similar to the one in Theorem 5.

10.6   Conclusions

We revealed the relationship between fuzzy controllers and PID controllers. We proved that a fuzzy controller is a piecewise PID controller. From the theorems in the paper, we showed that a fuzzy controller has stronger control capability than traditional PID controllers because a nonlinear term is included in the fuzzy controller. However, in the region around the origin, a fuzzy controller is functioned similar to a PID controller.

References

1. L. A. Zadeh, Fuzzy sets, *Information Control*, Vol. 8, pp. 338-353, 1965.

2. M. Sugeno, *Industrial Applications of Fuzzy Control*, North-Holland, Amsterdam, 1985.

3. R. R. Yagar and D. P. Filev, *Essentials of Fuzzy Modeling and Control,* John Wiley and Sons, New York, 1994.

4. R. M. Tong, Some properties of fuzzy feedback systems, *IEEE Transactions on Systems, Man, and Cybernetics*, Vol. 10, No. 6, pp. 327-330, 1980.

5. T. Terano, K. Asai, and M. Sugeno, *Applied Fuzzy Systems*, Academic Press, New York, 1994.

6. Z. Y. Zhao, M. Tomizuka, and S. Isaka, Fuzzy gain scheduling of PID controller, *IEEE Transactions on Systems, Man, and Cybernetics*, Vol. 23, No. 5, pp. 1392-1398, 1993.

7. H. Ying, Constructing nonlinear variable gain controllers via Takagi-Sugeno fuzzy control, *IEEE Transactions on Fuzzy Systems*, Vol. 6, No. 2, pp. 226-234, May 1998.

8. R. Palm, Sliding mode fuzzy control, *Proc. 1st IEEE International Conference on Fuzzy Systems*, San Diego, CA, pp. 519-526, 1992.

9. W. Pydrycz, *Fuzzy Control and Fuzzy Systems*, John Wiley and Sons, New York, 1993.

10. H. X. Li, The mathematical essence of fuzzy controls and fine fuzzy controllers, in *Advances in Machine Intelligence and Soft-Computing*, Paul Wang, ed., Vol. IV, Bookwrights Press, Durham, pp. 55-74, 1997.

11. L. X. Wang, *Adaptive Fuzzy Systems and Control*, Prentice-Hall, Englewood Cliffs, 1994.

12. L. X. Wang, *A Course in Fuzzy Systems and Control*, Prentice-Hall, Englewood Cliffs, 1997.

13. D. Driankov, H. Hellendoorn, and M. Reinfrank, *An Introduction to Fuzzy Control,* 2nd Ed., Springer, Berlin, 1996.

14. L. Reznix, *Fuzzy Controllers,* Newnes Publishers, Oxford, 1997.

15. K. M. Passino and S. Yurkovich, *Fuzzy Control*, Addison-Wesley, NY, 1998.

16. H. X. Li and C. L. P. Chen, The interpolation mechanism of fuzzy control and its relationship to PID control, *International Journal of Fuzzy Systems*, Vol. 2, No. 1, March 2000.

Chapter 11

Adaptive Fuzzy Controllers Based on Variable Universes

In this chapter, we introduce variable universes-based adaptive fuzzy controllers. The concept comes from interpolation forms of fuzzy control introduced in Chapter 8. First, we define monotonicity of control rules, and we prove that the monotonicity of interpolation functions of fuzzy control is equivalent to the monotonicity of control rules. This means that there is no contradiction among the control rules under the condition for the control rules being monotonic. Then the structure of the contraction-expansion factor is discussed. At last, based on variable universes, we present three models of adaptive fuzzy control, namely, an adaptive fuzzy control model with *potential heredity*, adaptive fuzzy control model with *obvious heredity* and adaptive fuzzy control model with *successively obvious heredity*.

11.1 The Monotonicity of Control Rules and The Monotonicity of Control Functions

There are two different types of fuzzy controllers: the trial-and-error approach and the theoretical approach [1, 2]. A set of "If-Then" rules is collected from experience-based knowledge for the trial-and-error approach. In theoretical approach, the structure and parameters of fuzzy controller are designed such that a certain performance criteria are guaranteed. In adaptive fuzzy control, the structure or parameters of the controller change during the operation. Several popular adaptive fuzzy controllers and their applications have been extensively studied in references [3-7]. In this chapter, we shall take a different approach to discuss a "variable universe-based" adaptive fuzzy control.

Without loss of generality, we consider fuzzy controllers with two inputs and one output. Let X and Y be the universes of input variables and Z the universe of output variables. The families of all unimodal and normal fuzzy sets on these universes are, respectively, denoted by $\mathcal{F}_0(X)$, $\mathcal{F}_0(Y)$, and $\mathcal{F}_0(Z)$. If a certain order relation $\leq$ is defined in $\mathcal{F}_0(X)$, $\mathcal{F}_0(Y)$, and $\mathcal{F}_0(Z)$, the ordered sets $(\mathcal{F}_0(X), \leq)$, $(\mathcal{F}_0(Y), \leq)$ and $(\mathcal{F}_0(Z), \leq)$ are formed. Taking three ordered subsets $\mathcal{A} \subset \mathcal{F}_0(X)$, $\mathcal{B} \subset \mathcal{F}_0(Y)$, and $\mathcal{C} \subset \mathcal{F}_0(Z)$, we know that the control rules of fuzzy control can be described by a mapping R:

$$R: \ \mathcal{A} \times \mathcal{B} \longrightarrow \mathcal{C}, \quad (A, B) \longmapsto R(A, B) \overset{\triangle}{=} C ,$$

where R is called the rules or rule base of a controller.

Definition 1   The rule $R(A, B)$ is said to be monotonic increasing (decreasing) with respect to A, if $R(A, B)$ is isotonic (anti-isotonic), i.e., $\forall \ A', A'' \in \mathcal{A}$,

$$A' \leq A'' \Longrightarrow R(A', B) \leq R(A'', B) , \quad (R(A', B) \geq R(A'', B)) . \tag{11.1}$$

Similarly, the monotonicity of $R(A, B)$ with respect to B can be defined. When $R(A, B)$ is monotonic increasing (decreasing) not only with respect to A but also with respect to B, $R(A, B)$ is called completely monotonic increasing (decreasing). And when $R(A, B)$ is monotonic increasing (decreasing) with respect to A but monotonic decreasing (increasing) with respect to B, $R(A, B)$ is said to be mixedly monotonic.

Note 1   When a fuzzy controller is of multi-output, R should be a vector-valued function. It is easy to define the monotonicity of R.

Here we assume that universes are all real number intervals. Let $U \in \{X, Y, Z\}$. For any $A \in \mathcal{F}_0(U)$, if the peak points of A are not unique, then a representative point among them should be taken by means of a certain way. This representative point (including the unique peak point) is called the normal peak point of A. Now we define a commonly used order relation "$\leq$" :

$$(\forall A_1, A_2 \in \mathcal{F}_0(U))(A_1 \leq A_2 \Longleftrightarrow x_1 \leq x_2) , \tag{11.2}$$

where x_1 and x_2 are, respectively, the normal peak points of A_1 and A_2.

So the linguistic values on a symmetric interval $[-a, \ a]$, NB, NM, NS, ZO, PS, PM, PB (negative big, negative medium, etc.), can be sequenced as the following:

$$NB \leq NM \leq NS \leq ZO \leq PS \leq PM \leq PB .$$

The control rule base, R, of a fuzzy controller is usually written as follows:

$$\text{if } x \text{ is } A_i \text{ and } y \text{ is } B_j , \quad \text{then } z \text{ is } C_{ij} \tag{11.3}$$

where $i = 1, 2, \cdots, p$, $j = 1, 2, \cdots, q$. From Chapter 8, we know that a fuzzy controller is approximately a binary piecewise interpolation function:

$$F(x, y) = \sum_{i=1}^{p} \sum_{j=1}^{q} \mu_{A_i}(x) \mu_{B_j}(y) z_{ij} . \tag{11.4}$$

If the control function of the control system is written as $f : \ X \times Y \longrightarrow Z$, $(x, y) \longmapsto z = f(x, y)$, then $F(x, y)$ approximates to $f(x, y)$, i.e.,

$$(\forall \ \varepsilon > 0)(\exists N)(n \geq N \Longrightarrow \sup_{(x,y) \in X \times Y} |F(x, y) - f(x, y)| \leq \varepsilon) . \tag{11.5}$$

Equation (11.5) indicates that we can regard $F(x, y)$ and $f(x, y)$ as the same. Clearly, $f(x, y)$ is usually a nonlinear function, which means that a fuzzy controller is a nonlinear approximator. In order to discuss the monotonicity of control action, we can define the monotonicity of the control function $f(x, y)$ with respect to (x, y), similarly to Definition 1.

Theorem 1 For a given fuzzy control system where its rule base is shown as Expression (8.1) and $\mathcal{A}$ and $\mathcal{B}$ are two-phased groups of base elements, then $R(A, B)$ is monotonic increasing (decreasing) with respect to A or B, or completely monotonic increasing (decreasing), or mixedly monotonic, if and only if $F(A, B)$ has corresponding monotonicity.

Proof We only give the proof of the mixedly monotonic case, and the proofs of other cases are similar to it.

Necessity: Let $R(A, B)$ be monotonic increasing with respect to A but monotonic decreasing with respect to B . First we prove that $F(x, y)$ is monotonic increasing with respect to x. In fact, $\forall\ (x', y_0),\ (x'', y_0),\ x' < x''$, by the two-phased property of $\mathcal{A}$ and $\mathcal{B}$, $\exists i_1, i_2 \in \{1, 2, \cdots, p\}$ and $\exists j_0 \in \{1, 2, \cdots, q\}$, such that

$$F(x', y_0) = \mu_{A_{i_1}}(x')\mu_{B_{j_0}}(y_0)z_{i_1\ j_0} + \mu_{A_{i_1}}(x')\mu_{B_{j_0+1}}(y_0)z_{i_1, j_0+1}$$
$$+ \mu_{A_{i_1}+1}(x')\mu_{B_{j_0}}(y_0)z_{i_1+1, j_0} + \mu_{A_{i_1}+1}(x')\mu_{B_{j_0+1}}(y_0)z_{i_1+1, j_0+1}$$

and

$$F(x'', y_0) = \mu_{A_{i_2}}(x'')\mu_{B_{j_0}}(y_0)z_{i_2 j_0} + \mu_{A_{i_2}}(x'')\mu_{B_{j_0+1}}(y_0)z_{i_2, j_0+1}$$
$$+ \mu_{A_{i_2}+1}(x'')\mu_{B_{j_0}}(y_0)z_{i_2+1, j_0} + \mu_{A_{i_2}+1}(x'')\mu_{B_{j_0+1}}(y_0)z_{i_2+1, j_0+1}\ ,$$

where

$$1 = \mu_{A_{i_1}}(x') + \mu_{A_{i_1}+1}(x') = \mu_{A_{i_2}}(x'') + \mu_{A_{i_2}+1}(x')$$
$$= \mu_{B_{j_0}}(y_0) + \mu_{B_{j_0+1}}(y_0).$$

Based on the property of monotonicity of $R(A, B)$, it is easy to know that $(\forall\ i)(z_{i1} \geq z_{i2} \geq \cdots \geq z_{iq})$. Because $(\forall\ j)(z_{1j} \leq z_{2j} \leq \cdots \leq z_{pj})$, $x_1 < x_2$ and $i_1 \leq i_2$, so when $i_1 = i_2$, we have

$$F(x', y_0) - F(x'', y_0)$$
$$= (\mu_{A_{i_1}}(x') - \mu_{A_{i_1}}(x''))(\mu_{B_{j_0}}(y_0)z_{i_1\ j_0} + \mu_{B_{j_0+1}}(y_0)z_{i_1, j_0+1})$$
$$+ (\mu_{A_{i_1}+1}(x') - \mu_{A_{i_1}+1}(x''))(\mu_{B_{j_0}}(y_0)z_{i_1+1, j_0} + \mu_{B_{j_0+1}}(y_0)z_{i_1+1, j_0+1})$$
$$\leq (\mu_{A_{i_1}}(x') - \mu_{A_{i_1}}(x''))(\mu_{B_{j_0}}(y_0) + \mu_{B_{j_0+1}}(y_0))z_{i_1 j_0}$$
$$+ (\mu_{A_{i_1}+1}(x') - \mu_{A_{i_1}+1}(x''))(\mu_{B_{j_0}}(y_0) + \mu_{B_{j_0+1}}(y_0))z_{i_1+1, j_0}$$
$$= (\mu_{A_{i_1}}(x') - \mu_{A_{i_1}}(x''))z_{i_1 j_0} + (\mu_{A_{i_1}+1}(x') - \mu_{A_{i_1}+1}(x''))z_{i_1+1, j_0}$$
$$\leq (\mu_{A_{i_1}}(x') - \mu_{A_{i_1}}(x'') + \mu_{A_{i_1}+1}(x') - \mu_{A_{i_1}+1}(x''))z_{i_1+1, j_0}$$
$$= 0$$

therefore $F(x', y_0) \leq F(x'', y_0)$. When $i_1 < i_2$, $i_1 + 1 \leq i_2$, we have,

$$F(x', y_0) \leq (\mu_{A_{i_1}}(x') + \mu_{A_{i_1+1}}(x'))(\mu_{B_{j_0}}(y_0)z_{i_1+1,j_0} + \mu_{B_{j_0+1}}(y_0)z_{i_1+1,j_0+1})$$
$$= \mu_{B_{j_0}}(y_0)z_{i_1+1,j_0} + \mu_{B_{j_0+1}}(y_0)z_{i_1+1,j_0+1}$$

$$F(x'', y_0) \geq \mu_{B_{j_0}}(y_0)z_{i_2 j_0} + \mu_{B_{j_0+1}}(y_0)z_{i_2,j_0+1}$$

and

$$F(x', y_0) - F(x'', y_0) \leq \mu_{B_{j_0}}(y_0)(z_{i_1+1,j_0} - z_{i_2 j_0}) + \mu_{B_{j_0+1}}(y_0)(z_{i_1+1,j_0+1} - z_{i_2,j_0+1})$$
$$\leq 0$$

therefore $F(x', y_0) \leq F(x'', y_0)$. From these two cases we know that $F(x, y)$ is monotonic increasing with respect to x. In the same way, we can prove that $F(x, y)$ is monotonic decreasing with respect to y.

Sufficiency :   If $F(x, y)$ is monotonic increasing with respect to y, and x_i and y_j are, respectively, the peak points of A_i and B_j, then we have $(x_i, y_j) = z_{ij}$. As the order relation in $\mathcal{A}$, $\mathcal{B}$, and $\mathcal{C}$ is defined by using the peak points, the monotonicity of $R(A, B)$ depends on these peak points. Hence, it is easy to know that $R(A, B)$ is monotonic increasing with respect to A but monotonic decreasing with respect to B based on $F(x_i, y_j) = z_{ij}$.   **Q.E.D.**

The theorem shows that there exists an important relation between rule bases and control functions.

11.2   The Contraction-expansion Factors of Variable Universes

11.2.1   The Contraction-expansion Factors of Adaptive Fuzzy Controllers with One Input and One Output

Given a fuzzy controller, the universe of input variables and the universe of output variables are, respectively, $X = [-E, E]$ and $Y = [-U, U]$, where E and U are real numbers. X and Y can be called initial universes being relative to variable universes.

Definition 2   A function $\alpha : \ X \longrightarrow [0, 1]$, $\ x \longmapsto \alpha(x)$, is called a contraction-expansion factor on universe X, if it satisfies the following conditions: (1) Evenness: $(\forall x \in X)(\alpha(x) = \alpha(-x))$; (2) zero-preserving: $\alpha(0) = 0$; (3) monotonicity: $\alpha(x)$ is strictly monotone increasing on $[0, E]$; and (4) compatibility: $(\forall x \in X)(|x| \leq \alpha(x)E)$.

For any $x \in X$, a variable universe on $X(x)$ is defined below:

$$X(x) \overset{\triangle}{=} \alpha(x)X \overset{\triangle}{=} [-\alpha(x)E, \alpha(x)E] \overset{\triangle}{=} \{\alpha(x)x' | x' \in X\} \ .$$

Figure 1 illustrates the idea of variable universes. Moreover, from the compatibility of Definition 2, it is easy to know that contraction-expansion factors satisfy the following condition:

(5) Normality: $\alpha(\pm E) = 1$, $\beta(\pm U) = 1$.

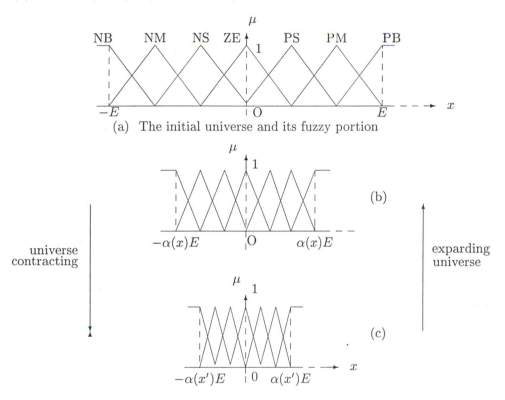

(a) The initial universe and its fuzzy portion

(b)

universe contracting

exparding universe

(c)

Figure 1 Contracting/expanding universe

Let $0 < \tau < 1$, and take

$$\alpha(x) = \left(\frac{|x|}{E}\right)^{\tau} ; \qquad (11.6)$$

then $\alpha(x)$ is a contraction-expansion factor satisfying Definition 1.

11.2.2   The Contraction-expansion Factors of Adaptive Fuzzy Controllers with Two Inputs and One Output

Let $X = [-E, E]$ and $Y = [-D, D]$ be the universes of input variables and $Z = [-U, U]$ be the universe of output variable. When Y is relatively independent from X, we can obtain the contraction-expansion factors $\alpha(x)$ of X, and $\beta(y)$ of Y and $\gamma(z)$ of Z. In some cases, Y may not be independent from X. Then β should be defined on $X \times Y$, i.e., $\beta = \beta(x, y)$. For example, denoting $D = EC$, and $Y = [-EC, EC]$, we can use one of the following two expressions:

$$\beta(x,y) = \left(\frac{|x|}{E}\right)^{\tau_1}\left(\frac{|y|}{EC}\right)^{\tau_2} \tag{11.7}$$

and

$$\beta(x,y) = \frac{1}{2}\left[\left(\frac{|x|}{E}\right)^{\tau_1} + \left(\frac{|y|}{EC}\right)^{\tau_2}\right], \tag{11.8}$$

where $0 < \tau_1, \ \tau_2 < 1$.

Note 2   The rate of change of error depends on error, in this case we can take $\beta = \beta(y)$, but not $\beta = \beta(x,y)$.

11.3   The Structure of Adaptive Fuzzy Controllers Based on Variable Universes

To consider the structure of variable universe-based adaptive fuzzy controllers, we use a fuzzy controller with two inputs and one output shown in Figure 2 as an example.

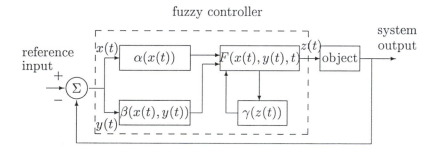

Figure 2   A variable universe-based adaptive fuzzy controller

As a fuzzy control system is a dynamic system, its base variables x, y, and z should depend on time t, denoted by $x(t)$, $y(t)$, and $z(t)$. So the universes should also be denoted by $X(x(t))$, $Y(y(t))$, and $Z(z(t))$. Then the "shapes or forms" of membership functions A_i, B_j, and C_{ij} change according to the change of the universes. It is easy to understand that they should be denoted by $\mu_{A_i(t)}(x(t))$, $\mu_{B_j(t)}(y(t))$, and $\mu_{C_{ij}(t)}(z(t))$. This makes the rule base in (11.3) a group of dynamic rules, $R(t)$:

$$\text{if } x(t) \text{ is } A_i(t) \text{ and } y(t) \text{ is } B_j(t), \text{ then } z(t) \text{ is } C_{ij}(t) \tag{11.9}$$

Because Expression (11.9) equals Expression (11.3) when $t = 0$, Expression (11.3) is called initial rules. Also the control function becomes dynamic, denote it by $F(x(t), y(t), t)$, i.e.,

$$F(x(t), y(t), t) = \sum_{i=1}^{p} \sum_{j=1}^{q} \mu_{A_i(t)}(x(t)) \mu_{B_j(t)}(y(t)) z_{ij}(t) \ . \qquad (11.10)$$

From the definition of variable universe, we know that the monotonicity of initial rule base, $R = R(0)$, ensures the monotonicity of $R(t)$ $(t > 0)$. It means that there exists no contradiction among the rules when we process the contracting/expanding universes. So it ensures that control function $F(x(t), y(t), t)$ is significant.

Figure 3 illustrates the change of control function with one input and one output $F(x(t), t)$.

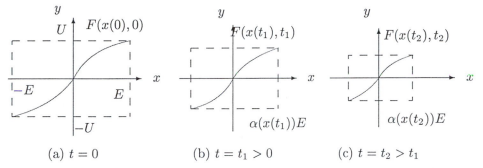

(a) $t = 0$ (b) $t = t_1 > 0$ (c) $t = t_2 > t_1$

Figure 3   The change of control function as time goes on

Figure 3 also indicates that the initial control function at t_{k+1} inherits the initial control function at t_k.

Without loss of generality, from now on, we only consider discrete-time case.

11.4 Adaptive Fuzzy Controllers with One Input and One Output

11.4.1 Adaptive Fuzzy Controllers with Potential Heredity

Let the initial control rule base, $R(0) = R$, be "if x is A_i then y is B_i," $i = 1, 2, \cdots, n$, where $\{A_i\}_{(1 \le i \le n)}$ and $\{B_i\}_{(1 \le i \le n)}$ are, respectively, a group of linear base elements on initial universes $X = [-E, E]$ and $Y = [-U, U]$, where their peak point sets $\{x_i\}_{(1 \le i \le n)}$ and $\{y_i\}_{(1 \le i \le n)}$ satisfy $-E = x_1 < x_2 < \cdots < x_n = E$; and $-U = y_1$ and $y_n = U$. For all $k = 0, 1, 2, \cdots$, take $x_i(0) = x_i$ and $y_i(0) = y_i$; and define a group of linear base elements $\mu_{A_i(k)}(x(k))$ as follows:

$$\mu_{A_1(k)}(x(k)) = \begin{cases} \dfrac{[x(k) - x_2(k)]}{[x_1(k) - x_2(k)]}, & x_1(k) \le x(k) \le x_2(k); \\ 0, & x_2(k) \le x(k) \le x_n(k), \end{cases} \qquad (11.11)$$

$$\mu_{A_i(k)}(x(k)) = \begin{cases} \dfrac{[x(k) - x_{i-1}(k)]}{[x_i(k) - x_{i-1}(k)]}, & x_{i-1}(k) \le x(k) \le x_i(k); \\[2mm] \dfrac{[x(k) - x_{i+1}(k)]}{[x_i(k) - x_{i+1}(k)]}, & x_i(k) < x(k) \le x_{i+1}(k); \\[2mm] 0, & \text{otherwise} \end{cases} \qquad (11.12)$$

$$i = 2, 3, \cdots, n - 1,$$

$$\mu_{A_n(k)}(x(k)) = \begin{cases} 0, & x_1(k) \le x(k) < x_{n-1}(k); \\[2mm] \dfrac{[x(k) - x_{n-1}(k)]}{[x_n(k) - x_{n-1}(k)]}, & x_{n-1}(k) \le x(k) \le x_n(k). \end{cases} \qquad (11.13)$$

Clearly, the group of base elements is a group of two-phased base elements.

Next we discuss a potentially hereditary adaptive fuzzy control algorithm with one input and one output:

Step 0: For an arbitrarily given initial input value, $x(0) \in X$, calculate the output value of the controller $y(1)$:

$$y(1) = F(x(0), 0) = \sum_{i=1}^{n} \mu_{A_i(0)}(x(0))y_i(0);$$

Step 1: The output of the system emerges after $y(1)$ acts on the object. Obtain the input of the controller, $x(1)$, after the output compares with the reference input, and calculate

$$x_i(1) = \alpha(x(1))x_i(0), \qquad y_i(1) = \beta(y(1))y_i(0),$$
$$y(2) = F(x(1), 1) = \sum_{i=1}^{n} \mu_{A_i(1)}(x(1))y_i(1);$$

Step 2: The output of the system emerges after $y(2)$ acts on the object. Obtain the input of the controller, $x(2)$, after the output compares with the reference input, and calculate

$$x_i(2) = \alpha(x(2))x_i(0), \qquad y_i(2) = \beta(y(2))y_i(0),$$
$$y(3) = F(x(2), 2) = \sum_{i=1}^{n} \mu_{A_i(2)}(x(2))y_i(2);$$
$$\cdots\cdots$$

Step k: The output of the system emerges after $y(k)$ acts on the object. Obtain the input of the controller, $x(k)$, after the output compares with the reference input, and calculate

$$x_i(k) = \alpha(x(k))x_i(0), \qquad y_i(k) = \beta(y(k))y_i(0), \tag{11.14}$$

$$y(k+1) = F(x(k), k) = \sum_{i=1}^{n} \mu_{A_i(k)}(x(k))y_i(k); \tag{11.15}$$

$$\cdots\cdots$$

For the sake of convenience, for any membership function $\mu_A(x)$, we denote $\mu_{A(0)}(x) \triangleq \mu_A(x)$.

Theorem 2. The control algorithm mentioned above, $(\forall k)(x(k) \neq 0)$, has the following form:

$$y(k+1) = \beta(y(k))F\left(x(k)\Big/\alpha(x(k))\right)$$

$$= \beta(y(k))\sum_{i=1}^{n} \mu_{A_i}\left(x(k)\Big/\alpha(x(k))\right)y_i(0), \tag{11.16}$$

where $F(x,0) \triangleq F(x)$. If $\alpha(x(k))$ and $\beta(y(k))$ satisfy the stationary property:

$$\lim_{k\to+\infty} x(k) = 0 \Longrightarrow \left(\lim_{k\to+\infty} \frac{\alpha(x(k-1))}{\alpha(x(k))} = 1, \quad \lim_{k\to+\infty} \frac{\beta(y(k))}{\beta(y(k-1))} = 1\right)$$

then $y(k+1)$ must converge as $x(k) \to 0$. Denote $a = \lim_{k\to+\infty} y(k+1)$. If $x(k)$ is an infinitesimal of higher order compared with $\alpha(x(k))$, then $a = 0$.

Proof   We can obtain Expression (11.16) directly from Expressions (11.14) and (11.15).

To prove the convergence of $y(k+1)$, note the following equation:

$$|y(k+1) - y(k)| = |F(x(k), k) - F(x(k-1), k-1)|$$
$$= |F(x(k), k) - F(x(k), k-1) + F(x(k), k-1) - F(x(k-1), k-1)|$$
$$\leq |F(x(k), k) - F(x(k), k-1)| + |F(x(k), k-1) - F(x(k-1), k-1)| \; .$$

For $F(x, k-1)$ is a continuous function, $\forall\, \varepsilon > 0, \exists\, \delta > 0$, when $|x(k) - x(k-1)| < \delta$

$$|F(x(k), k-1) - F(x(k-1), k-1)| < \frac{\varepsilon}{3} \; .$$

As $x(k) \to 0$, $\exists\, K_1$, when $k > K_1$, $|x(k) - x(k-1)| < \delta$, the following equation is also true:

$$F((x(k), k) = \frac{\beta(y(k))}{\beta(y(k-1))} F\left(x(k)\frac{\alpha(x(k-1))}{\alpha(x(k))}, k-1\right) \; .$$

Let M be an upper bound of $|F(x, k-1)|$, then we have

$$|F(x(k), k) - F(x(k), k-1)|$$

$$= \left| \frac{\beta(y(k))}{\beta(y(k-1))} F\left(x(k)\frac{\alpha(x(k-1))}{\alpha(x(k))}, k-1\right) - F(x(k), k-1) \right|$$

$$\leq \left| \frac{\beta(y(k))}{\beta(y(k-1))} F\left(x(k)\frac{\alpha(x(k-1))}{\alpha(x(k))}, k-1\right) - F\left(x(k)\frac{\alpha(x(k-1))}{\alpha(x(k))}, k-1\right) \right|$$

$$+ \left| F\left(x(k)\frac{\alpha(x(k-1))}{\alpha(x(k))}, k-1\right) - F(x(k), k-1) \right|$$

$$\leq M \left| \frac{\beta(y(k))}{\beta(y(k-1))} - 1 \right| + \left| F\left(x(k)\frac{\alpha(x(k-1))}{\alpha(x(k))}, k-1\right) - F(x(k), k-1) \right|.$$

Since $\beta(y(k))/\beta(y(k-1)) \to 1$ and $\alpha(x(k-1))/\alpha(x(k)) \to 1$, $\exists K_2$, when $k > K_2$,

$$|\alpha(x(k-1))/\alpha(x(k)) - 1| < \sqrt{\delta}.$$

For $x(k) \to 0$, $\exists K_3$, when $k > K_3$, $|x(k)| < \sqrt{\delta}$. So when $k > \max\{K_2, K_3\}$,

$$\left| x(k)\frac{\alpha(x(k-1))}{\alpha(x(k))} - x(k) \right| = |x(k)| \left| \frac{\alpha(x(k-1))}{\alpha(x(k))} - 1 \right| < \sqrt{\delta}\sqrt{\delta} = \delta .$$

And now

$$\left| F\left(x(k)\frac{\alpha(x(k-1))}{\alpha(x(k))}, k-1\right) - F(x(k), k-1) \right| < \varepsilon/3.$$

Moreover, $\exists K_4$, when $k > K_4$,

$$|\beta(y(k))/\beta(y(k-1)) - 1| < \varepsilon/3M.$$

Taking $K = \max\{K_1, K_2, K_3, K_4\}$, when $k > K$,

$$|y(k+1) - y(k)| < \varepsilon/3 + \varepsilon/3 + \varepsilon/3 = \varepsilon,$$

i.e., $y(k+1) \to a$.

At last, when $x(k)$ is an infinitesimal of higher order compared with $\alpha(x(k))$, i.e., $x(k)/\alpha(x(k)) \to 0$ $(x(k) \to 0)$, we easily obtain $a = 0$ by Expression (11.16). **Q.E.D**

Note 3   The contraction-expansion factor $\alpha(x)$ satisfies $x(k)/\alpha(x(k)) \to 0$ $(x(k) \to 0)$. For example, $\alpha(x) = (|x|/E)^\tau$ $(0 < \tau < 1)$ holds the condition. In fact, when $x(k) \to 0$, we have

$$|x(k)/\alpha(x(k))| = |x(k)|E^\tau/|x(k)|^\tau = |x(k)|^{1-\tau}E^\tau \to 0.$$

Note 4   Here, the so-called "potential heredity" means that data $y_i(k)$ (see Expression (11.14)) do not come from $F(x(k-1), k-1)$ directly, but indirectly come from $y(k) = F(x(k-1), k-1)$ through $\beta(k)$.

11.4.2 Adaptive Fuzzy Controllers with Obvious Heredity

In this section, we discuss the algorithm for adaptive fuzzy controller with obvious heredity. The approach is similar to the previous section, except the difference in equation $y_i(k)$, $k = 1, 2, \cdots$.

Step 0: For an arbitrarily given initial input value $x(0) \in X$, calculate the output value of the controller $y(1)$:

$$y(1) = F(x(0), 0) = \sum_{i=1}^{n} \mu_{A_i(0)}(x(0))y_i(0); \tag{11.17}$$

Step 1: The output of the system emerges after $y(1)$ acts on the object. Obtain the input of the controller, $x(1)$, after the output compares with the reference input. Take $x_i(1) = \alpha(x(1))x_i(0)$, and calculate

$$y_i(1) = F(x_i(1), 0) = \sum_{s=1}^{n} \mu_{A_s(0)}(x_i(1))y_s(0) \tag{11.18}$$

$$y(2) = F(x(1), 1) = \sum_{i=1}^{n} \mu_{A_i(1)}(x(1))y_i(1) . \tag{11.19}$$

$$\cdots \cdots$$

Step k: The output of the system emerges after $y(k)$ acts on the object. Obtain the input of the controller, $x(k)$, after the output compares with the reference input. Take $x_i(k) = \alpha(x(k))x_i(0)$, and calculate

$$y_i(k) = F(x_i(k), 0) = \sum_{s=1}^{n} \mu A_s(k-1)(x_i(k))y_s(0) \tag{11.20}$$

$$y(k+1) = F(x(k), k) = \sum_{i=1}^{n} \mu_{A_i(k)}(x(k))y_i(k) \tag{11.21}$$

$$\cdots \cdots$$

Theorem 3 The control algorithm mentioned above has the following calculating formula:

$$y(k+1) = \sum_{s=1}^{n} \sum_{i=1}^{n} \mu_{A_i}\left(x(k) \Big/ \alpha(x(k))\right) \mu_{A_s}(\alpha(x(k))x_i(0))y_s(0) , \tag{11.22}$$

where $k = 0, 1, 2, \cdots$ and we stipulate $\alpha(x(0)) = 1$. Moreover, when $x(k) \to 0$, $y(k+1) \to 0$.

Proof   We can obtain Expression (11.22) directly from Expressions (11.20) and (11.21).   Next we prove the convergence of $y(k + 1)$.   Write $E(0) \triangleq E, E(k) \triangleq \alpha(x(k))E$ $(k = 1, 2, \cdots)$ and $X(k) \triangleq [-E(k), E(k)]$ $(k = 0, 1, 2, \cdots)$, and let

$$M(k) \triangleq \sup\{|F(x, 0)| \| x \in X(k)\}, \ k = 0, 1, 2, \cdots$$

Because $F(x, 0)$ is monotonic and even $F(0, 0) = 0$, so when $x(k) \rightarrow 0$, $M(k) \rightarrow 0$. From $x_i(k) \in X(k)$, we know that $|F(x_i(k), 0)| \leq M(k)$. Thus,

$$|y(k + 1)| = \left| \sum_{i=1}^{n} \mu_{A_i(k)}(x(k))y_i(k) \right| \leq \max_i \{|y_i(k)|\}$$
$$= \max_i \{|F(x_i(k), 0)|\} \leq M(k) \rightarrow 0 \ (k \rightarrow \infty) \ .$$

Therefore, $y(k + 1) \rightarrow 0 \ (k \rightarrow \infty)$. **Q.E.D.**

11.4.3   Adaptive Fuzzy Controllers with Successively Obvious Heredity

In this section, we discuss the algorithm for an adaptive fuzzy controller with successively obvious heredity. The approach is similar to the previous section, except the difference in equation $y_i(k)$, $k = 2, 3, \cdots$.

Step 0: See (11.17);
Step 1: See (11.18) and (11.19);
Step 2: The output of the system emerges after $y(2)$ acts on the object. Then get the input of the controller $x(2)$, after the output compares with the reference input. Take $x_i(2) = \alpha(x(2))x_i(0)$ and calculate

$$y_i(2) = \begin{cases} F(x_i(2), 1) = \sum_{s=1}^{n} \mu_{A_s(1)}(x_i(2))y_s(1), & x(2) \in X(1); \\ F(x_i(2), 0) = \sum_{s=1}^{n} \mu_{A_s(0)}(x_i(2))y_s(0), & \text{otherwise}; \end{cases} \tag{11.23}$$

$$y(3) = F(x(2), 2) = \sum_{i=1}^{n} \mu_{A_i(2)}(x(2))y_i(2) \ . \tag{11.24}$$

$$\cdots \cdots$$

Step k: The output of the system emerges after $y(k)$ acts on the object. Then get the input of the controller $x(k)$ after the output compares with the reference input. Take $x_i(k) = \alpha(x(k))x_i(0)$ and calculate

$$y_i(k) = \begin{cases} F(x_i(k), k-1) = \sum_{s=1}^{n} \mu_{A_s(k-1)}(x_i(k))y_s(k-1), & x(k) \in X(k-1); \quad (11.25) \\ F(x_i(k), 0) = \sum_{s=1}^{n} \mu_{A_s(0)}(x_i(k))y_s(0), & \text{otherwise}; \quad\quad\quad (11.26) \end{cases}$$

$$y(k+1) = F(x(k), k) = \sum_{i=1}^{n} \mu_{A_i(k)}(x(k))y_i(k) . \qquad (11.27)$$

. . . . . .

The following theorem shows the convergence of the above algorithm.

Theorem 4 $\lim\limits_{k \to +\infty} x(k) = 0 \Longrightarrow \lim\limits_{k \to +\infty} y(k+1) = 0.$

Proof. Denote $J \triangleq \{k_1, k_2, k_3, \cdots\} = \{k | x(k) \in X(k-1)\}$ and

$$J' \triangleq \{k_1', k_2', k_3', \cdots\} = \{1, 2, 3, \cdots\} \setminus J .$$

Now we consider the following several cases:

Case 1: J' is a finite set. Then $\exists K$, when $k \geq K$, $y_i(k)$ are calculated by Expression (11.25). Denote

$$N(k) \triangleq \sup\{|F(x, k-1)| | x \in X(k-1)\} .$$

From the structure of $F(x, k)$, we know $k \geq K$, $N(k)$ is monotonic decreasing and $|X(k-1)| \to 0$ as $x(k-1) \to 0$, where $|X(k-1)|$ is a certain measure of the interval $X(k-1)$. So we have $N(k) \to 0$. Thus,

$$|y(k+1)| = |F(x(k), k)| \leq N(k) \to 0 \ (k \to \infty).$$

Case 2: J is a finite set. There exists K, when $k \geq K$, $y_i(k)$ are calculated by Expression (11.26). From Theorem 3 we know $x(k) \to 0$ as $y(k+1) \to 0$.

Case 3: J' and J are all infinite sets. The sequence $\{x(k)\}$ is divided into two subsequences $\{x(k_j')\}$ and $\{x(k_j)\}$. Given $x(k) \to 0$ and $x(k_j') \to 0$, as $x(k_j) \to 0$, $\forall \varepsilon > 0$, there exists $J_1 > 0$, when $j > J_1$, $|y(k_{j+1}')| < \varepsilon$; and there exists J_2, when $j > J_2, |y(k_{j+1})| < \varepsilon$. Denoting $K = \max\{k_{J_1}, k_{J_2}\}$, when $k > K$, we have

$$|y(k+1)| \leq \max\{y(k_{J_1}), y(k_{J_2})\} < \varepsilon.$$

So $y(k+1) \to 0 \ (k \to +\infty)$. **Q.E.D.**

11.5 Adaptive Fuzzy Controllers with Two Inputs and One Output

Given an initial control rule base, $R(0) = R$, $\{A_i\}_{(1 \leq i \leq p)}$, $\{B_j\}_{(1 \leq j \leq q)}$, and $\{C_{ij}\}_{(1 \leq i \leq p, 1 \leq j \leq q)}$, are, respectively, a group of linear base elements on initial universes $X = [-E, E]$, $Y = [-EC, EC]$, and $Z = [-U, U]$, where their peak points satisfy that $-E = x_1 < x_2 < \cdots < x_p = E$, $-EC = y_1 < y_2 < \cdots < y_q = EC$. Taking $x_i(0) = x_i$, $y_j(0) = y_j$, and $z_{ij}(0) = z_{ij}$, the membership functions $\mu_{A_i(k)}(x(k))$

are defined as Expressions (11.11), (11.12), and (11.13), where n is replaced by p. And $\mu_{B_j(k)}(y(k))$ are defined as follows:

$$\mu_{B_1(k)}(y(k)) = \begin{cases} \dfrac{[y(k) - y_2(k)]}{[y_1(k) - y_2(k)]}, & y_1(k) \le y(k) \le y_2(k); \\ 0, & y_2(k) \le y(k) \le y_q(k), \end{cases} \tag{11.28}$$

$$\mu_{B_j(k)}(y(k)) = \begin{cases} \dfrac{[y(k) - y_{j-1}(k)]}{[y_j(k) - y_{j-1}(k)]}, & y_{j-1}(k) \le y(k) \le y_j(k); \\ \dfrac{[y(k) - y_{j+1}(k)]}{[y_j(k) - y_{j+1}(k)]}, & y_j(k) < y(k) \le y_{j+1}(k); \\ 0, & \text{otherwise} \end{cases} \tag{11.29}$$

$$j = 2, 3, \cdots, q-1,$$

$$\mu_{B_q(k)}(y(k)) = \begin{cases} 0, & y_1(k) \le y(k) < y_{q-1}(k); \\ \dfrac{[y(k) - y_{q-1}(k)]}{[y_q(k) - y_{q-1}(k)]}, & y_{q-1}(k) \le y(k) \le y_q(k). \end{cases} \tag{11.30}$$

The three algorithms presented in the former section can be easily generalized to this case. For the time being, we take the adaptive fuzzy control algorithm with obvious heredity as an example.

Step 0: For arbitrarily given initial input values $x(0) \in X$ and $y(0) \in Y$, calculate the output value of the controller $z(1)$:

$$z(1) = F(x(0), y(0), 0) = \sum_{i=1}^{p} \sum_{j=1}^{q} \mu_{A_i(0)}(x(0)) \mu_{B_j(0)}(y(0)) z_{ij}(0);$$

Step 1: The output of the system emerges after $z(1)$ acts on the object. Obtain the inputs of the controller $x(1)$ and $y(1)$ after the output compares with the reference input. Take $x_i(1) = \alpha(x(1)) x_i(0)$ and $y_j(1) = \beta(x(1), y(1)) y_j(0)$, and calculate

$$z_{ij}(2) = F(x_i(1), y_j(1), 0) = \sum_{s=1}^{p} \sum_{t=1}^{q} \mu_{A_s(0)}(x_i(1)) \mu_{B_t(0)}(y_j(1)) z_{st}(0);$$

$$z(2) = F(x(1), y(1), 1) = \sum_{i=1}^{p} \sum_{j=1}^{q} \mu_{A_i(1)}(x(1)) \mu_{B_j(1)}(y(1)) z_{ij}(1);$$

$\cdots\cdots$

Step k: The output of the system emerges after $z(k)$ acts on the object. Obtain the inputs of the controller $x(k)$ and $y(k)$ after the output compares with the reference input. Take $x_i(k) = \alpha(x(k)) x_i(0)$ and $y_j(k) = \beta(x(k), y(k)) y_j(0)$, and calculate

$$z_{ij}(k) = F(x_i(k), y_j(k), 0)$$

$$= \sum_{s=1}^{p} \sum_{t=1}^{q} \mu_{A_s(0)}(x_i(k)) \mu_{B_t(0)}(y_j(k)) z_{st}(0); \quad (11.31)$$

$$z(k+1) = F(x(k), y(k), k)$$

$$= \sum_{i=1}^{p} \sum_{j=1}^{q} \mu_{A_i(k)}(x(k)) \mu_{B_j(k)}(y(k)) z_{ij}(k); \quad (11.32)$$

$$\cdots \cdots$$

Similar to Theorem 3, it is easy to prove the following theorem.

Theorem 5 The algorithm mentioned above has the following formula:

$$z(k+1) = \sum_{s=1}^{p} \sum_{t=1}^{q} \sum_{i=1}^{p} \sum_{j=1}^{q} \mu_{A_i}\left(x(k)\Big/\alpha(x(k))\right) \mu_{A_s}\left(\alpha(x(k)) x_i(0)\right)$$

$$\cdot \mu_{B_j}\left(y(k)\Big/\beta(x(k), y(k))\right) \mu_{B_t}\left(\beta(x(k), y(k)) y_j(0)\right) z_{st}(0) . \quad (11.33)$$

Moreover, $x(k) \to 0$ as $y(k) \to 0$ and $z(k+1) \to 0$. We neglect the proof, since it is similar to the proof in Theorem 3.

11.6 Conclusions

The variable universe is proposed in this chapter. Based on this idea and the interpolation mechanism of fuzzy controllers (see Chapter 8), we presented an innovative variable-universe-based adaptive fuzzy controller. This type of controller can achieve high precision and can be applied to many fuzzy controls that need higher precision. We also proved several theorems to show the convergence of the algorithms.

References

1. L. X. Wang, *A Course in Fuzzy Systems and Control*, Prentice-Hall, Englewood Cliffs, 1997.

2. L. X. Wang, *Adaptive Fuzzy Systems and Control*, Prentice-Hall, Englewood Cliffs, 1994.

3. C. T. Lin and C. S. G. Lee, Fuzzy adaptive learning control network with on-line neural learning, *Fuzzy Sets and Systems*, Vol. 71, pp. 25-45, 1995.

4. C. C. Lee, Fuzzy logic control systems, Fuzzy logic controller, Parts I and II, *IEEE Transactions on Systems, Man, and Cybernetics*, Vol. 20, No. 2, pp. 404-435, 1990.

5. M. Sugeno, ed., *Industrial Applications of Fuzzy Control*, North-Holland, Amsterdam, 1985.

6. R. R. Yager and L. A. Zadeh, ed., *Fuzzy Sets, Neural Networks and Soft Computing*, Van Nostrand Reinhold, New York, 1994.

7. H. X. Li, The mathematical essence of fuzzy controls and fine fuzzy controllers, in *Advances in Machine Intelligence and Soft-Computing*, Paul P. Wang, ed., Vol. IV, Bookwrights Press, Durham, pp. 55-74, 1997.

Chapter 12

The Basics of Factor Spaces

The original definition of "factor spaces" was proposed by Peizhuang Wang [1]. He used factor spaces to explain the source of randomness and the essence of probability laws. In 1982, he gave an axiomatic definition of factor spaces [2]. Since then he has applied factor spaces to the study of artificial intelligence [3-5]. Several applications in the area of fuzzy information processing have been discussed [6]. This chapter provides an introduction to the basic concepts and methods of applications of factor spaces.

12.1   What are "Factors"?

The word "factor" is a primary term in the factor spaces theory. Although it is not simple to provide a concise definition, nevertheless, we can illustrate its meaning from the following three distinctive viewpoints.

Attributional: When we have a bumper harvest in a good year we want to know what causes it or what contributed to it. For example, the appropriate amount of the rainfall could be a key contributor to the bumper harvest. The rainfall is referred to as a factor; the amount of rainfall, say 50 milliliters, is referred to as a state; and the concept "appropriate" is referred to as a characteristic (not a factor!). In dealing with factors, it is important that we distinguish states from characteristics.

In general, a factor is designated by a noun, a state by a numeral, and a characteristic by an adjective. The factor "temperature", for instance, is a noun; 36°C, 100°C, $\cdots$, are states; and "hot", "cold", etc., are characteristics. A factor may be viewed as a common symbol, sign, or code of its states and characteristics. A state is a sign or symbol that represents an instance of a factor and a characteristic is a sign or symbol that roughly describes the factor.

When a state/characteristic of a factor leads to certain effects or results, we will attribute the effects or results to the factor, not the state/characteristic. This is because "factor" is more essential and pertinent than its state or characteristics. Continuing from the "harvest" example, we could not conclude that the harvest was attributable to the *amount* of rainfall. However, from past experience, we learned that an appropriate amount of rainfall was a key factor to the harvest; insufficient level of rainfall would result in poor harvest. Apparently, there is a causal relationship between the harvest and the rainfall.

There are two steps involved in the process of establishing a causal relationship. The first step is to find possible factors that have significant relationships with the effects, and the second step is to identify the states or characteristics of a given factor on a specified level of effects. The first step is considered as basic whereas the second step is more essential, although abstract.

Analytical: Thought is closely related to concepts. Concepts are often formed by means of comparisons among a set of objects. However, comparison requires objects to have both the differences and the common features; otherwise different objects can not always be compared. For example, the concept of men and women can be formed by comparing under the common feature–sex. Red, green, yellow, etc., have the common feature called "color" that helps to form concepts of each type of color. These common features are just factors and they are the common labels of a class of states or a group of characteristics. Terms such as age, height, and profession are examples of factors. Thus, factors can be viewed as an analytical way of recognizing the real world.

Describable: Like a point in Cartesian space, an object can be regarded as residing in a space constructed by factors. For example, the personal data of John Doe are: age 26 years old, sex male, and profession engineering, etc. In a sense, John is being determined by the states of each factor relevant to him. By specifying a sufficient number of factors, John can be uniquely determined and described by them. In other words, John can be viewed simply as a point in abstract coordinate space, and that space is capable of representing any object just like the Cartesian coordinates for the physical systems. One of the important tasks in constructing such an abstract coordinate system is to specify the factors like age, sex, title, characteristics, interest, ..., etc. We can view this abstract coordinate space as the generalized coordinate space whose dimensions are factors.

12.2   The State Space of Factors

A factor may not be significant to an object. For example, it is meaningless to talk about the sex of a stone. An object u is called relevant to a factor f (in the direction from u to f) if there exists a state $f(u)$ of f corresponding to u.

Let U be a set of objects and V be a set of factors; we call $(U, V]$ a *left matched pair* if it satisfies the condition that for any $u \in U$, V contains all factors relevant to u.

For a given left matched pair $(U, V]$, a *relation* R between U and V is defined as follows:

$R(u, f) = 1$ iff u is relevant to f. For simplicity, we only consider R as an ordinary (non-fuzzy) relation in this book. Define

$$D(f) = \{u \in U \mid R(u, f) = 1\}$$
$$V(u) = \{f \in U \mid R(u, f) = 1\} \ .$$

A factor $f \in V$ can be regarded as a mapping, acting on an object u and resulting in $f(u)$. That is,

$$f : \ D(f) \longrightarrow X(f), \qquad u \longmapsto f(u) \ ,$$

where $X(f) = \{f(u) \mid u \in U\}$ is called the *state space* of f and any element in $X(f)$ is called a state of f.

According to the nature of state spaces, factors can be classified into four major categories:

Measurable Factors: Factors such as time, length, mass, height, weight, etc. are measurable factors. Usually their state spaces are subsets, such as intervals, of one dimensional or n-dimensional Euclidean spaces. For example, X (height) $= [0, 200]$ where the unit is in centimeters. Sometimes, for practical purposes, the state space of a measurable factor is represented as a discrete subset of an interval. Thus, height is given with the range $\{1,2,3,\cdots,200\}$.

Nominal Factors: Factors like profession, nationality, race, religion, etc. are qualitative in nature. These factors are called nominal factors whose state spaces are sets of terms. For instance,

$$X(\text{title}) = \{\text{professor, engineer, lawyer}, \cdots\} \ .$$

Degree (/Ordinal) Factors: Factors such as ability, feasibility, quality, degree of satisfaction, degree of necessity, etc. are called degree factors. Their state space usually is the interval $[0,1]$. For example,

$$X(\text{degree of satisfaction}) = [0, 1] \ .$$

Switch (/Binary) Factors: Factors such as success, eligibility, permission, etc. whose state space simply contains two values $\{yes, no\}$ are called switch factors. In general, the values of a state space may be represented by any two appropriate symbols relevant to the context. For instance,

$$X(\text{lifeness}) = \{\text{life, lifeless}\} \ .$$

Although the name of a switch factor is a noun, for convenience's sake, other forms of naming such as (1) a name with a question mark, and (2) two state values linked with a hyphen, are accepted. For example, $X(\text{life?}) = \{\text{yes, no}\}$, and $X(\text{life-lifeless}) = \{\text{life, lifeless}\}$.

12.3   Relations and Operations of Factors

Some basic relations and operations on factors are defined and illustrated as follows.

12.3.1   The Zero Factor

The symbol θ is called the *empty state* if for any state x the set containing x and θ, and the ordered pair formed by x and θ, equal to x. That is,

$$\{x, \theta\} = \{x\}, \quad \text{and} \quad (x, \theta) = x = (\theta, x) . \tag{12.1}$$

The symbol $\mathbf{0}$ is called the zero factor if

$$X(\mathbf{0}) = \{\theta\}. \tag{12.2}$$

That is, the only state of the zero factor is the empty state. The importance of the zero factor in the factor space theory is similar to the empty set in set theory. From the above properties, we have, for any factor f,

$$X(f) \times X(\mathbf{0}) = X(f) = X(\mathbf{0}) \times X(f) . \tag{12.3}$$

From here on, we will assume the zero factor is always a member of V in any left-matched pair $(U, V]$, i.e., $\mathbf{0} \in V$.

Recall that for a given left-matched pair $(U, V]$, a factor $f \in V$ is the same as a mapping $f : D(f) \longrightarrow X(f)$. For convenience of discussion in subsequent sections, we extend the domain of f to the whole set U as shown below.

$$f : U \longrightarrow X(f), \quad u \longmapsto \begin{cases} f(u), & u \in D(f) \\ \theta, & u \in U \setminus D(f) . \end{cases}$$

12.3.2   Equality of Factors

A factor f is equal to a factor g if f and g are equal in the sense of mappings; that is, $D(f) = D(g)$, $X(f) = X(g)$, and $f(u) = g(u)$ for any $u \in D(f)$.

12.3.3   Subfactors

There are instances where the states of a factor g depend completely on the states of another factor f. For example, define f to be the plane coordinates and g the abscissas (or ordinates). Then a state of f is a point (x, y) in the plane that determines the state of g, an x on the abscissa (or y on the ordinate). Here, the state space of g is represented as a "subspace" or "bifurcation" of the state space of f.

A factor g is called a *proper subfactor* of f, denoted by $f > g$, if there exists a set Y satisfying $Y \neq \emptyset$ and $Y \neq \{\theta\}$, and

$$X(f) = X(g) \times Y . \tag{12.4}$$

A factor g is called a subfactor of f, denoted by $f \geq g$ if $f > g$ or $f = g$.

Note 1   The condition $Y \neq \{\theta\}$ in the above definition is indispensable; otherwise by (12.4) we can derive $f < g$ for any f which is inconceivable.

Note 2   It is obvious that the zero factor is a subfactor of any factor according to (12.3).

Note 3   Generally, the order of the direct product of the state spaces in the factor space theory is immaterial, that is, $X(f) \times X(g)$ and $X(g) \times X(f)$ are equivalent. Hence, (12.4) can also be written as

$$X(f) = Y \times X(g).$$

12.3.4   Conjunction of Factors

A factor h is called the *conjunction of factors* f and g, denoted by

$$h = f \wedge g \tag{12.5}$$

if h is the greatest common subfactor of f and g. That is, if $f \geq h$ and $g \geq h$, and for any factor e such that $f \geq e$ and $g \geq e$, then $h \geq e$.

A factor g is the *conjunctive factor* of a family of factors $\{f_t\}_{(t \in T)}$, denoted by $g = \bigwedge_{t \in T} f_t$, if g is the greatest common subfactor of f_t $(t \in T)$. In other words, $(\forall\, t \in T)(f_t \geq g)$ and for any factor h, $(\forall\, t \in T)(f_t \geq h)$ implies $g \geq h$.

Example 1   Let f be the length and the width of cuboids and g be the width and the height of cuboids, then $h = f \wedge g$ is the width of cuboids.

12.3.5   Disjunction of Factors

A factor h is called the *disjunction of factors* f and g, denoted by

$$h = f \vee g \tag{12.6}$$

if $h \geq g$ and for any factor e such that $e \geq f$ and $e \geq g$ implies $e \geq h$.

A factor $g = \bigvee_{t \in T} f_t$ is the *disjunction factor* of a family of factors $\{f_t\}_{(t \in T)}$, if $(\forall\, t \in T)(g \geq f_t)$ and for any factor h such that $(\forall\, t \in T)(h \geq f_t)$, then $h \geq g$.

Note that for a given left pair $(U, V]$, disjunctive factors can be determined by conjunctive factors. For example, $h = f \vee g$ if and only if

$$h = \wedge\{e \in V \mid e \geq f, \ e \geq g\} . \tag{12.7}$$

Conversely, conjunctive factors can also be determined by distinctive factors:

$$f \wedge g = \vee \{e \in V \mid e \leq f, \ e \leq g\} \ . \tag{12.8}$$

Example 2   Let f be the abscissas of all points in a plane and g be the ordinates of these points, then $h = f \vee g$ is the set of coordinates of all points in the plane.

12.3.6   Independent Factors

A family of factors $\{f_t\}_{(t \in T)}$ is called *independent* if it satisfies the following condition:

$$(\forall \ s, t \in T)(f_s \wedge f_t = 0) \ . \tag{12.9}$$

Obviously the subfactors of independent factors are independent and the zero factor is independent of any factors.

12.3.7   Difference of Factors

A factor h is called the *difference factor* between factor f and factor g, denoted by $h = f - g$, if

$$(f \wedge g) \vee h = f \quad \text{and} \quad h \wedge g = 0 \ . \tag{12.10}$$

Example 3   Let f be coordinates of a plane and g be abscissas of points of that plane, then $h = f - g$ are ordinates of points of the plane.

12.3.8   Complement of a Factor

Let F be a class of factors in a problem domain. Define $\mathbf{1}$ to be the *complete factor* with respect to F if every factor in F is a subfactor of it. For any factor $f \in F$, define

$$f^c = \mathbf{1} - f \tag{12.11}$$

to be the *complementary factor* of f with respect to $\mathbf{1}$.

12.3.9   Atomic Factors

A factor f is called an *atomic factor* if f does not have proper subfactors except the zero factor.

Let F be the class of factors in a problem domain. The set of all atomic factors in F is called the family of atomic factors which is denoted by π. Clearly, π is independent. Also, we can easily prove that if a family of factors $\{f_t\}_{(t \in T)}$ is independent, then

$$X\left(\bigvee_{t \in T} f_t \right) = \prod_{t \in T} X(f_t) \ . \tag{12.12}$$

If the family of atomic factors exists in the class of factors F then any factor f in F can be viewed as a disjunction of some subset of π. In other words, the set of all factors in F is equivalent to the power set of π. In notation, we have

$$F = \mathcal{P}(\pi) = \{S \mid S \subset \pi\}$$

$$X(f) = \prod_{g \in f} X(g), \quad \text{and} \quad X(\mathbf{1}) = \prod_{f \in \pi} X(f).$$

The power set of π is a Boolean algebra whose interesting properties have inspired us to develop an axiomatic approach to the factor space theory.

12.4   Axiomatic Definition of Factor Spaces

For a given left-matched pair $(U, V]$ and $F \subset V$, the family $\{X(f)\}_{(f \in F)}$ is called a factor space on U if it satisfies the following axioms:

(f.1) $F = F(\vee, \wedge, c, \mathbf{1}, \mathbf{0})$ is a complete Boolean algebra;

(f.2) $X(\mathbf{0}) = \{\theta\}$; and

(f.3)   If $\forall\, T \subset F$ and $(\forall\, s, t \in T)\ (s \neq t \Longrightarrow s \wedge t = \mathbf{0})$, then

$$\bigvee_{f \in T} f = \prod_{f \in T} f , \tag{12.13}$$

where the right-hand side of the equality is the direct product of the mappings of f in T (since factors can be regarded as mappings).

We call F the set of factors, $f\ (\in F)$ a factor, $X(f)$ the state space of f, $\mathbf{1}$ the complete factor, and $X(\mathbf{1})$ the complete space.

Notice that, for an independent T, $X(\prod_{f \in T} f) = \prod_{f \in T} X(f)$, hence we have $X(\bigvee_{f \in T} f) = \prod_{f \in T} f$, by axiom (f.3).

Example 4   Let n be a natural number. Define $I_n = \{1, 2, \cdots, n\}$, and $F = \mathcal{P}(I_n) = \{f \mid f \subset I_n\}$. For any $f \in F$, set

$$X(f) = \prod_{i \in f} X(i) ,$$

where $X(i)$ is a set for all $i \in f$. Further define $\prod_{i \in \emptyset} X(i) = \{\emptyset\}$, $\mathbf{0} = \emptyset$, and $\theta = \emptyset$. It readily follows that $\{X(f)\}_{(f \in F)}$ is a factor space. In particular, when $X(i) = \Re$ (the set of real numbers) then $\{X(f)\}_{(f \in F)}$ is a family of Euclidean spaces with dimensions n or less; for $n = 3$, $\{X(f)\}_{(f \in F)}$ can be viewed as a Cartesian coodinate system with variable dimensions.

Example 5   For a given left-matched pair $(U, V]$, let F be a subset of V and F is closed with respect to the infinite disjunction, infinite conjunction, and the difference operations of factors. If we put

$$\mathbf{1} = \bigvee_{f \in F} f, \quad \text{and} \quad f^c = \mathbf{1} - f \quad (f \in F)$$

then $\{X(f)\}_{(f\in F)}$ is a factor space.

Example 6   Let S be a set and $F = \{S, \emptyset\}$. Thus, F is a complete Boolean algebra with $\mathbf{1} = S$ and $\mathbf{0} = \emptyset$. If the state space $X(S)$ can be determined then $\{X(S), \{\emptyset\}\}$ forms a factor space with $\theta = \emptyset$. And the factor space degenerates to the state space $X(S)$ since $\{\emptyset\}$ is unnecessary.

The concept of the state space here is a generalization of the same concept used in control theory, the "characteristics space" or "parameter space" in pattern recognition, and the "phase space" in physics, and so forth. The last example shows that a state space can be viewed as a factor space, or a special case of a factor space. The use of "factor spaces" has merits over the other terminologies because a factor space exists not only with a fixed state space, but also with a family of state spaces of "variable dimensions"— a key idea in factor spaces.

Proposition 1   Let $\{X(f)\}_{(f\in F)}$ be a factor space. For any $f, g \in F$,

$$X(f \vee g) = X(f - g) \times X(f \wedge g) \times X(g - f) . \tag{12.14}$$

Proof   For F is a Boolean algebra, we have

$$f \vee g = (f - g) \vee (f \wedge g) \vee (g - f).$$

Since $(f - g)$, $(f \wedge g)$, and $(g - f)$ can easily be shown to be independent from each other, the result follows from (12.13).   **Q.E.D.**

12.5   A Note on The Definition of Factor Spaces

Since it is known that factors, in mathematics, may be considered as mappings from the universe (U) to their state spaces, we can give another definition of factor spaces based on the concept of such mappings.

Definition 1   For a given universe U, let $F = \{f \mid f : U \longrightarrow X(f)\}$ be a family of mappings. The family of sets, $\{X(f)\}_{(f\in F)}$, is called a factor space on U, if it satisfies the following axioms:

(f.1')   There exists an algebraic structure such that $F = F(\vee, \wedge, c, \mathbf{1}, \mathbf{0})$ is a complete Boolean algebra.

(f.2') $(\forall T \subset F)\Big((\forall f, g \in T)(f \wedge g = \mathbf{0}) \Longrightarrow \bigvee_{f\in T} f = \prod_{f\in T} f\Big).$

Under this definition, clearly $\mathbf{0} = \bigvee_{f\in\phi} f$. For any $T \subset F$, if $(\forall f, g \in T)(f \wedge g = \mathbf{0})$ then $\bigvee_{f\in T} f$ is a mapping; namely, $\bigvee_{f\in T} f : U \longrightarrow \prod_{f\in T} f$. Thus,

$$X\Big(\bigvee_{f\in T} f\Big) = \prod_{f\in T} X(f) .$$

In particular, for the zero factor $\mathbf{0}$,

$$X(\mathbf{0}) = \prod_{f \in \phi} X(f) = \{\theta \mid \theta : \phi \longrightarrow \phi\}.$$

Here θ is called the empty state, and $\theta = \phi$. Furthermore, for any $f \in F$,

$$X(\mathbf{0}) \times X(f) = X(f) \times X(\mathbf{0}) = \prod_{g \in \phi \cup \{f\}} X(g) = X(f).$$

The operations and relations between factors, such as equalities, subfactors, conjunctions, disjunctions, independent factors, difference factors, complement factors, and atomic factors are dealt with similarly as in the earlier sections. We illustrate the point with an example on subfactors.

Let f and g be factors of F and $f \neq g$. If there exists a non-zero factor $h \in F$ such that the mapping of the direct product of f and g is simply h, then g is called a proper subfactor of f and is denoted by $f > g$. We call g a subfactor of f, denoted by $f \geq g$, if $f > g$ or $f = g$.

12.6   Concept Description in a Factor Space

"Concept" is one of the most important bases for thinking and knowledge building in human reasoning. Generally, concepts may be classified in three forms:

Form 1. Intension: Indicating the essence and attributes of a concept.

Form 2. Extension: Indicating the aggregate of objects according to a concept.

Form 3. Conceptual structure: Using relation between concepts to illustrate a concept.

Traditional set theory can represent a crisp concept in extension form. Fuzzy set theory can express general concepts (either crisp or fuzzy) in extension form. Both theories, however, have not considered the question of how to "select and transform" the universes of interest so that concepts could be described and analyzed. In other words, the open question is: How can we use mathematical methods to represent the intension of a concept? Our research on this question is based upon the factor space theory.

Assume $\mathcal{C} = \{\alpha, \beta, \gamma, \cdots\}$ is a group of concepts under the common universe U. Let V be a family of factors such that U and V form a left-matched pair $(U, V]$. Also, let F be a set of factors of V such that F is *sufficient*; i.e., satisfying

$$(\forall\, u_1,\ u_2 \in U)(\exists\, f \in F)(f(u_1) \neq f(u_2)) . \tag{12.15}$$

The triple $(U, \mathcal{C}, F]$ or $(U, \mathcal{C}, \{X(f)\}_{(f \in F)}]$ is called a *description frame* of $\mathcal{C}$. Two properties follow immediately:

1) $(\forall\, f, g \in F)(f \geq g \Longrightarrow (f = g \times (f - g),\ g \wedge (f - g) = \mathbf{0}))$;
2) $(\forall\, f \in F)(\mathbf{1} = f \times f^c)$.

Proposition 2   For a given description frame $(U, \mathcal{C}, F]$, the complete factor **1** must be an injection.

Proof   Since $(U, \mathcal{C}, F]$ is a descriptive frame, for any $u_1, u_2 \in U$, there exists a factor $f \in F$ such that $f(u_1) \neq f(u_2)$. By the properties stated above, we have

$$\mathbf{1}(u_1) = (f(u_1), f^c(u_1)) \neq (f(u_2), f^c(u_2)) = \mathbf{1}(u_2).$$

This means the **1** is an injection.   **Q.E.D.**

Note 4   Since a factor $f : U \longrightarrow X(f) = \{f(u) \mid u \in U\}$ is always a surjection that implies the complete factor **1** is a bijection.

Note 5   The "sufficiency" with respect to F mentioned above means that for any two distinctive objects u_1 and u_2, there exists at least one factor f in F, such that their state values are different in f.

Let $(U, \mathcal{C}, F]$ be a description frame and $\alpha \in \mathcal{C}$. The extension of α in U is a fuzzy set $A\ (\in \mathcal{F}(U))$ on U, where A is a mapping:

$$A : \ U \longrightarrow [0, 1], \qquad u \longmapsto A(u).$$

$A(u)$ is called the degree of membership of u with respect to α or A. When $A(u) = 1$, we say that u definitely accords with α or u completely belongs to A; while $A(u) = 0$, u definitely does not accord with α or u does not belong to A. When $A(U) = \{0, 1\}$, A degenerates to a crisp set and its α is called a *crisp concept*. For a description frame $(U, \mathcal{C}, F]$ and $f \in F$, $X(f)$ are called the *representation universe* of $\mathcal{C}$ with respect to f; in particular, $X(\mathbf{1})$ are called the **complete representation universe**. A factor space is, therefore, just a family of representation universes of $\mathcal{C}$.

Note 6 $A(u)$ **and** $\mu_A(u)$ **are of the same sense.**

According to Zadeh's extension principle, we can extend any $f \in F$ by

$$f : \ \mathcal{F}(U) \longrightarrow \mathcal{F}(X(f)), \qquad A \longmapsto f(A)$$

where $f(A)(x) = \bigvee_{f(u)=x} A(u)$ and $x = X(f)$. We call $f(A)$ the representation extension of α in the representation universe $X(f)$.

This means that a concept or its extension on U can be transformed to the representation universe $X(f)$, $f \in F$. In other words, a concept can be decomposed and modeled by $\{X(f)\}_{(f \in F)}$.

12.7   The Projection and Cylindrical Extension of the Representation Extension

Let $(U, \mathcal{C}, \{X(f)\}_{(f \in F)}]$ be a description frame and $f, g \in F$ such that $f \geq g$. Define

$$\downarrow_g^f \colon X(f) \longrightarrow X(g), \qquad (x, y) \longmapsto \downarrow_g^f (x, y) = x \,,$$

where $X(f) = X(g) \times X(f - g)$, $x \in X(g)$, and $y \in X(f - g)$. We call $\downarrow_g^f$ the *projection* from f to g.

For any concept $\alpha \in \mathcal{C}$, let $A \in \mathcal{F}(U)$ be the extension of α on U. For any $f, g \in F$, and $f \geq g$, if the representation extension $B = f(A) \in \mathcal{F}(X(f))$ of α with respect to f is known then the representation extension of α with respect to g can be derived by the "projection".

Using the extension principle, we can extend $\downarrow_g^f$ to:

$$\downarrow_g^f \colon \mathcal{F}(X(f)) \longrightarrow \mathcal{F}(X(g)), \qquad B \longmapsto \downarrow_g^f (B) = \downarrow_g^f B$$

$$x \longmapsto (\downarrow_g^f B)(x) = \bigvee_{\downarrow_g^f (x,y)=x} B(x, y) = \bigvee_{g \in X(f-g)} B(x, y). \qquad (12.16)$$

We call $\downarrow_g^f B$ the projection of B from f to g.

Let us consider a natural question. For a concept $\alpha \in \mathcal{C}$, let A be the extension of α on U and $B = f(A)$ is the known representation extension of α with respect to g. If we have derived $\downarrow_g^f B$, the projection of B from f to g, does that equal to $B' = g(A) \in \mathcal{F}(X(g))$, the representation extension of α with respect to g? The answer is positive.

Lemma 1   Let X, Y, and Z be three universes. Given two mappings $f : X \longrightarrow Y$, and $g : Y \longrightarrow Z$, and an arbitrary fuzzy set A of $\mathcal{F}(X)$,

$$g(f(A)) = (g \circ f)(A) \,. \qquad (12.17)$$

Proof   For any $z \in Z$, we have

$$g(f(A))(z) = \bigvee_{g(y)=z} f(A)(y) = \bigvee_{g(y)=z} \left(\bigvee_{f(x)=y} A(x) \right)$$

$$= \vee \{A(x) \mid f(x) = y, g(y) = z\} = \bigvee_{g(f(x))=z} A(x)$$

$$= \bigvee_{(g\circ f)(x)=z} A(x) = (g \circ f)(A)(z) \,.$$

So (12.17) is true.   **Q.E.D.**

Proposition 3   Let $(U, \mathcal{C}, F]$ be a description frame. For any $f, g \in F$, and $f \geq g$, if A is the extension of $\alpha \in \mathcal{C}$, then

$$\downarrow_g^f f(A) = g(A) \qquad (12.18)$$

Proof   Since $g = \downarrow_g^f f$ is the composite mapping of f and $\downarrow_g^f$, the conclusion holds by Lemma 1.   **Q.E.D.**

Let $(U, \mathcal{C}, \{X(f)\}_{(f \in F)}]$ be a description frame. Assuming $f, g \in F$ with $f \geq g$, define

$$\uparrow_g^f \colon X(g) \longrightarrow \mathcal{P}(X(g)), \qquad x \longmapsto \uparrow_g^f (x) = \{x\} \times X(f - g) \,,$$

where $X(f) = X(g) \times X(f - g)$, $x \in X(g)$, and $y \in X(f - g)$. We call $\uparrow_g^f$ the cylindrical extension of g to f.

Using the cylindrical extension method, the representation extension $B = g(A) \in \mathcal{F}(X(g))$ of α with respect to g, may generate a rough representation extension of α with respect to f.

Similar to the extension principle, we have

$$\uparrow_g^f \colon \mathcal{F}(X(g)) \longrightarrow \mathcal{F}(X(f)), \quad B \longmapsto \uparrow_g^f (B)$$

$$\uparrow_g^f B \colon X(f) \longrightarrow [0,1], \quad (x, y) \longmapsto (\uparrow_g^f B)(x, y) = B(x),$$

where $X(f) = X(g) \times X(f - g)$, $x \in X(g)$, and $y \in X(f - g)$. We call $\uparrow_g^f B$ the cylindrical extension of B from g to f.

Consider a similar question discussed earlier. For a concept $\alpha \in \mathcal{C}$, let A be the extension of α on U and $B = g(A)$ is the known representation extension of α with respect to g. If we have derived $\uparrow_g^f B$, the cylindrical extension of B from g to f, does that equal to $B' = f(A) \in \mathcal{F}(X(f))$, the representation extension of α with respect to f? A simple counter example will show the answer is negative. However, a weaker result is in

Proposition 4   For a given description frame $(U, \mathcal{C}, F]$, and for any $f, g \in F$ with $f \geq g$, if $A \in \mathcal{F}(U)$ is the extension of $\alpha \in \mathcal{C}$, then

$$\uparrow_g^f g(A) \supset f(A) \,. \tag{12.19}$$

Proof   For $f = g \vee (f - g)$ and $g \wedge (f - g) = \mathbf{0}$ implies $f = g \times (f - g)$. Thus, for any $(x, y) \in X(f) = X(g) \times X(f - g)$, we have

$$
\begin{aligned}
f(A)(x, y) &= \bigvee_{f(u)=(x,y)} A(u) = \vee\{A(u) \mid f(u) = (x, y)\} \\
&= \vee\{A(u) \mid (g(u), (f - g)(u)) = (x, y)\} \\
&= \vee\{A(u) \mid (g(u) = x, (f - g)(u) = y\} \\
&\leq \vee\{A(u) \mid (g(u) = x\} = \bigvee_{g(u)=x} A(u) = g(A)(x) \,.
\end{aligned}
$$

This proves (12.19).   **Q.E.D.**

12.8 Some Properties of the Projection and Cylindrical Extension

Proposition 5 For any factor space $\{X(f)\}_{(f\in F)}$, and any $f,g,h\in F$ such that $f\geq g\geq h$, we have

1) if $B\in\mathcal{F}(X(f))$, then

$$\downarrow_h^g (\downarrow_g^f B) = \downarrow_h^f B\ ; \tag{12.20}$$

2) if $B\in\mathcal{F}(X(h))$, then

$$\uparrow_g^f (\uparrow_h^g B) = \uparrow_h^f B\ ; \tag{12.21}$$

3) if $B\in\mathcal{F}(X(g))$, then

$$\downarrow_g^f (\uparrow_g^f B) = B\ ; \tag{12.22}$$

4) if $B\in\mathcal{F}(X(f))$, then

$$\uparrow_g^f (\downarrow_g^f B) \supset B\ . \tag{12.23}$$

Proof 1) Notice that $\downarrow_h^f = \downarrow_h^g \circ \downarrow_g^f$. By the extension principle on the composition of mappings, (see Lemma 1), we have (12.20).

2) Since $f\geq g\geq h$ this implies

$$X(f) = X(f-g) \times X(g-h) \times X(h).$$

For any

$$(x,y,z)\in X(f-g) \times X(g-h) \times X(h),$$

we have

$$(\uparrow_g^f (\uparrow_h^g B))(x,y,z) = (\uparrow_h^g B)(y,z) = B(x)$$
$$(\uparrow_h^f B)(x,y,z) = B(x)\ .$$

Hence, (12.21) is proved.

3) For any $x\in X(g)$, we have

$$(\downarrow_g^f (\uparrow_g^f B))(x) = \bigvee_{y\in X(f-g)} (\uparrow_g^f B)(x,y) = \bigvee_{y\in X(f-g)} B(x) = B(x)\ .$$

This proves (12.22).

4) Since $X(f) = X(g) \times X(f-g)$. For any $(x,y)\in X(f)$, we have

$$(\uparrow_g^f (\downarrow_g^f B))(x,y) = (\downarrow_g^f B)(x) = \bigvee_{y'\in X(f-g)} B(x,y') \geq B(x,y)\ .$$

Hence, (12.23) follows.   **Q.E.D.**

The next proposition establishes an equality condition for Expression (12.23).

Proposition 6 Let $\{X(f)\}_{(f\in F)}$ be a factor space on U. Assume $f,g\in F$ such that $f\geq g$, and $B\in\mathcal{F}(X(f))$. Then, $\uparrow_g^f (\downarrow_g^f B) = B$ if and only if

$$(\forall\,(x,y)\in X(g) \times X(f-g))(B(x,y) = B(x))\ . \tag{12.24}$$

Proof   Since the "if" part is clearly valid we proved the "only if" part.

Suppose (12.24) is not true; then there exists y_1 and $y_2 \in X(f - g)$ such that $B(x, y_1) \geq B(x, y_2)$. Thus, for any $y \in X(f - g)$,

$$B(x, y_2) \leq B(x, y_1) \leq \bigvee_{y \in X(f-g)} B(x, y) = (\downarrow_g^f B)(x) = (\uparrow_g^f (\downarrow_g^f B))(x, y) .$$

In particular, when $y = y_2$ we deduce a contradiction. This completes the proof. **Q.E.D.**

Return to Expression (12.19) in Proposition (12.17). It is interesting to study the conditions under which Expression (12.19) is an equality. The following corollary provides an answer.

Corollary 1   Under the conditions of Proposition 12.17, $\uparrow_g^f g(A) = f(A)$ if and only if

$$(\forall\, (x, y) \in X(f) = X(g) \times X(f - g))(f(A)(x, y) = f(A)(x)) . \qquad (12.25)$$

Proof   From Proposition 3, we know $\downarrow_g^f f(A) = g(A)$. After the substitution of $g(A)$ by $\downarrow_g^f f(A)$ in Expression (12.19), we obtain $(\uparrow_g^f (\downarrow_g^f f(A))) \supset f(A)$. Set $B = f(A)$, we have $(\uparrow_g^f (\downarrow_g^f B)) \supset B$. By Proposition 6, the corollary is proved. **Q.E.D.**

The significance of the corollary is that both Expressions (12.19) and (12.23) are essentially the same. That is,

$$\uparrow_g^f g(A) = f(A), \quad \text{and} \quad (\uparrow_g^f (\downarrow_g^f B)) = B , \qquad (12.26)$$

in essence are the same.

A factor space has many properties with respect to the projection and the cylindrical extension of factors. Their properties are given in the following propositions.

Proposition 7   For a given factor space $\{X(f)\}_{(f \in F)}$ and for any $f, g \in F$ with $f \geq g$, we have

1)   if $\{B_t\}_{(t \in T)}$ is a family of fuzzy subsets in $X(g)$, then

$$\left(\uparrow_g^f \left(\bigcup_{t \in T} B_t \right) \right) = \left(\bigcup_{t \in T} (\uparrow_g^f B_t) \right), \quad \left(\uparrow_g^f \left(\bigcap_{t \in T} B_t \right) \right) = \left(\bigcap_{t \in T} (\uparrow_g^f B_t) \right) ; \qquad (12.27)$$

2)   if $B \in \mathcal{F}(X(g))$ and B^c, the complementary set of B, in $X(g)$, then

$$\uparrow_g^f B^c = (\uparrow_g^f B)^c ; \qquad (12.28)$$

3)   if $B \in \mathcal{F}(X(f))$ and B^c, the complementary set of B, in $X(g)$, then

$$\downarrow_g^f B^c \supset (\downarrow_g^f B)^c ; \qquad (12.29)$$

4)   if $\{B_t\}_{(t \in T)}$ is a family of fuzzy subsets in $X(f)$, then

$$\left(\downarrow_g^f \left(\bigcup_{t \in T} B_t \right) \right) = \left(\bigcup_{t \in T} (\downarrow_g^f B_t) \right), \quad \left(\downarrow_g^f \left(\bigcap_{t \in T} B_t \right) \right) = \left(\bigcap_{t \in T} (\downarrow_g^f B_t) \right) \quad (12.30)$$

5)   Let $B, B_n \in \mathcal{F}(X(f))$, and $n = 1, 2, 3, \cdots$. Then

$$B_n \uparrow B \Longrightarrow (\downarrow_g^f B^n) \uparrow (\downarrow_g^f B) \quad (12.31)$$

where $B_n \uparrow B$ if and only if $\bigcup\limits_{n=1}^{\infty} B_n = B$.

Proof   1)   For any $(x, y) \in X(f) = X(g) \times X(f - g)$, we have

$$\left(\uparrow_g^f \left(\bigcup_{t \in T} B_t \right) \right)(x, y) = \left(\bigcup_{t \in T} B_t \right)(x)$$

$$= \bigvee_{t \in T} B_t(x) = \bigvee_{t \in T} (\uparrow_g^f B_t)(x, y)$$

$$= \left(\bigcup_{t \in T} (\uparrow_g^f B_t) \right)(x, y) .$$

So 1) is true. Similarly, we can prove the other identity.

2)   For any $(x, y) \in X(f) = X(g) \times X(f - g)$, we have

$$(\uparrow_g^f B^c)(x, y) = B^c(x) = 1 - B(x) = 1 - (\uparrow_g^f B)(x, y) = (\uparrow_g^f B)^c(x, y).$$

This completes the proof of 2).

3)   For any $(x, y) \in X(f) = X(g) \times X(f - g)$, we have

$$(\downarrow_g^f B)^c(x) = 1 - (\downarrow_g^f B)(x) = 1 - \bigvee_{y \in X(f-g)} B(x, y)$$

$$(\downarrow_g^f B^c)(x) = \bigvee_{y \in X(f-g)} B^c(x, y) = \bigvee_{y \in X(f-g)} (1 - B(x, y)) .$$

By comparing the right-hand sides of the above equations, we can conclude that

$$(\downarrow_g^f B)^c(x) \le (\downarrow_g^f B^c)(x).$$

This proves 3).

4)   The first identity. For any $(x, y) \in X(f) = X(g) \times X(f - g)$, we have

$$\left(\downarrow_g^f \left(\bigcup_{t \in T} B_t \right) \right)(x) = \bigvee_{y \in X(f-g)} \left(\bigcup_{t \in T} B_t \right)(x, y)$$

$$= \bigvee_{y \in X(f-g)} \left(\bigvee_{t \in T} B_t(x, y) \right) = \bigvee_{t \in T} \left(\bigvee_{y \in X(f-g)} B_t(x, y) \right)$$

$$= \bigvee_{t \in T} (\downarrow_g^f B_t)(x) = \left(\bigcup_{t \in T} (\downarrow_g^f B_t) \right)(x) .$$

Hence, the first part is proved.

In the second identity, notice that

$$\left(\downarrow_g^f \left(\bigcap_{t\in T} B_t\right)\right)(x) = \bigvee_{y\in X(f-g)} \left(\bigcap_{t\in T} B_t\right)(x, y) = \bigvee_{y\in X(f-g)} \left(\bigwedge_{t\in T} B_t(x, y)\right).$$

$$\left(\bigcap_{t\in T} (\downarrow_g^f B_t)\right)(x) = \bigwedge_{t\in T} (\downarrow_g^f B_t)(x) = \bigwedge_{t\in T} \left(\bigvee_{y\in X(f-g)} B_t(x, y)\right).$$

Also, for any $y \in X(f - g)$, y satisfies

$$\bigwedge_{t\in T} B_t(x, y) \leq \bigwedge_{t\in T} \left(\bigvee_{y\in X(f-g)} B_t(x, y)\right).$$

Hence, we have the following inequality:

$$\bigvee_{y\in X(f-g)} \left(\bigwedge_{t\in T} B_t(x, y)\right) \leq \bigwedge_{t\in T} \left(\bigvee_{y\in X(f-g)} B_t(x, y)\right).$$

This completes the proof of the second identity.

5)   Since $B_n \uparrow B \Longrightarrow \bigcup_{n=1}^{\infty} B_n = B$ which, in turn, implies $\downarrow_g^f B = \downarrow_g^f \left(\bigcup_{n=1}^{\infty} B_n\right) = \bigcup_{n=1}^{\infty} (\downarrow_g^f B_n)$, and that implies $(\downarrow_g^f B^n) \uparrow (\downarrow_g^f B)$.   **Q.E.D.**

Proposition 8   Let $\{X(f)\}_{(f\in F)}$ be a factor space and $\{f_t\}_{(t\in T)}$ be an independent family of factors in F. For any $B_t \in \mathcal{F}(X(f_t))$, $t \in T$, and $f = \bigvee_{t\in T} f_t$, the following equality holds:

$$\bigcap_{t\in T} (\uparrow_{f_t}^f B_t) = \prod_{t\in T} B_t . \tag{12.32}$$

Proof   For any w,

$$w \in \prod_{t\in T} X(f_t) = \left\{w \mid w : T \longrightarrow \bigcup_{t\in T} X(f_t),\ w(t) \in X(f_t),\ t \in T\right\},$$

we have

$$\left(\bigcap_{t\in T} (\uparrow_{f_t}^f B_t)\right)(w) = \bigwedge_{t\in T} (\uparrow_{f_t}^f B_t)(w) = \bigwedge_{t\in T} B_t(w(t)) = \left(\prod_{t\in T} B_t\right)(w).$$

This completes the proof.   **Q.E.D.**

12.9   Factor Sufficiency

Let $(U, \mathcal{C}, \{X(f)\}_{(f\in F)}]$ be a description frame. For any $\alpha \in \mathcal{C}$, let $A \in \mathcal{F}(U)$ be its extension. Since the complete factor **1** is a bijection, **1** can transform A

without distortion to the representation extension of α, $\mathbf{1}(A) \in \mathcal{F}(X(\mathbf{1}))$, and vice versa. That is, in case the extension of α on U is unknown but $B \in \mathcal{F}(X(\mathbf{1}))$, the representation extension of α on $X(\mathbf{1})$ can be found; we can apply the complete factor $\mathbf{1}$ to transform B back to A ($= \mathbf{1}^{-1}(B)$) on U. Generally, the following equality holds:

$$\mathbf{1}^{-1}(\mathbf{1}(A)) = \mathbf{1} . \tag{12.33}$$

Given a problem, if we were able to master its complete factor then the problem would be solved in essence. However, the complete factor $\mathbf{1}$ is hard to master because it is quite complicated and implicit.

The complete factor $\mathbf{1}$ holds the "overall situation" which means that A, the extension of α for any $\alpha \in \mathcal{C}$, is equivalent to $B(\mathbf{1})$, the representation extension of α, and $B(\mathbf{1}) = \mathbf{1}(A)$. Here, "equivalent" means a level of "sufficiency". This led us to the idea that we should pursue "parts of the situation" instead of the "overall situation". The approach is, for every fixed concept $\alpha \in \mathcal{C}$, find a factor f that is "simpler" than the complete factor $\mathbf{1}$ and $B(f) = f(A) \in \mathcal{F}(X(f))$ that is the representation extension of α with respect to f. Then, we construct $\uparrow_f^1 f(A)$, the cylindrical extension of $f(A)$ from f to $\mathbf{1}$. If $\uparrow_f^1 f(A) = \mathbf{1}(A)$, then f is considered "sufficient" with respect to the concept α since f can represent α, without distortion, in the complete space $X(\mathbf{1})$. Thus, we have the following:

Definition 2   Let $(U, \mathcal{C}, F]$ be a description frame. For any concept $\alpha \in \mathcal{C}$ and its extension $A \in \mathcal{F}(U)$, when $\mathbf{1}(A) \neq X(\mathbf{1})$ and ϕ, we call the factor $f \in F$ sufficient with respect to α if it satisfies

$$\uparrow_f^1 f(A) = \mathbf{1}(A) . \tag{12.34}$$

Every factor that is independent of f is called a surplus with respect to α.

Proposition 9   Let $(U, \mathcal{C}, F]$ be a description frame. For any concept $\alpha \in \mathcal{C}$ and its extension $A \in \mathcal{F}(U)$, if $\mathbf{1}(A) = X(\mathbf{1})$ or $\mathbf{1}(A) = \phi$, then every factor $f \in F$ is both simultaneously sufficient and surplus with respect to α.

Proof   First we state the following fact: For any $f, g \in F$, if $f \geq g$, then

$$\downarrow_g^f X(f) = X(g), \quad \uparrow_g^f X(g) = X(f), \quad \downarrow_g^f \phi = \phi = \uparrow_g^f \phi, \tag{12.35}$$

The fact is easy to prove. The proposition follows the proof of the two cases below.
Case 1   When $\mathbf{1}(A) = X(\mathbf{1})$, for any $f \in F$, we have

$$\uparrow_f^1 f(A) = \uparrow_f^1 (\downarrow_f^1 \mathbf{1}(A)) = \uparrow_f^1 (\downarrow_f^1 X(\mathbf{1})) = \uparrow_f^1 X(f) = X(\mathbf{1}) = \mathbf{1}(A).$$

This means that every factor $f \in F$ is sufficient with respect to α. Since every factor $f \in F$ is independent of $\mathbf{0}$, f is also a surplus with respect to α.
Case 2   When $\mathbf{1}(A) = \phi$, for any $f \in F$, we have

$$\uparrow_f^1 f(A) = \uparrow_f^1 (\downarrow_f^1 \mathbf{1}(A)) = \uparrow_f^1 (\downarrow_f^1 \phi) = \uparrow_f^1 \phi = \phi = \mathbf{1}(A).$$

This means that every factor $f \in F$ is sufficient with respect to α. Similarly, we can say that every factor $f \in F$ is a surplus with respect to α.   **Q.E.D.**

Proposition 10   Let $(U, \mathcal{C}, \{X(f)\}_{(f \in F)}]$ be a description frame. For any $\alpha \in \mathcal{C}$, let $A \in \mathcal{F}(U)$ be its extension. Then for any $f \in F$, we have
   1)   The following conditions are equivalent to each other:
      (a) f is sufficient with respect to α;
      (b)

$$\uparrow^1_f (\downarrow^1_f \mathbf{1}(A)) = \mathbf{1}(A) \ . \tag{12.36}$$

      (c)   For any $(x, y) \in X(\mathbf{1}) = X(f) \times X(f^c)$,

$$\mathbf{1}(A)(x, y) = (\downarrow^1_f \mathbf{1}(A))(x) \ . \tag{12.37}$$

      (d)   For any $(x_1, y_1), (x_2, y_2) \in X(\mathbf{1}) = X(f) \times X(f^c)$,

$$x_1 = x_2 \Longrightarrow \mathbf{1}(A)(x_1, y_1) = \mathbf{1}(A)(x_2, y_2) \ . \tag{12.38}$$

   2)   If f is a surplus with respect to α, then $(\downarrow^1_f \mathbf{1}(A))(x)$ is independent of x; i.e.,

$$(\downarrow^1_f \mathbf{1}(A))(x) = \text{constant}, \quad x \in X(f). \tag{12.39}$$

Proof   1)   We omit the proof since it is straightforward.
   2)   For f to be a surplus with respect to α, there exists a sufficient factor $g \ (\in F)$ which is sufficient with respect to α such that $f \wedge g = \mathbf{0}$. Since

$$\mathbf{1} = (f \vee g) \vee (f \vee g)^c = (f \vee g) \vee (f^c \vee g^c) \ ,$$

we have

$$X(\mathbf{1}) = X(f) \times X(g) \times X(f^c \wedge g^c).$$

Also, since

$$\mathbf{1} = f \vee f^c, \quad \mathbf{1} = (f \vee g) \vee (f \vee g)^c = f \vee [g \vee (f^c \vee g^c)]$$
$$f \wedge [g \vee (f^c \vee g^c)] = f \wedge g \vee [f \wedge (f^c \wedge g^c)] = \mathbf{0} \vee \mathbf{0} = \mathbf{0} \ ,$$

we have $f^c = g \vee (f^c \wedge g^c)$. Thus, for any $(x, y, z) \in X(\mathbf{1}) = X(f) \times X(g) \times X(f^c \wedge g^c)$

$$(\downarrow^1_f \mathbf{1}(A))(x) = \bigvee_{(y,z) \in X(f^c)} \mathbf{1}(A)(x, y, z) = \bigvee_{y \in X(g)} ((\downarrow^1_g \mathbf{1}(A))(y)).$$

This means that $(\downarrow^1_f \mathbf{1}(A))(x)$ is independent of x.   **Q.E.D.**

Proposition 11   Let $(U, \mathcal{C}, \{X(f)\}_{(f \in F)}]$ be a description frame. For any $\alpha \in \mathcal{C}$, let $A \in \mathcal{F}(U)$ be its extension. If $f, g \in F$ and $f \geq g$, then
   1)   if g is sufficient with respect to α, so does f.
   2)   if f is surplus with respect to α, so does g.

Proof   1)   Since $f \geq g$, this implies $f = g \vee (f - g)$. Thus,

$$1 = f \vee f^c = g \vee (f - g) \vee f^c.$$

Clearly, g, $f - g$, and f^c are independent from each other. Hence, $g^c = (f - g) \vee f^c$. Observe the following equalities:

$$X(\mathbf{1}) = X(f) \times X(f^c) = X(g) \times X(g^c) = X(g) \times X(f - g) \times X(f^c),$$

$$X(g^c) = X(f - g) \times X(f^c), \quad X(f) = X(g) \times X(f - g).$$

For any (x_1, y_1), $(x_2, y_2) \in X(\mathbf{1}) = X(f) \times X(f^c)$, if $x_1 = x_2$ we can prove $\mathbf{1}(A)(x_1, y_1) = \mathbf{1}(A)(x_2, y_2)$. In fact, we can write x_1 and x_2 as follows:

$$x_1 = (x_{11}, x_{12}), \quad x_2 = (x_{21}, x_{22}) \in X(\mathbf{1}) = X(f) \times X(f^c).$$

For $x_1 = x_2$, then $x_{11} = x_{21}$. Since g is sufficient with respect to α, by 1) (d) of the last proposition, we have

$$\mathbf{1}(A)(x_1, y_1) = \mathbf{1}(A)((x_{11}, x_{12}), y_1) = \mathbf{1}(A)(x_{11}, (x_{12}, y_1))$$
$$= \mathbf{1}(A)(x_{21}, (x_{22}, y_2)) = \mathbf{1}(A)((x_{21}, x_{22}), y_2) = \mathbf{1}(A)(x_2, y_2).$$

Therefore, f is sufficient with respect to α.

2)   The proof is simple and hence omitted.   **Q.E.D.**

12.10   The Rank of a Concept

Definition 3   Let $(U, C, F]$ be a description frame. Let $\alpha \in C$, whose extension is A and $\mathbf{1}(A) \notin \{X(\mathbf{1}), \emptyset\}$. Define

$$r(\alpha) = r(A) = \wedge \{f \in F \mid f \text{ is sufficient with respect to } \alpha\}$$

to be the rank of the concept α. When $\mathbf{1}(A) = X(\mathbf{1})$, or $\emptyset$, we set $r(\alpha) = r(A) = 0$ (the zero factor).

Note 7   The rank of a concept α is a factor which is the greatest lower bound of the sufficient factors (in the sense of the conjunction operator). Those factors that are greater than $r(\alpha)$ are sufficient with respect to α, but those factors that are less than $r(\alpha)$ are not sufficient. The complementary factor of $r(\alpha)$ is the least upper bound of the surplus factors (in the sense of the disjunction operator), which means that the factors less than it are surpluses with respect to α but the factors greater than it are not surpluses.

We now ask: Is the rank of a concept itself sufficient with respect to that concept?

Theorem 1   Let $(U, C, F]$ be a description frame and $\alpha \in C$, whose extension is $A \in \mathcal{F}(U)$. For any $g \in F$, if g is sufficient with respect to α, then for any $f \in F$ with $f \geq g$ we have the following:

$$\uparrow_g^f g(A) = f(A) . \tag{12.40}$$

Conversely, for any $g \in F$, if there exists a sufficient factor $f \in F$ with respect to α, and $f \geq g$ such that the Equality (12.40) holds, then g is also sufficient with respect to α.

Proof   First we prove the Equality (12.40). Since g is sufficient with respect to α, by the definition of "sufficiency", $\uparrow_g^1 g(A) = \mathbf{1}(A)$ holds. Also, by Proposition 11, f is sufficient with respect to α. Thus, we have

$$\uparrow_f^1 (\uparrow_g^f g(A)) = \uparrow_g^1 g(A) = \mathbf{1}(A) = \uparrow_f^1 f(A) . \tag{12.41}$$

The second half of the theorem is proved by observing:

$$\uparrow_g^1 g(A) = \uparrow_f^1 (\uparrow_g^f g(A)) = \uparrow_f^1 f(A) = \mathbf{1}(A).$$

Q.E.D.

The corollary below provides a partial answer to our question above.

Corollary 2   Let $(U, \mathcal{C}, F]$ be a description frame and $\alpha \in \mathcal{C}$, whose extension is $A \in \mathcal{F}(U)$. $\gamma(\alpha)$ is sufficient with respect to α if and only if there exists a sufficient factor $f \in F$ with respect to α such that

$$\uparrow_{\gamma(\alpha)}^f \gamma(\alpha)(A) = f(A) . \tag{12.42}$$

12.11   Atomic Factor Spaces

Definition 4   Let $\{X(f)\}_{(f \in F)}$ be a factor space. When F is an atomic lattice, F is called an *atomic set of factors* and $\{X(f)\}_{(f \in F)}$ is called an atomic factor space.

Let π be the family of all atomic factors in F. Then F is the same as $\mathcal{P}(\pi)$, the power set of π, i.e., $F = \mathcal{P}(\pi)$.

Proposition 12   Let $(U, \mathcal{C}, F]$ be a description frame, $F = \mathcal{P}(\pi)$ be an atomic set of factors, and $\alpha \in \mathcal{C}$. If $\pi_1 = \pi \setminus \pi_2$, and $\pi_2 = \{f \in \pi \mid f$ is surplus with respect to $\alpha\}$, then

$$\gamma(\alpha) = \vee\{f \mid f \in \pi_1\} = \pi_1 . \tag{12.43}$$

Proof   From the properties of atomic lattices, for any $f \in F$,

$$f = \vee\{g \in \pi \mid f \geq g \quad (\text{i.e., } g \in f)\} .$$

Take an atomic factor $e \in \pi_1$ that is not a surplus with respect to α. Then for an arbitrary sufficient factor g with respect to α, we have $e \wedge g \neq 0$ because e is an atomic factor and $e \leq g$. Thus,

$$e \leq \gamma(\alpha) = \wedge\{f \in F \mid f \text{ is sufficient with respect to } \alpha\}.$$

This implies $\pi_1 = \vee\{f \mid \in \pi_1\} \leq \gamma(\alpha)$.

We now prove $\pi_1 = \gamma(\alpha)$. If this is not the case, then there exists $h \in \pi_2$ such that $h \leq \gamma(\alpha)$. That is, for an arbitrary sufficient factor g with respect to α, we have $h \leq \gamma(\alpha) \leq g$ which is contradictory to the assumption $h \in \pi_2$. Therefore, $\pi_1 = \gamma(\alpha)$.   **Q.E.D.**

The proposition means that when F is an atomic set of factors the rank of a concept is formed by all of the non-surplus atomic factors. Under the same conditions of the proposition, we have the following

Corollary 3   1)   For any $f \in F$, f is sufficient with respect to α if and only if $f \geq r(\alpha)$; and

2)   For any $f \in F$, f is a surplus with respect to α if and only if $f \wedge r(\alpha) = \mathbf{0}$.

12.12   Conclusions

Factor spaces can be regarded as a kind of mathematical frame on artificial intelligence, especially on fuzzy information processing. This chapter introduced some basic notions about factor spaces. First of all, factors were interpreted by non-mathematical language. And several operations between factors were defined. Then an axiomation definition of factor spaces was given mathematically. Secondly, concept description in a factor space was discussed in detail. At last, the sufficiency of a factor and the rank of a factor are proposed, which are very important means for representing concepts.

References

1. P. Z. Wang, Stochastic differential equations, in *Advances in Statistical Physics,* B. L. Hao and L. Yu, eds., Science Press, Beijing, 1981.

2. P. Z. Wang and M. Sugeno, The factor fields and background structure for fuzzy subsets, *Fuzzy Mathematics,* 2(2), pp. 45-54, 1982.

3. P. Z. Wang, A factor space approach to knowledge representation, *Fuzzy Sets and Systems,* Vol. 36, pp. 113-124, 1990.

4. P. Z. Wang, Factor space and fuzzy tables, *Proceedings of Fifth IFSA World Congress,* Korea, pp. 683-686, 1993.

5. A. Kandel, X. T. Peng, Z. Q. Cao, and P. Z. Wang, Representation of concept by factor spaces, *Cybernetics and Systems,* Vol. 21, 1990.

6. H. X. Li and V. C. Yen, Factor spaces and fuzzy decision-making, *Journal of Beijing Normal University,* Vol. 30, No. 1, pp. 15-21, 1994.

Chapter 13

Neuron Models Based on Factor Spaces Theory and Factor Space Canes

This chapter discusses several neuron models based on factor space discussed in previous chapter. Factor space offers a mathematical frame of describing objects and concepts. However, it has some limitations for certain applications. Here, we extend the factor space concept to factor space canes. We introduce several factor space canes and switch factors and their growth relation. Finally, we study class partition for multifactorial fuzzy decision-making using factor space canes.

13.1   Neuron Mechanism of Factor Spaces

Given an atomic factor space $\{X(f)\}_{(f \in F)}$, where the family of all atomic factors in F, $\pi = \{f_1, f_2, \cdots, f_m\}$, is a finite set, for an object u, which its states in the state spaces $X(f_j)$ $(j = 1, 2, \cdots, m)$ are $x_j = f_j(u)$ $(j = 1, 2, \cdots, m)$, according to Chapter 6, we can obtain its state in the complete space $X(1)$:

$$x = \mathbf{1}(u) \approx M_m(f_1(u), f_2(u), \cdots, f_m(u)) = M_m(x_1, x_2, \cdots, x_m) ,$$

where $M_m : [0,1]^m \longrightarrow [0,1]$ is an ASM_m-func. Especially M_m is usually taken as Σ, i.e.,

$$x = M_m(x_1, x_2, \cdots, x_m) = \sum_{j=1}^{m} w_j x_j , \tag{13.1}$$

where w_j $(j = 1, 2, \cdots, m)$ is a group of constant weights, i.e., $w_j \in [0,1]$ $(j = 1, 2, \cdots, m)$ and $\sum_{j=1}^{m} w_j = 1$.

A factor space can be regarded as a "transformer". If input in a set of states, $x_1, x_2, \cdots, x_m$, then it outputs only one state $x = M_m(x_1, x_2, \cdots, x_m)$ by means of the composition function of M_m or the factor space (see Figure 1).

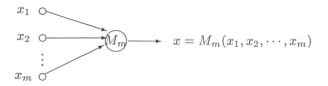

Figure 1   Composition function of a factor space

Where $M_m = \Sigma$, the atomic factors f_j $(j = 1, 2, \cdots, m)$ can be regarded as m input channels, and the weights w_j $(j = 1, 2, \cdots, m)$ regarded, respectively, as the damping coefficients of the channels f_j $(j = 1, 2, \cdots, m)$. The complete factor $\mathbf{1}$ is regarded as one output channel. If a set of input data $x_1, x_2, \cdots, x_m$ is given, an output datum $x = \sum_{j=1}^{m} w_j x_j$ can be obtained. Thus, we have a neuron model shown in Figure 2.

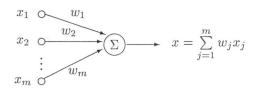

Figure 2   A kind of model of neurons based on factor spaces

13.2   The Models of Neurons without Respect to Time

13.2.1   Threshold Models of Neurons

If a "step" (threshold) $\theta \in [0, 1]$ is set in the output channel of the model of neurons, then the result of composition by Σ can be output only if it is more than the threshold θ, else the output is regarded as zero. So the output y of a neuron can be represented as follows (see Figure 3):

$$y = \varphi\left(\sum_{j=1}^{m} w_j x_j - \theta \right) , \tag{13.2}$$

where $\varphi(x)$ is a piecewise linear function defined as follows:

$$\varphi(x) = \begin{cases} x, & x \geq 0 \\ 0, & x < 0 . \end{cases} \tag{13.3}$$

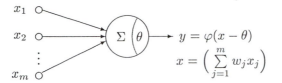

Figure 3   The model of neurons with threshold

13.2.2   Linear Model of Neurons

When the threshold $\theta = 0$, Expression (13.2) is simplified as a linear function:

$$x = \sum_{j=1}^{m} w_j x_j \, . \tag{13.4}$$

This is the linear model of neurons, which is a special case of the threshold models of neurons.

13.2.3   General Threshold Model of Neurons

In Expression (13.2), if the multifactorial function Σ is replaced by general multifactorial function M_m, then we have a general threshold model of neurons as follows:

$$y = \varphi(M_m(x_1, x_2, \cdots, x_m) - \theta) \, . \tag{13.5}$$

Note 1   Since the multifactorial function M_m is a basic tool of the composition of states in factor spaces, based on the viewpoint of factor spaces, Expression (13.5) is regarded as a general threshold model of neurons.

When $\theta = 0$, Expression (13.5) also has a simple form:

$$x = M_m(x_1, x_2, \cdots, x_m) \, . \tag{13.6}$$

According to different expressions of M_m, (13.6) has different special examples (of course, these can suit Expression (13.5)):

$$x = M_m(x_1, x_2, \cdots, x_m) = \bigwedge_{j=1}^{m} x_j, \tag{13.7}$$

$$x = M_m(x_1, x_2, \cdots, x_m) = \bigvee_{j=1}^{m} x_j, \tag{13.8}$$

$$x = M_m(x_1, x_2, \cdots, x_m) = \sum_{j=1}^{m} x_j, \tag{13.9}$$

where $w_j \in [0,1]$ $(j = 1, 2, \cdots, m)$ and $\sum\limits_{j=1}^{m} w_j = 1$, which is just Expression (13.4).

$$x = \sum_{j=1}^{m} w_j(x_j)x_j, \tag{13.10}$$

where $w_j : [0,1] \longrightarrow [0,1]$, $t \longmapsto w_j(t)$, is a continuous function, and satisfies the normalized condition: $\sum\limits_{j=1}^{m} w_j(x_j) = 1$, which is a kind of variable weights with one variable.

$$x = \bigvee_{j=1}^{m} w_j x_j, \tag{13.11}$$

where $w_j \in [0,1]$ $(j = 1, 2, \cdots, m)$ and $\bigvee\limits_{j=1}^{m} w_j = 1$.

$$x = \bigvee_{j=1}^{m} w_j(x_j)x_j, \tag{13.12}$$

where $w_j : [0,1] \longrightarrow [0,1]$, $t \longmapsto w_j(t)$, is a continuous function, and satisfies the condition: $\bigvee\limits_{j=1}^{m} w_j(x_j) = 1$.

$$x = \bigvee_{j=1}^{m} (w_j \wedge x_j), \tag{13.13}$$

where $w_j \in [0,1]$ $(j = 1, 2, \cdots, m)$ and $\bigvee\limits_{j=1}^{m} w_j = 1$.

$$\bigvee_{j=1}^{m} [w_j(x_j) \wedge x_j], \tag{13.14}$$

where $w_j(x_j)$ are the same as the ones in (13.12).

$$x = \left(\prod_{j=1}^{m} x_j \right)^{1/m}, \tag{13.15}$$

$$x = \left(\frac{1}{m} \sum_{j=1}^{m} x_j^p \right)^{1/p}, \quad p > 0, \tag{13.16}$$

$$x = \left(\sum_{j=1}^{m} w_j x_j^p \right)^{1/p}, \tag{13.17}$$

where $p > 0$, $w_j \in [0,1]$ $(j = 1, 2, \cdots, m)$ and $\sum\limits_{j=1}^{m} w_j = 1$.

$$x = \left(\sum_{j=1}^{m} w_j(x_j)x_j^p \right)^{1/p}, \tag{13.18}$$

where $p > 0$ and $w_j(x_j)$ are same as the ones in (13.10).

According to the generations of ASM_m-funcs, by using the models mentioned above, we can generate many more complicated models of neurons.

13.2.4   The Models of Neurons Based on Weber-Fechner's Law

In the 19th century, G. T. Fechner, a German psychologist, discussed the problem of the reaction of people receiving some outside stimulation, based on E. H. Weber's work. Let r be the reaction of people receiving some stimulation, and s be the real intensity of the stimulation. They have the following result:

$$r = k \ln s + c, \tag{13.19}$$

where k and c are constants. This is the celebrated Weber-Fechner's law. Although the law is in connection with some sense organs of the human body (i.e., the law is of some macro-properties), the reaction of the organs is based on the reaction of the neurons in the human body. Therefore, we hold that the reaction of a neuron receiving the stimulation from its synapses also follows Weber-Fechner's law, i.e.,

$$y_j = k_j \ln x_j + c_j, \tag{13.20}$$

where x_j is the intensity of stimulation from jth synapse, and y_j is the reaction of the neuron with respect to x_j based on Weber-Fechner's law. So Expression (13.5) becomes the following expression:

$$y = \varphi(M_m(k_1 \ln x_1 + c_1, k_2 \ln x_2 + c_2, , \cdots, k_m \ln x_m + c_m) - \theta). \tag{13.21}$$

When $M_m = \Sigma$, we have the following expression:

$$y = \varphi\left(\sum_{j=1}^{m} w_j(k_j \ln x_j + c_j) - \theta \right). \tag{13.22}$$

This is a kind of model of neurons based on Weber-Fechner's law. In order to determine k_j and c_j, $\ln x_j$ is first developed as the following power series:

$$\ln x_j = (x_j - 1) - \frac{1}{2}(x_j - 1)^2 + \frac{1}{3}(x_j - 1)^3 - \cdots$$

Then y_j is taken approximately as the two former terms of the power series:

$$y_j = k_j[(x_j - 1) - \frac{1}{2}(x_j - 1)^2] + c_j = k_j(2x_j - \frac{1}{2}x_j^2 - 2) + c_j. \tag{13.23}$$

If y_j is regarded as a function of x_j, i.e., $y_j = y_j(x_j)$ by using the following boundary value condition:

$$y_j(0) = 0, \quad y_j(1) = 1, \tag{13.24}$$

then it is easy to determine that $k_j = \frac{2}{3}$ and $c_j = \frac{4}{3}$. Thus,

$$y_j = \frac{2}{3}(2x_j - \frac{1}{2}x_j^2 - 2) + \frac{4}{3} = \frac{1}{3}(4x_j - x_j^2). \tag{13.25}$$

Expression (13.25) is substituted into Expression (13.22) and we have

$$y = \varphi\left(\frac{1}{3}\sum_{j=1}^{m} w_j(4x_j - x_j^2) - \theta\right).$$

(13.26)

This is a simplified model of neurons based on Weber-Fechner's law. Although Expression (13.26) is little more complicated than Expression (13.2), it may take an interesting stride forward on the way of approximating real neurons.

13.3   The Models of Neurons Concerned with Time

For a given description frame $(U, \mathcal{C}, \{X(f)\}_{(f \in F)})$, the state of the frame can be regarded as to be not concerned with time. If for any $u \in U$, at any instant, u is concerned with time, we have $u = u(t)$. However, once the object u separates itself from the description frame, the change for it depending on time will be not shown explicitly. In other words, the change is with respect to the factor space $\{X(f)\}_{(f \in F)}$. For example, let U be the set of some people, say $U = \{$John, Kate, Lucy, $\cdots\}$, and the set of factors F be taken as $F = \{f_1, f_2, f_3, f_4, \cdots\} = \{$height, weight, age, sex, $\cdots\}$. For a $u \in U$, say $u =$ John, he was born in 1970. If we set $t_0 = 1970$, $t_1 = 1971$, $t_2 = 1972, \cdots$, then $x_j(x) \triangleq f_j(u(t))$, the state for u being with respect to $f_j \in F$, changes depending on time $t \in T = \{t_0, t_1, t_2, \cdots\}$. Of course, sometimes for a fixed $u \in U$, there exist some factors such that u does not change depending on time with respect to these factors. For instance, in the above example, u can not change with respect to f_4 (i.e., sex).

When a factor space $\{X(f)\}_{(f \in F)}$ is used to represent an optic neuron, and u is an object in motion or itself is in change, u is concerned with time: $u = u(t)$, and the stimulation from the synapse, brought by u, x_j is also concerned with time t : $x_j = x_j(t)$. For general neurons, there are also such similar "perceptions" concerned with time.

Generally speaking, if every synapse changes its state every τ, where τ is a given interval of time, then we have a model of neurons concerned with time as follows:

$$y(t + \tau) = \varphi\left(\sum_{j=1}^{m} w_j x_j(t) - \theta\right).$$

(13.27)

If we consider the non-response period of neurons (including absolute non-response period and relative non-response period), a nonlinear first order differential equation as follows is often used to simulate the change of the membrane potential of biological neurons:

$$\tau\frac{du(t)}{dt} = -u(t) + \sum_{j=1}^{m} w_j x_j(t) - \theta,$$

(13.28)

$$y(t) = \varphi(u(t)).$$

(13.29)

13.4   The Models of Neurons Based on Variable Weights

Up to now, in the models of neural network we have known, connection weights play a key role. Clearly, these weights are constants. Of course, in the process of learning, these weights are continually adjusted such that they converge on their stable values. But they are in essence constants. On the other side, people have known that connection weights are of plasticity. This means that connection weights should change depending on the change of the intensity of synapses to be stimulated. The change of connection weights is not arbitrary but of some tendency showing some relation of functions.

It is well known that the plasticity of connection weights is a key basic of the learning in the neural network. We shall illustrate our idea of displaying the plasticity in the following.

13.4.1   The Excitatory and Inhibitory Mechanism of Neurons

A neuron is basically made up of five parts: cell body, cell membrane, dendrite, axon, and synapse. An axon can transmit the electric impulse signals in its cell body to other neurons through its synapses. Dendrites receive the electric signals from other neurons. Because of cell bodies being different, there may be two kinds of properties between two neurons: excitatory and inhibitory (see Figure 4).

Figure 4   The excitation and inhibition of
the connection between neurons

A cell, if the membrane potential of another cell connected with the cell goes up as the cell generates electric impulses, is said to have excitatory connection with another cell; if not, it has an inhibitory connection. It is worthy of note that excitatory connection and inhibitory connection are relative, for a cell (neuron) itself cannot be designated as being excitatory or inhibitory. In other words, the excitation and inhibition are shown in a cell (neuron) only when the cell (neuron) has some connections with other cells (neurons). For example, though there is an excitatory connection between cell A and cell B, there may be an inhibitory connection between cell A and cell C (see Figure 5).

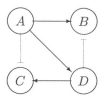

Figure 5   The relativity of excitation
and inhibition of neurons

The basic form of a neuron connecting with other neurons is relatively steady such that the electric signals which flow into its dendrites (the signals are received from the synapses of other neurons) should classify the whole "input channels" into two classes: one is excitatory and the other is inhibitory (see Figure 6).

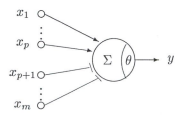

Figure 6   Two classes of input channels

13.4.2   The Negative Weights Description of the Inhibitory Mechanism

The excitatory and inhibitory mechanism of a neuron embody that the excitatory input will take "gain" effort and the inhibitory input will take "attenuation" effort, in the whole quantity after the composition. Such attenuation can be described by negative weights.

Given a set of constant weights, $w_j \in [0, 1]$ $(j = 1, 2, \cdots, m)$, with $\sum_{j=1}^{m} w_j = 1$, write

$$w_j' = \begin{cases} w_j, & 1 \leq j \leq p, \\ -w_j, & p+1 \leq j \leq m. \end{cases} \tag{13.30}$$

Then w_j' $(j = 1, 2, \cdots, m)$ is a set of constant weights with negative weights. Thus, we have a model of neurons with excitatory and inhibitory mechanism as follows:

$$y = \varphi\Big(\sum_{j=1}^{m} w_j x_j - \theta \Big) = \varphi\Big(\sum_{j=1}^{p} w_j x_j - \sum_{j=p+1}^{p} w_j x_j - \theta \Big). \tag{13.31}$$

Note 2   The constant weights with negative weights, mentioned above, lose the normality, i.e., dissatisfy the condition $\sum_{j=1}^{m} w_j' = 1$. But we can make them satisfy the normality by using the following two methods.

Method 1   A neuron is regarded as one piece formed by putting the excitatory part and the inhibitory part together, where each part has its own weights system (see Figure 7): w_j $(j = 1, 2, \cdots, m)$, $\sum_{j=1}^{m} w_j = 1$, and w_i' $(i = 1, 2, \cdots, n)$ $\sum_{i=1}^{n} w_i' = 1$

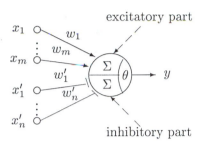

Figure 7   A neuron regarded as one piece
formed by putting the two together

Clearly, based on the idea, the model of neurons should be written as follows:

$$y = \varphi\left(\sum_{j=1}^{m} w_j x_j - \sum_{i=1}^{n} w_i' x_i' - \theta \right). \tag{13.32}$$

Method 2   We can propose generalized normality, i.e., about (13.30), set

$$a_j = w_j' \Big/ \left| \sum_{j=1}^{m} w_j' \right|, \quad \sum_{j=1}^{m} w_j' \neq 0. \tag{13.33}$$

Then such constant weights a_j $(j = 1, 2, \cdots, m)$ satisfy the following generalized normality:

$$\sum_{j=1}^{m} a_j = \begin{cases} 1, & \sum_{j=1}^{m} w_j' > 0, \\[2mm] -1, & \sum_{j=1}^{m} w_j' < 0. \end{cases} \tag{13.34}$$

From this, Expression (13.31) becomes the following form:

$$y = \varphi\left(\sum_{j=1}^{m} a_j x_j - \theta \right). \tag{13.35}$$

13.4.3   On Fukushima's Model

As mentioned above, the inhibitory input of a neuron takes attenuation effort in the whole quantity after the composition. Also, it can be regarded as taking shunt effect to the excitatory action. So in the model, the excitatory input and inhibitory input are no longer put together under Σ, but a relation between the Σ form of the excitatory input and the Σ form of the inhibitory input is given. For example, Fukushima gave a simple model (according to Figure 7, we have modified it a little):

$$y = \varphi \left(\frac{\varepsilon + \sum\limits_{j=1}^{m} w_j x_j}{\varepsilon + \sum\limits_{i=1}^{n} w'_i x'_i} - 1 \right), \tag{13.36}$$

where $\varepsilon \in (0,1)$. In order to consider the input-output property of the model, we set

$$e = \sum_{j=1}^{m} w_j x_j, \quad h = \sum_{i=1}^{n} w'_i x'_i. \tag{13.37}$$

Thus Expression (13.36) becomes the following simpler form:

$$y = \varphi \left(\frac{\varepsilon + e}{\varepsilon + h} - 1 \right) = \varphi \left(\frac{e - h}{\varepsilon + h} \right). \tag{13.38}$$

When $h \ll \varepsilon$, we approximately take the following expression:

$$y = \varphi \left(\frac{1}{\varepsilon} (e - h) \right). \tag{13.39}$$

This moment, the input-output property of the neuron is practically a linear threshold element.

When $e \gg \varepsilon$ and $h \gg \varepsilon$, we approximately have the following expression:

$$y = \varphi \left(\frac{e}{h} - 1 \right). \tag{13.40}$$

If assuming $e = \xi x$ and $h = \eta x$, we have

$$y = \varphi \left(\frac{\xi - \eta}{2\eta} \left(1 + \text{th} \left(\frac{1}{2} \log_2 \eta x \right) \right) \right), \tag{13.41}$$

where $\text{th}\, x$ is a hyperbolic tangent function of x. Clearly, Expression (13.41) is near to Weber-Fechner's law.

13.4.4   The Model of Neurons Based on Univariable Weights

Under constant weights, the excitatory and inhibitory mechanism coexisting in a neuron is shown by using negative weights. Now without negative weights, if for every j, w_j taken as a function of x_j, which means that w_j is a univariable weight, then the mechanism mentioned above will be shown in the changes of the weights.

Definition 1   A set of functions $w_j : [0,1] \longrightarrow [0,1]$ is called a set of univariable weights, if every $w_j(x)$ is a monotonically continuous function. If $w_j(x)$ is a monotonically increasing function, then it is called an excitatory weight. If $w_j(x)$ is a monotonically decreasing function, then it is called an inhibitory weight.

Note 3 The univariable weights defined above may not satisfy the normality: $\sum_{j=1}^{m} w_j(x_j) = 1$.

Not losing generality, $w_1(x), \cdots, w_p(x)$ are assumed to be excitatory weights, and $w_{p+1}(x), \cdots, w_m(x)$ to be inhibitory weights. Set

$$e = \sum_{j=1}^{p} w_j(x_j)x_j, \quad h = \sum_{j=p+1}^{m} w_j(x_j)x_j , \tag{13.42}$$

then we have Expression (13.38) again.

As a matter of fact, the input intensity x_j may change along with time t, i.e., $x_j = x_j(t)$ so that w_j may also change along with t: $w_j = w_j(x_j(t))$. Therefore,

$$y(t + \tau) = \varphi\left(\sum_{j=1}^{m} w_j(x_j(t))x_j(t) - \theta \right). \tag{13.43}$$

In addition, if we set

$$e(t) = \sum_{j=1}^{p} w_j(x_j(t))x_j(t), \quad h(t) = \sum_{j=p+1}^{m} w_j(x_j(t))x_j(t), \tag{13.44}$$

we get Fukushima's model with time variable:

$$y(t + \tau) = \varphi\left(\frac{e(t) - h(t)}{\varepsilon + h(t)}\right). \tag{13.45}$$

13.5 Naïve Thoughts of Factor Space Canes

Factor spaces can offer a mathematical frame of describing objective things and concepts. However, from the viewpoint of applications, as every mathematical tool has its limitation, factor spaces also have certain limitation. For example, for any factors f and g, f and g cannot always be permitted in the same factor space. Assuming f = lifeless and g = sex, if f and g can be put into the same factor space, then there is the disjunction operation between f and g. Clearly f and g can be regarded as independent. Let $h = f \vee g$. The state space of h is that,

$$\begin{aligned}
x(h) = X(f \vee g) &= X(f) \times X(g) = \{\text{life, lifeless}\} \times \{\text{male, female}\} \\
&= \{(\text{life, male}), (\text{life, female}), (\text{lifeless, male}), (\text{lifeless, female})\} \\
&= \{\text{life} \cdot \text{male}, \text{life} \cdot \text{female}, \text{lifeless} \cdot \text{male}, \text{lifeless} \cdot \text{female}\} , \tag{13.46}
\end{aligned}$$

where $x \cdot y$ (for example, x = life and y = male so that $x \cdot y$ = life · male) is a compound word that means that it is the meaning of both x and y. It is easy to recognize that lifeless · male and lifeless · female are meaningless. How can we talk about the sex of a stone? So we should not put such f and g into the same factor space. In fact, sex is a factor of biological matter while the hierarchy of the factor lifeless being is higher than the hierarchy of sex. Thus, "hierarchy" will be an important relation between factors, which is one of the backgrounds of factor space canes.

We start from an example to see what is a factor space cane. Now we consider the concept "people". Let $V(\text{people})$ be the set of all factors concerned with people, where there is a factor $f_0 = \text{sex} \in V(\text{people})$. Clearly, its state space $X(f_0) = \{\text{male, female}\}$. By using f_0, people can be classified into two classes: "men" and "women". Of course, the concepts "men" and "women" are "subconcepts" of the concept "people".

Let $V(\text{men})$ be the set of all factors concerned with men, and $V(\text{women})$ be the set of all factors concerned with women. $V(\text{men})$ and $V(\text{women})$ can be regarded as the families of factors induced by f_0 (see Figure 8).

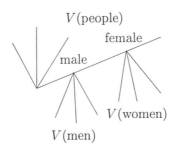

Figure 8   The structure-like "cane"

As men(women) are people, the factors concerned with people must be concerned with men(women). This means that $V(\text{people}) \subset V(\text{men})$ and $V(\text{people}) \subset V(\text{women})$. On the other side, people, with respect to sex, has just two classes: men and women. So $V(\text{men}) \cap V(\text{women}) = V(\text{people})$, i.e., the factors concerned with people are just the common factors concerned with men and women. Generally speaking, the more concrete a concept, the bigger the family of factors concerned with the concept.

Moreover, $V(\text{men}) \backslash V(\text{people})$ is the family of factors concerned only with men, and $V(\text{women}) \backslash V(\text{people})$ is the family of factors concerned only with women. Therefore,

$$(V(\text{men}) \backslash V(\text{people})) \cap (V(\text{women}) \backslash V(\text{people})) = \emptyset . \qquad (13.47)$$

As a concept is classified into some subconcepts, the factors concerned only with every subconcept are new factors based on the "old" factors concerned with the concept. For instance, if the concept "people" is classified into "men" and "women", then $V(\text{men})$ and $V(\text{women})$ are the families of new factors being the opposite of the factors in $V(\text{people})$ (see Figure 9).

We now consider the place of factor $f_0 = \text{sex}$. Because $f_0 \in V(\text{people})$, $f_0 \in V(\text{men}) \cap V(\text{women})$ so that $f_0 \notin (V(\text{men}) \backslash V(\text{people})) \cup (V(\text{women}) \backslash V(\text{people}))$, which means that there is no place of f_0 in the families of factors concerned only with men or women. It is easy to understand that f_0 has, respectively, only one state "male" or "female" corresponding to "men" or "women", i.e., there is no change in the state space of f_0. In other words, f_0 is a trivial factor this moment. Of course the states of f_0 also have no change in $V(\text{men})$ or $V(\text{women})$. However, for the convenience of expression, f_0 can be kept in $V(\text{men})$ and $V(\text{women})$. But if assuming $f_0 \notin V(\text{men}) \cup V(\text{women})$, the case is not very complicated. Then there is the expression:

$$V(\text{people}) \setminus \{f_0\} = V(\text{men}) \cap V(\text{women}). \tag{13.48}$$

So clearly $f_0 \notin (V(\text{men}) \setminus V(\text{people})) \cup (V(\text{women}) \setminus V(\text{people}))$. That is, the "stretch" $\{V(\text{men}), V(\text{women})\}$ on $V(\text{people})$, induced by f_0, will satisfy Expression (13.48), when f_0 is deleted.

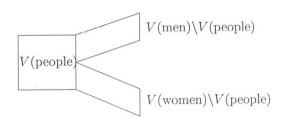

Figure 9   The growing of families of factors

There are two ideas to consider factor space canes. Thus, the next two sections will introduce two forms of factor space canes.

13.6   Melon-type Factor Space Canes

The families of factors such as $V(\text{people})$, $V(\text{men})$, $V(\text{women})$, $V(\text{men}) \setminus V(\text{people})$, $V(\text{women}) \setminus V(\text{people})$, etc., mentioned above, generally satisfy the definition of factor spaces (mainly are Boolean algebra structures), although they have not been proved to be "the sets of factors" defined in factor spaces. In fact, we always make use of the families of factors to generate the sets of factors. For instance, if we can find all atomic factors in $V(\text{people})$, denoted by $\pi(\text{people})$, and set $F(\text{people}) = \mathcal{P}(\pi(\text{people}))$, then $F(\text{people})$ is just a set of factors. Without loss of generality, we can assume that the families are the sets of factors. Thus, we can make use of them to form factor spaces.

Write $F_1 \triangleq V(\text{people})$ and we have a factor space with respect to "people", denoted by $[\text{people}] \triangleq \{X(f)\}_{(f \in F_1)}$. And setting $F_{11} \triangleq V(\text{men}) \setminus V(\text{people})$ and $F_{12} \triangleq V(\text{women}) \setminus V(\text{people})$, we also have two factor spaces, $[\text{men}] \triangleq \{X(f)\}_{(f \in F_{11})}$ and $[\text{women}] \triangleq \{X(f)\}_{(f \in F_{12})}$.

Let us suppose that there exists such a factor $f_* =$ childbirth in F_{12}, where $X(f_*) = \{\text{childbirthable(women)}, \text{childbirthless(women)}\}$.

The family of factors concerned with concept "childbirthable women" is denoted by V(childbirthable). Similarly we have V(childbirthless). Clearly, V(childbirthable) $\cap V$(childbirthless)$= V$(women). As mentioned above, V(childbirthable)$\backslash V$(women) is the family of factors concerned only with childbirthable women, and V(childbirthless) $\backslash V$(women) is the family of factors concerned only with childbirthless women. Naturally,

$$(V(\text{childbirthable})\backslash V(\text{women}))\cap(V(\text{childbirthless}) \backslash V(\text{women})) = \emptyset .$$

If setting

$$F_{121} \triangleq V(\text{childbirthable})\backslash V(\text{women}) \text{ and } F_{122} \triangleq V(\text{childbirthless}) \backslash V(\text{women}),$$

as shown in Figure 10, two families of factors F_{121} and F_{122} are stretched out from F_{12}. So we get two factor spaces:

$$[\text{childbirthable}] \triangleq \{X(f)\}_{(f\in F_{121})}, \quad [\text{childbirthless}] \triangleq \{X(f)\}_{(f\in F_{122})} .$$

If the symbol $[\cdot]$ is regarded as a "melon", where "$\cdot$" stands for a word or sentence, such as "men", "women", and "childbirthless", etc., then these melons can be linked together to form a cane (see Figure 11), which is why we call them factor space canes.

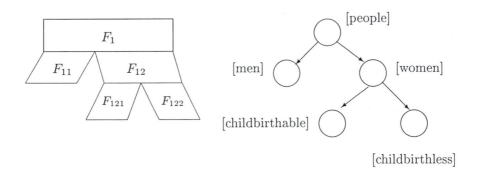

Figure 10   The stretching of   Figure 11   A melon-type factor
the set of factors   space cane

When we discuss general properties of people, we can make use of the factor space [people]; when we discuss the properties of women, we can make use of the factor space [women] based on [people];when we discuss the properties of childbirthable women, we can make use of the factor space [childbirthable] based on [women] and [people]; and so on. A melon-type factor space cane describes concepts or things from general to concrete by using a "step by step" approach, where every "melon" is of relative independence.

13.7 Chain-type Factor Space Canes

We all know that iron chains are such a kind of utensil that are formed by the means of one ring linked to another. They have an essential characteristic that only one ring can not be called a chain while, some rings to be linked together by means of a certain form such as one ring linked by another can be called a chain. Chain-type factor space canes also have such characteristic. However, they are more complicated than the chains mentioned above. They are a kind of tree-type chains.

Since we have made use of $[\,\cdot\,]$ to stand for a factor space in a melon-type factor space cane, for difference, a new symbol $<\,\cdot\,>$ will be used to stand for a factor space in a chain-type factor space cane.

Write $F'_1 \triangleq V(\text{people})\ (= F_1)$, $F'_{11} \triangleq V(\text{men})$, $F'_{12} \triangleq V(\text{women})$, $F'_{121} \triangleq V(\text{childbirthable})$, and $F'_{122} \triangleq V(\text{childbirthless})$. From the sets of factors, we can form the following factor spaces:

$$< \text{people} > \triangleq \{X(f)\}_{(f \in F'_1)}, \quad < \text{men} > \triangleq \{X(f)\}_{(f \in F'_{11})},$$
$$< \text{women} > \triangleq \{X(f)\}_{(f \in F'_{12})}, \quad < \text{childbirthable} > \triangleq \{X(f)\}_{(f \in F'_{121})},$$
$$< \text{childbirthless} > \triangleq \{X(f)\}_{(f \in F'_{122})}$$

These factor spaces can form a chain-type factor space cane, shown in Figure 12.

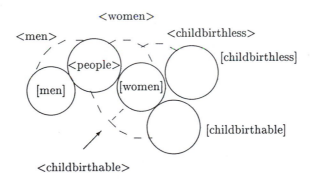

Figure 12   A chain-type factor space cane

In Figure 12, the dashed lines indicate that the rings form one body. Except "initial ring" <people>, every ring itself can not form a factor space, while a ring that is linked with other "predecessor" rings can be a factor space. Clearly, $[\,\cdot\,]$-type factor spaces and $< \cdot >$-type factor spaces have the following relations:

$$< \text{people} >= [\text{people}], \quad < \text{men} >= [\text{people}] \cup [\text{men}],$$
$$< \text{women} >= [\text{people}] \cup [\text{women}],$$
$$< \text{childbirthable} >=< \text{women} > \cup [\text{childbirthable}]$$
$$= [\text{people}] \cup [\text{women}] \cup [\text{childbirthable}]$$
$$< \text{childbirthless} >=< \text{women} > \cup [\text{childbirthless}]$$
$$= [\text{people}] \cup [\text{women}] \cup [\text{childbirthless}] .$$

From these relations, we know that a chain-type factor space cane seems to be a trunk of cactus, where every "path" of it (for example, [people]$\longrightarrow$[women]$\longrightarrow$ [childbirthable]) can form a factor space. There is a characteristic: the more concrete the concepts to be described (for instance, "people"$\longrightarrow$"women"$\longrightarrow$ "childbirthable"), the bigger the capacity (dimension) of the factor spaces describing these concepts (for example, <people>$\subset$<women> $\subset$<childbirthable>$\subset \cdots$).

13.8   Switch Factors and Growth Relation

The forming of a factor space cane depends on a kind of special factor such as $f_0 (= \text{sex})$ and $f_* (=\text{childbirth})$ that we have considered. Such a special factor can be regarded as a shunt switch (see Figure 13 and Figure 14).

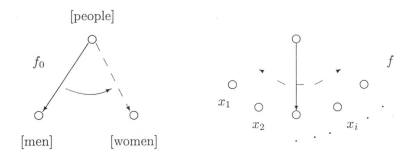

Figure 13 f_0 as a shunt switch   Figure 14   A general switch factor f

As a matter of fact, any factor f can be regarded a switch factor, where each state of it, $x \in X(f)$, is a contact point of the "switch". Of course, in applications, the switch factors with infinite contact points are almost useless. So we always assume that a switch factor has only finite states (contact points).

Let $f =$lifeless and $g = \text{sex}$, must be biological matter, $D(g) \subset D(f)$ and $f(D(g)) = \{\text{life}\}$. From this special case, we have a general growth relation about factors:

For a given left pair $(U, V]$, and for any $f, g \in V$, we call factor f growing factor g, denoted by $f \searrow g$, if $D(f) \supset D(g)$ and there exists $x \in f(D(f))$ such that $f(D(g)) = \{x\}$.

Example 1   Let factor f = clothes kind, and factor g = trousers length. Clearly,

$$X(f) = \{\text{jacket}, \text{trousers}, \text{overcoat}, \text{shirt}, \cdots\}\ ;$$
$$X(g) = \{30, 40, \cdots, 120\}\quad (\text{unit}:\ \text{cm})\ .$$

We take the universe as $U = \{\text{all clothes}\}$. Naturally, $D(f) = U$. It is easy to know that $D(f) \supset D(g) = \{\text{all trousers}\}$ and $f(D(g)) = \{\text{trousers}\}$. Hence $f \searrow g$.

Note 4   The zero factor $\mathbf{0}$ can not be grown by any factor, because $D(\mathbf{0}) = \emptyset$ so that $(\forall f \in V)(f(D(\mathbf{0})) = \emptyset)$. Conversely, it is easy to prove that the zero factor can not grow any factor.

Given a left pair $(U, V]$, an ordered relation $\supseteq$ in V is defined as follows:

$$f \supseteq g \iff f \searrow g \quad \text{or} \quad f = g\ .$$

For the convenience of discussion, we always assume that, for any non-zero factor f, it has at least two states, i.e., $|X(f)| \geq 2$.

Proposition 1 $(V, \supseteq)$ is a partial ordering set.
Proof   The reflexivity holds obviously. We prove the transitivity as follows.
In fact, let $f \supseteq g \supseteq h$. If $f = g$ or $g = h$, naturally $f \supseteq h$ holds. Assume $f \neq g \neq h$. Then $D(h) \subset D(g) \subset D(f)$, and there exists $x \in f(D(f))$, such that $f(D(g)) = \{x\}$. From the note mentioned above, we know $h \neq \mathbf{0}$ so that $D(h) \neq \emptyset$. Thus, $f(D(h)) \neq \emptyset$. But $f(D(h)) \subset f(D(g))$. Therefore $f(D(h)) = \{x\}$, i.e., $f \supseteq h$. This proves the transitivity.
Now we prove the antisymmetry. Let $f, g \in V$ with $f \neq g$. We only need to prove the following expression:

$$f \supseteq \neg(g \supseteq f) \tag{13.49}$$

because it means that the antecedent (premise or condition) of the proposition does not hold so that the consequent (consequence) is always true. In fact, from $f \supseteq g$, we know

$$D(g) \subset D(f) \quad \text{and} \quad (\exists x \in f(D(f)))(f(D(g)) = \{x\})\ .$$

Since $|X(f)| \geq 2$, there exists $y \in X(f)$ with $y \neq x$ such that $(\exists u \in D(f))(f(u) = y)$. Clearly u is not in relation to g, i.e., $R(u, g) = 0$. This means that $D(f) \not\subset D(g)$. Hence, $g \supseteq f$ does not hold.   **Q.E.D.**

Now we make use of the growth relation to define switch factors.
Given a left pair $(U, V]$, a factor $f \in V$ is called a switch factor, if $X(f) = \{x_1, x_2, \cdots, x_n\}$ and there exist $g_i \in V$ $(i = 1, 2, \cdots, n)$, such that

$$f \supseteq g_i,\ f(D(g_i)) = \{x_i\}, \quad i = 1, 2, \cdots, n$$

denoted by $f : \{g_i\}_{(1 \leq i \leq n)}$. When $n = 2$, the switch factor f is called a simple switch factor. The set of all switch factors in V is denoted by W.

Clearly $(W, \supseteq)$ is a partial ordering subset of $(V, \supseteq)$, which plays an important role in the discussion of factor space canes.

13.9   Class Partition and Class Concepts

Given a left pair $(U, V]$, for any switch factor $f \in W$, where $f : \{g_i\}_{(1 \leq i \leq n)}$ and $X = \{x_1, x_2, \cdots, x_n\}$, clearly $f^{-1}(x_i) = \{u \in U \mid f(u) = x_i\}$ $(i = 1, 2, \cdots, n)$ are a partition of $D(f)$, called f-class partition. Every $f^{-1}(x_i)$ is called a x_i-class, and the concept for its extension to be such a class is called a class concept.

Example 2   Let the factor $f = $ lifeless which is a switch factor for $X(f) = \{$life, lifeless$\}$. Thus $f^{-1}($life$)$ and $f^{-1}($lifeless$)$ forms, respectively, "life-class" and "lifeless-class", i.e., biological class and non-biological class.

Note 5   We have pointed out a viewpoint that a factor $f \in V$ can be regarded as a mapping $f : D(f) \longrightarrow X(f)$, and for convenience, f is always extended as

$$f : U \longrightarrow X(f), \quad u \longmapsto \begin{cases} f(u), & u \in D(f) \\ \theta, & u \in U \setminus D(f) , \end{cases}$$

where θ is the empty state. Based on the properties of θ (see [1]), we have

$$X(f) = \{x_1, x_2, \cdots, x_n\} = \{\theta, x_1, x_2, \cdots, x_n\} .$$

If set $x_0 \triangleq \theta$, then $f^{-1}(x_i)$ $(i = 0, 1, \cdots, n)$ also forms a partition of U, where $f^{-1}(x_0) = f^{-1}(\theta)$. In fact, θ-class is useless in the partition of U with respect to f.

According to the above wording, every switch factor $f : \{g_i\}_{(1 \leq i \leq n)}$ decides a partition of U. And the partition decides an equivalent relation, denoted still by f. In other words, a switch factor can be regarded an equivalent relation on U so that we can get a quotient set $U/f = \{[u]_f \mid u \in U\}$, where $[u]_f$ is an equivalent class for u being its representative element. Of course,

$$C \in U/f \iff (\exists i \in \{0, 1, \cdots, n\})(C = f^{-1}(x_i)).$$

If we set $\mathcal{A}(U, W) = \bigcup_{f \in W} U/f$, then we clearly have

$$\mathcal{A}(U, W) = \{C \mid (\exists f \in W, f : \{g_i\}_{(1 \leq i \leq n)})(C = f^{-1}(x_i))\},$$

which is the all classes decided by W, called W-class family.

For any class $C \in \mathcal{A}(U, W)$, write

$$V(C) \triangleq \{f \in V \mid D(f) \supset C\}, \quad E(C) \triangleq \{f \in V \mid C \cap D(f) \neq \emptyset\}.$$

It is easy to prove that the following three expressions hold:

$$V(C) = \bigcap_{u \in C} V(u), \quad E(C) = \bigcup_{u \in C} V(u), \quad V(C) \subset E(C),$$

where $V(u) \triangleq V(\{u\})$. In fact, V is just the family of common factors with respect to the objects in the class C.

In order to analyze V, we first study the structure of a left pair $(U, V]$.

Proposition 2   Let $(U, V]$ be a left pair. Then
1)   For any factor g, if there exists $f \in V$ such that $f \geq g$, then $g \in V$.
2)   For any factor f and g, if $f \in V$, then $f \wedge g \in V$.
3) $f, g \in V \Longrightarrow f - g \in V$.
4) $f, g \in V \Longrightarrow f \vee g \in V$.

Proof   1)   If $g = f$ or $g = \mathbf{0}$, then naturally $g \in V$. If $f > g$, there exists a nonempty set $Y \neq \{\theta\}$, such that $X(f) = X(g) \times Y$. Thus, for any $u \in D(f)$, we have $f(u) = (x, y) \in X(g) \times Y$ which means that u is in relation to g, i.e., $R(u, g) = 1$. Hence $g \in V$.

2)   This is a direct conclusion of 1).

3)   Based on the definition of the difference of two factors: $h = f - g$ if and only if $h \wedge g = \mathbf{0}$ and $(f \wedge g) \vee h = f$, we easily know $f \geq f - g$. Then from 1), we get $f - g \in V$.

4)   By means of $X(f \vee g) = X(f - g) \times X(f \wedge g) \times X(g - f)$, we know that the states of $f \vee g$ are in relation to the states of $f - g$, $f \wedge g$, and $g - f$. Then from the clear fact for the states of $f - g$, $f \wedge g$, and $g - f$ being in relation to the states of f and g, we have $f \vee g \in V$.   **Q.E.D.**

Note 6   The cases 2) and 4) of Proposition 2 can be easily extended into infinite cases as follows:

$$\{f_t\}_{(t \in T)} \subset V \Longrightarrow \bigwedge_{t \in T} f_t, \ \bigvee_{t \in T} f_t \in V \ .$$

Furthermore, if set $\mathbf{1} \triangleq \bigvee_{f \in V} f$ and $(\forall f \in V)(f^c \triangleq \mathbf{1} - f)$, then we have the following corollary.

Corollary $(V, \vee, \wedge, c, \mathbf{1}, \mathbf{0})$ is a complete Boolean algebra.

It is worthy to point out that, for a left pair $(U, V]$, $\bigvee_{t \in T} f_t$ may be meaningless relative to applications for a family $\{f_t\}_{(t \in T)} \subset V$. For example, when $f = $ lifeless and $g = $ sex, from the discussion of Section 5, $f \vee g$ is meaningless. But the "meaningless" in applications may not hinder the formal work in mathematics. However, for the need of applications, we should give the following definition:

A left pair $(U, V]$ is called normal, if it satisfies the condition: For any family of factors $\{f_t\}_{(t \in T)} \subset V$,

$$\bigvee_{t \in T} f_t \text{ being meaningless} \Longrightarrow (\exists s, t \in T)(f_s \searrow f_t). \tag{13.50}$$

We can always assume that any left pair here is normal.
For $V(C)$, it is easy to prove the following proposition.

Proposition 3   For any class $C \in \mathcal{A}(U, W)$, we have
1) $(\forall f \in V(C))(\forall g \in V \setminus \{0\})(f \geq g \Longrightarrow g \in V(C))$.
2) $\{f_t\}_{(t \in T)} \subset V(C) \Longrightarrow \bigwedge\limits_{t \in T} f_t, \bigvee\limits_{t \in T} f_t \in V(C)$.
3) $f, g \in V(C) \Longrightarrow f - g \in V(C)$.

For need, we now introduce a result in set theory:

Given a set G, take a family of subsets of G, $\{G_h\}_{(h \in H)}$, and take a family of subsets of H, $\{H(t)\}_{(t \in H)}$. The following expressions hold:

$$\cup \{G_h \mid h \in \bigcup\limits_{t \in T} H(t)\} = \bigcup\limits_{t \in T} \left(\bigcup\limits_{h \in H(t)} G_h \right), \tag{13.51}$$

$$\cup \{G_h \mid h \in \bigcap\limits_{t \in T} H(t)\} \subset \bigcap\limits_{t \in T} \left(\bigcup\limits_{h \in H(t)} G_h \right), \tag{13.52}$$

$$\cap \{G_h \mid h \in \bigcup\limits_{t \in T} H(t)\} = \bigcap\limits_{t \in T} \left(\bigcap\limits_{h \in H(t)} G_h \right), \tag{13.53}$$

$$\cap \{G_h \mid h \in \bigcap\limits_{t \in T} H(t)\} \supset \bigcup\limits_{t \in T} \left(\bigcap\limits_{h \in H(t)} G_h \right) \supset \bigcap\limits_{t \in T} \left(\bigcap\limits_{h \in H(t)} G_h \right). \tag{13.54}$$

From the above four expressions, we easily know the following result.

Proposition 4   For any family of classes $\{C_t\}_{(t \in T)} \subset \mathcal{A}$, we have:
1) $E\left(\bigcup\limits_{t \in T} C_t \right) = \bigcup\limits_{t \in T} E(C_t)$;
2) $E\left(\bigcap\limits_{t \in T} C_t \right) \subset \bigcap\limits_{t \in T} E(C_t)$;
3) $V\left(\bigcup\limits_{t \in T} C_t \right) = \bigcap\limits_{t \in T} V(C_t)$;
4) $V\left(\bigcap\limits_{t \in T} C_t \right) \supset \bigcup\limits_{t \in T} V(C_t)$.

Note 6   A class $C \in \mathcal{A}(U, W)$ is a subset of U, which is often the extension of a certain concept α. The class regarded as the extension of a concept α may be denoted by C_α. If $C' \in \mathcal{A}(U, W)$ is the extension of another concept β, it should be denoted by C'_β. However, for convenience, C'_β is also denoted by C_β. If C_α is regarded as a whole body, we may not confuse "C" of C_α and "C" of C_β.

Given classes C_α, $C_\beta \in \mathcal{A}(U, W)$, if $C_\alpha \supset C_\beta$, then C_α is called an upper class of C_β, and C_β is called a subclass of C_α. From these, we can call concept α an upper concept of concept β, and call β a subconcept of α.

Proposition 5   For any classes C_α, C_β, $C_\gamma \in \mathcal{A}(U, W)$, if β is a subconcept of α, then
1) $V(C_\alpha) \subset V(C_\beta)$.
2)   If $C_\alpha = C_\beta \cup C_\gamma$, then $V(C_\alpha) = V(C_\beta) \cap V(C_\gamma)$, and satisfies the following condition:

$$(V(C_\beta) \setminus V(C_\alpha)) \cap (V(C_\gamma) \setminus V(C_\alpha)) = \emptyset .$$

Proof   Obviously 1) is true. And 2) is a direct result of the 3) in Proposition 4.
Q.E.D.

The 1) item in this proposition means that the common factors of an upper concept
must be the common factors of its subconcept, conversely it is not true.

Example 3   Let α, β, and γ stand, respectively, for the concepts "people",
"men", and "women". Clearly α is an upper concept of β and γ. If we take factor
$f =$ "age", then f is a common factor of C_α, and also a common factor of C_β and
C_γ. But if we take factor $g =$ "big beard", then g is a common factor of C_β but not
a common factor of C_α. Therefore, factors may appear to be "fractal phenomenon"
(see Figure 10).

Clearly, we may not describe such fractal phenomenon in one factor space, but
we can make use of a different factor space in a factor space cane to deal with a
different class. This is one of the basic tasks of factor space canes.

13.10   Conclusions

First, we presented a neuron mechanism based on factor spaces. That is, a factor
space can be viewed as a neuron with multi-input and single output. We can also
extend this mechanism to form a neural network or a fuzzy neural network. Second,
several models of such neurons were discussed in detail, such as the models without
respect to time, the ones with respect to time, the ones based on Weber-Fechner's
Law, and the ones based on variable weights. At last, in order to constitute more
complicated neurons and neural networks, factor space cane was introduced, and
some important properties for factor space canes were given. Several properties
and propositions indicate that the factor space cane is a useful tool for knowledge
representation.

References

1. B. Kosko, *Neural Networks and Fuzzy Systems*, Prentice-Hall, Englewood Cliffs, 1992.

2. K. Fukushima, Cognitron: A self-organizing multilayeral neural network, *Biological Cybernetics*, Vol. 20, pp. 121-136, 1975.

3. H. X. Li and V. C. Yen, *Fuzzy Sets and Fuzzy Decision–Making*, CRC Press, Boca Raton, 1995.

4. P. Z. Wang and H. X. Li, *Fuzzy Systems Theory and Fuzzy Computer*, Science Press, Beijing, 1995.

5. H. J. Zimmermann, *Fuzzy Sets Theory and Its Applications*, Kluwer Academic Publishers, Hingham, 1984.

6. P. Z. Wang and H. X. Li, *A Mathematical Theory on Knowledge Representation*, Tianjin Scientific and Technical Press, Tianjin, 1994.

7. H. X. Li, and V. C. Yen, Factor spaces and fuzzy decision–making, *Journal of Beijing Normal University*, Vol. 30, No. 1, pp. 15-21, 1994.

Chapter 14

Foundation of Neuro-Fuzzy Systems and an Engineering Application

This chapter discusses the foundation of neuro-fuzzy systems. First, we introduce Takagi, Sugeno, and Kang (TSK) fuzzy model [1,2] and its difference from the Mamdani model. Under the idea of TSK fuzzy model, we discuss a neuro-fuzzy system architecture: Adaptive Network-based Fuzzy Inference System (ANFIS) that is developed by Jang [3]. This model allows the fuzzy systems to learn the parameters adaptively. By using a hybrid learning algorithm, the ANFIS can construct an input-output mapping based on both human knowledge and numerical data. Finally, the ANFIS architecture is employed for an engineering example – an IC fabrication time estimation. The result is compared with other different algorithms: Gauss-Newton-based Levenberg-Marquardt algorithm (GN algorithm), and back-propagation of neural network (BPNN) algorithm. Comparing these two methods, the ANFIS algorithm gives the most accurate prediction result at the expense of the highest computation cost. Besides, the adaptation of fuzzy inference system provides more physical insights for engineers to understand the relationship between the parameters.

14.1 Introduction

During the past decades, we have witnessed a rapid growth of interest and experienced a variety of applications of fuzzy logic and neural networks systems [4-6]. Success in these applications has resulted in the emergence of neuro-fuzzy computing as a major framework for the design and analysis of complex intelligent systems [7,8]. Neuro-fuzzy systems are multi-layer feedforward adaptive networks that realize the basic elements and functions of traditional fuzzy logic systems. Since it has been shown that fuzzy logic systems are universal approximators, neuro-fuzzy control systems, which are isomorphic to traditional fuzzy logic control systems in terms of their functions, are also universal approximators [9,10]. Utilizing their network architectures and associated learning algorithms, neuro-fuzzy systems have been used successfully for modeling and controlling various complex systems.

Currently, several neuro-fuzzy networks exist in the literature. Most notable are Adaptive Network-based Fuzzy Inference System (ANFIS) developed by Jang [3], Fuzzy Adaptive Learning Control Network (FALCON) developed by Lin and Lee [11], NEuro-Fuzzy CONtrol (NEFCON) proposed by Nauck, Klawonn, and Kruse [12], GARIC developed by Berenji [13], and other variations from these developments [7,8]. These neuro-fuzzy structures and systems establish the foundation of neuro-fuzzy computing. Most neuro-fuzzy systems are developed based on the concept of neural methods on fuzzy systems. The idea is to learn the shape of membership functions for the fuzzy system efficiently by taking the advantage of adaptive property of the neural methods. Takagi, Sugeno, and Kang [1,2] are known as the first to utilize this approach. Later, Jang [3] elaborated upon this idea and developed a systematic approach for the adaptation with illustrations of several successful applications. In this chapter, we start with the Sugeno and Takagi fuzzy model followed by the Adaptive Network-based Fuzzy Inference System (ANFIS) proposed by Jang. Finally, we present an engineering application example that shows the effectiveness of the ANFIS and comparison with other learning approaches.

14.2 Takagi, Sugeno, and Kang Fuzzy Model

The Takagi, Sugeno, and Kang [1,2] (TSK) fuzzy model was known as the first fuzzy model that was developed to generate fuzzy rules from a given input-output data set. A typical fuzzy rule in their model has the form

$$\text{if } x \text{ is } A \text{ and } y \text{ is } B \text{ then } z = f(x,y),$$

where A and B are fuzzy sets in the antecedent, while $z = f(x,y)$ is a crisp function in the consequent. Usually $z = f(x,y)$ is a polynomial in the input variables x and y. It can be any function that describes the output of the model within the fuzzy region specified by the antecedent of the rule. The function can be a zero-degree Sugeno model in which the example, a two-input single-output model, has "If x is small and y is small then $z = 3$". If the function is an one-degree of polynomial, then it is called a first-order Sugeno model. For example, "If x is small and y is small then $z = -x + 2y + 4$". The zero-order Sugeno model is known as a special case of Mamdani fuzzy inference system. Note that only the antecedent part of the TSK model has the "fuzzyness", the consequent part is a crisp function.

In the TSK fuzzy model, the output is obtained through weighted average of consequents. This gives us a "smooth" effect and it is a nature and efficient gain scheduler [3]. This effect avoids the time-consuming process of defuzzification in a Mamdani model. Also, the TSK model cannot follow the compositional rule of inference (CRI) that exists in Mamdani fuzzy reasoning mechanism. The advantages of the TSK model are [3]: (1) it represents computational efficiency; (2) it works very well with linear techniques and optimization and adaptive techniques; (3) it guarantees continuity of the output surface; and (4) it is better suited for mathematical analysis. On the other hand, the Mamdani model is more intuitive and is better suited for human input.

14.3 Adaptive Network-based Fuzzy Inference System (ANFIS)

ANFIS, developed by Jang [3], is an extension of the TSK fuzzy model. This model allows the fuzzy systems to learn the parameters using adaptive backpropagation learning algorithm. In general ANFIS is much more complicated than fuzzy inference systems. For simplicity, in the following discussion we assume that the fuzzy inference system under consideration has four inputs x_1, x_2, x_3, and x_4 and each input has two linguistic terms, for example, $\{A_1, A_2\}$ for input x_1. Therefore, there are 16 fuzzy if-then rules. In implementing this algorithm, a five-layer network has to be constructed as shown in Figure 1. Figure 1(a) illustrates the reasoning mechanism for this TSK model (only one rule has been shown in this figure) and Figure 1(b) is the corresponding ANFIS architecture. Here we will give a brief explanation for the reasoning mechanism. For more detailed information, please refer to [1].

Without loss of generality, we use the first rule as an example. For a first order, four inputs TSK fuzzy model, the common fuzzy if-then rule has the following type:

$$\textit{Rule 1 :If } x_1 \textit{ is } A_1 \textit{ and } x_2 \textit{ is } B_1 \textit{ and } x_3 \textit{ is } C_1 \textit{ and } x_3 \textit{ is } D_1,$$
$$\textit{then } f_1 = p^1{}_1 x_1 + p^1{}_2 x_2 + p^1{}_3 x_3 + p^1{}_4 x_4 + p^1{}_5 ,$$

where superscript denotes the rule number.

We denote the output of the ith node in layer k as $O_{k,i}$. Every node in Layer1 can be any parameterized membership function, such as the generalized bell shape function:

$$\mu_A(x) = \frac{1}{1 + \left| \frac{x - c_i}{a_i} \right|^{2b_i}} \tag{14.1}$$

where a_i, b_i, and c_i are called "premise parameters", which are nonlinear coefficients. The function of the fixed node in Layer2 is to output the product of all inputs, and the output stands for the firing strength of the corresponding rule.

$$O_{2,i} = w_i = \mu_{A_j}(x_1)\mu_{B_j}(x_2) \cdots \mu_{D_j}(x_4), \quad i = 1, 2, \cdots, 16; \quad j = 1, 2 . \tag{14.2}$$

In Layer3, the function of the fixed node is used to normalize the input firing strengths.

$$O_{3,i} = \overline{w}_i = \frac{w_i}{\sum_{i=1}^{16} w_i}, \quad i = 1, 2, \cdots, 16 . \tag{14.3}$$

Every node in Layer4 is a parameterized function, and the adaptive parameters are called "consequent parameters". The node function is given by:

$$O_{4,i} = \overline{w}_i f_i = \overline{w}_i (p^i{}_1 x_1 + p^i{}_2 x_2 + p^i{}_3 x_3 + p^i{}_4 x_4 + p^i{}_5), \quad i = 1, 2, \cdots, 16 . \tag{14.4}$$

There is only one node in Layer5 with a simple summing function.

$$O_{5,1} = \sum_i \overline{w}_i f_i \tag{14.5}$$

Thus, the ANFIS network is constructed according to the TSK fuzzy model. This ANFIS architecture can then update its parameters according to the backpropagation algorithm.

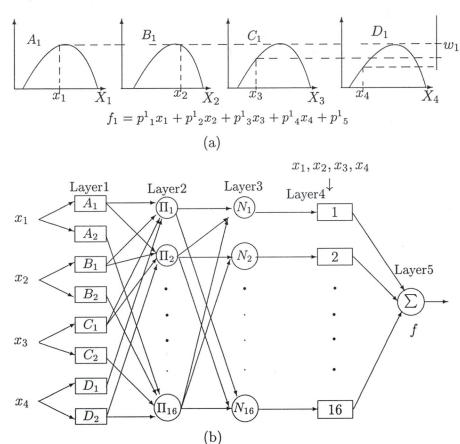

$$f_1 = p^1{}_1 x_1 + p^1{}_2 x_2 + p^1{}_3 x_3 + p^1{}_4 x_4 + p^1{}_5$$

(a)

(b)

Figure 1   A four-input ANFIS network

14.4 Hybrid Learning Algorithm for ANFIS

From Equations (14.4) and (14.5), we notice that given the values of premise parameters, the overall output is a linear combination of the consequence parameters. More precisely, rewrite Equation (14.4), and the output f in Figure 1(b) is:

$$
\begin{aligned}
f &= \sum_i \overline{w}_i f_i \\
&= \sum_i (\overline{w}_i x_1) p^i{}_1 + \sum_i (\overline{w}_i x_2) p^i{}_2 + \sum_i (\overline{w}_i x_3) p^i{}_3 + \sum_i (\overline{w}_i x_4) p^i{}_4 + \sum_i (\overline{w}_i) p^i{}_5 .
\end{aligned}
\tag{14.6}
$$

The learning can be divided to forward and backward passes. As indicated in Table 1, the forward pass of the learning algorithm stop at nodes at Layer4 and the consequent parameters are identified by the least squares method. In the backward pass, the error signals propagate backward and the premise parameters are undated by gradient descent.

Table 1 Hybrid Learning Algorithm for ANFIS

	Forward Pass	Backward Pass
Premise parameters	Fixed	Gradient descent
Consequent parameters	Least-squares	Fixed
Signals	Node outputs	Error signals

As noted by Jang [1], if the membership function is fixed and only the consequent part is adjusted, the ANFIS can be viewed as a functional-link network discussed in Chapter 5, where the enhancement node functions of the input variables are achieved by the membership functions. The only difference here is that the enhancement function in this fixed ANFIS takes advantage of the human knowledge, which is more revealing of the problem, than function expansion or random generated functions.

14.5 Estimation of Lot Processing Time in an IC Fabrication

Strong demand for IC products continually forces the fab managers to increase equipment utilization and productivity. The performance of on-time delivery is a key index to track both productivity improvement and customer service level in a fab. In order to remain competitive, it is necessary for manufacturing managers to achieve higher on-time delivery and meet monthly wafer out target. Not only on-time delivery performance but also maximum utilization of tools can be accomplished by constructing an overall cycle time model of a fab.

Cycle time plays an important role in an IC fab. It contains processing time and waiting time. Intuitively, the processing time should be deterministic. It is determined only by the quantity of wafers per lot, the tool type, and the recipe of the lot. In reality, at least two difficulties should be overcome. First, the modern IC fab is rather complex. It contains hundreds of machines, process steps and is highly re-entrant [21,25,27]; i.e., as wafers move from step-to-step, they often return to a given tool. As a consequence, it is almost impossible to construct all the matches of tool, recipe, etc. Second, the processing time includes in-process waiting time, which is influenced by many parameters, such as quantity of lots in fab, priority class of a lot, technology, stage group, etc. In other words, it is not easy to forecast remaining cycle time precisely under the condition of complex multi-product and various processes.

Ehteshami et al. [17] committed to delivery dates for a product by using historical average cycle time and adding some safety margin. He designed a safety margin to compensate for variability in the cycle time and to meet the desired service level. This is one kind of well-known SLACK policy. Several issues based on SLACK policy were also addressed in [16, 20, 22]. Adl et al. [14] proposed a hierarchical modeling and control technique. A low-level tracking policy was presented and integrated with a high-level state variable feedback policy. The low-level tracking policy was able to track low-frequency commands generated by the high-level controller. Such issues were also addressed in [18,25-27]. Raddon et al. [24] developed a model for forecasting throughput time. Another approach which has been addressed by many researchers is flow model [15,21]. Flow models predict the fab's behavior over a long time scale. Sattler [22] used queueing curve approximation in a fab to determine productivity improvements.

We use ANFIS algorithm as well as other algorithms to construct the processing time model of a single tool by using the historical measured data sets. This approach is reasonable under the assumption that the processing time pattern of tools will be stationary or not varying too much during a period of time. Similarly, the waiting time model of a tool can be obtained by this approach but with more parameters and complications.

After all the tool models are obtained, we can construct an overall cycle time model for a fab by combining all the individual tool models. It should be noted that here we focus on the interpretation of the development of one specific tool in a real IC fab by the three algorithms. In the future, a product cycle time estimation mechanism could be constructed in accordance with this tool model concept.

A tool model can be regarded as a highly nonlinear system, i.e., the relationship between the output and several inputs of the tool model is highly nonlinear. Therefore, the processing time model of the tool is also nonlinear in general. To solve the modeling problem previously described, three mathematical techniques are adopted to construct the processing time model of the tool. In this section, the model architecture and algorithms will be presented. The results and comparison of these three approaches will be described in the next section.

14.5.1 Algorithm 1: Gauss-Newton-based Levenberg-Marquardt Method

The Gauss-Newton-based Levenberg-Marquardt method (GN-based LM method) [19, 23] becomes popular in dealing with the multi-input nonlinear modeling problem. The operation principle is explained below (refer to Figure 2). At first, a specified explicit function ($y = f(x, \diamond)$) is expanded to obtain a linear approximation with some unknown coefficients. Through iterative calculation (gradient descent) about the expansion center, the desired model ($\mathbf{B}$, coefficient matrix) is obtained. While coding the system, several parameters, including input/output data pairs (x, t), the explicit function (y), and initial coefficient matrix ($\mathbf{B_0}$) must be given in advance.

In Figure 2 where x is the input vector of size t; y is the model's scalar output; $\diamond$ is the parameter vector of size m; $E(\diamond)$ is the sum of squared errors; $r(\diamond)$ is the difference between t_p and y; $\mathbf{J}$ is the Jacobian matrix of r; λ is a real value; $\mathbf{D}$ is a diagonal matrix, which can be determined by

$$\mathbf{D} = \text{diag}(J^T J) + I$$

Model: $Y = \mathbf{xB}$

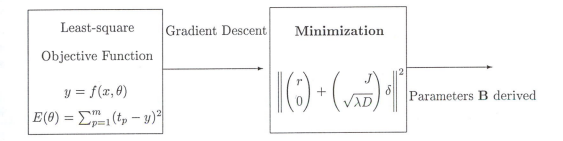

Figure 2   Gauss-Newton-based Levenberg-Marquardt method

14.5.2 Algorithm 2: Backpropagation Neural Network

In the recent decades, the backpropagation neural network (BPNN) has been used extensively and proven useful in several engineering fields. As shown in Figure 3, this technique can also be used to construct the processing time model. The input layer is responsible for receiving input data, and the number of input nodes depends on the number of inputs.

The hidden layer serves as the central part in neural network, and there are no standard ways to decide neither the node number nor the layer number. The node transfer function also varies case by case, and sigmoid function is a popular nonlinear transfer function. The output layer is used to output the result.

In the BPNN, the link of nodes between two layers is characterized by a weighting value. To model an unknown system by BPNN, the weighting value must be extracted from the well-trained neural network.

14.5.3 Algorithm 3: ANFIS Algorithm

ANFIS architecture is an TSK-based fuzzy model whose parameters can be identified by using the hybrid learning algorithm. This model is discussed in the previous section.

Output vector

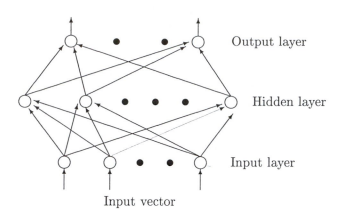

Figure 3   A back propagation neural network

14.5.4 Simulation Result

As mentioned earlier, the problem encountered here is to construct the processing time model of a specific tool by using 1520 actual data pairs acquired from an actual fab. The processing time model of the tool consists of 6 inputs and 1 output. The input parameters should be determined by expert engineers, and they are wafer quantity, priority class, technology group, route, stage number, and stage group number, respectively. In fact, the choice of input parameters is not straightforward, instead it is usually determined heuristically or by using fuzzy inference system. Of course, the output parameter is the processing time of the tool.

The distribution of raw data is described as follows. The inputs range from 0 to 30 except the stage number, which ranges from 0 to 5515. The output data range from 0 to 7279. Since there is significant variation on the values of input and output data, preprocessing of the data is required.

To obtain a good model, it is important to preprocess the raw data. Since the range of the six input data value is very different, all the input values are transformed linearly into a common range. Among those data sets, 1360 sets of data pairs are used as learning data to construct the model, while the remaining 160 sets are selected as the test data. The test output is then compared with the actual output. Finally, the percentage error of individual compared result and the mean absolute percentage error (APE) are calculated.

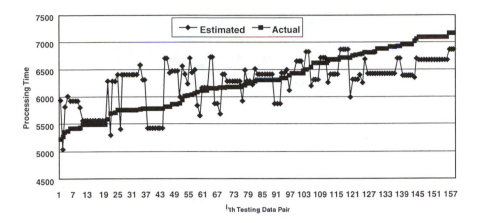

Figure 4 Result of Gauss-Newton-based LM model

14.5.4.1 Gauss-Newton-based LM Model Construction

If the explicit function is specified as the first order equation, $\mathbf{Y}=\mathbf{X}*\mathbf{B}$ ($\mathbf{X}$ is the input vector, $\mathbf{Y}$ is the output, and $\mathbf{B}$ is the unknown coefficient), then the model coefficient $\mathbf{B}=[139.85; 386.5; 39.47; 333.33; -152.68; 89.59]$, and APE=8.7%. If the second order explicit function is specified as $\mathbf{Y}=[\ \mathbf{X}*\mathbf{X},\ \mathbf{X},\ 1]^{*}\ \mathbf{B}$, then the model coefficient $\mathbf{B} = [-0.1;\ 19.7;\ -1.4;\ -12.1;\ 6.7;\ -10;\ 18.4;\ -214.9;\ 37.8;\ 195.1;\ -11.4;\ -90.9;\ 5857]$, and APE=7.8%, as shown in Figure 4. Note that the comparison data have been sorted by the actual output.

14.5.4.2 BP Neural Network Model Construction

Input data preprocessing is also done as the previous case. In this case the training data consists of 1360 data sets, which are extracted randomly from the raw data, and the test data are the remaining 160 sets of data. The BPNN algorithm is implemented in a neural network with the following specified parameters: 6 input nodes, 6 hidden nodes, 1 output node, 1 hidden layer, Delta learn rule, and sigmoid-type transfer function The test output is compared with the actual output, as shown in Figure 5, and APE is also calculated (APE=9.31%).

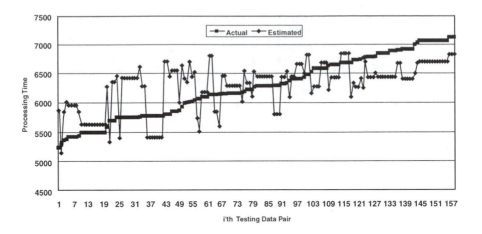

Figure 5 Result of BPNN

14.5.4.3 ANFIS Model Construction

By examining the input-data dependence for the output, we found that two inputs
are relatively unimportant in this case because the output shows little dependence on
them. In order to accelerate the computing speed, four key parameters are chosen as
inputs to implement this algorithm. The number of the training and test pairs is the
same as the previous case. Other design parameters are described as follows: The
number of nodes: Layer1 has 8 nodes; Layer2 has 16 nodes; Layer3 has 16 nodes;
Layer4 has 16 nodes and Layer5 has 1 node. In short, it has 57 nodes. The number
of fuzzy if-then rules: Since each of the four input variables has two linguistic terms,
the maximum number of fuzzy rules is then equal to 2*2*2*2=16. The number
of linear parameters: Linear parameters are consequent parameters. There are 5
linear parameters for each rule, hence 80 for the whole system. The number of
nonlinear parameters: Nonlinear parameters are premise parameters. The term
set of this ANFIS is $\{A_1, A_2, B_1, B_2, C_1, C_2, D_1, D_2\}$. Since each term has 3 premise
parameters, there are 24 nonlinear parameters. Those parameters can be determined
by the hybrid learning algorithm. Upon finding all parameters, the processing time
model is obtained. Based on this model, the test error can be calculated. The APE
of test error is 3.44%, and comparison result between the estimated output and the
actual output is shown in Figure 6.

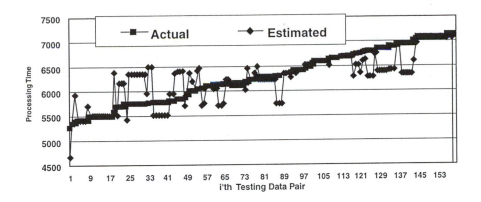

Figure 6 Comparison result of ANFIS

The absolute percentage errors (APE) as derived from the above three algorithms are listed in Table 2. Note that ANFIS has the highest accuracy but it takes more computational effort. On the contrary, BPNN can gain the result rapidly but the APE is up to 10%.

Table 2 Modeling Results of Three Different Algorithms

	APE of the Test Error(%)	Training Data Size	Test Data Size
GN(first-order)	8.66	1360	160
GN(second-order)	5.88	1360	160
BPNN	9.31	1360	160
ANFIS	3.44	1360	160

14.5 Conclusions

This chapter addressed the foundation of neuro-fuzzy systems. We discussed the TSK model first and then the Adaptive Network-based Fuzzy Inference System (ANFIS). ANFIS architecture is a TSK-based fuzzy model whose parameters can be identified by using the hybrid learning algorithm. We also discussed an engineering application using ANFIS and other learning approaches. Analysis and comparison of them are discussed. In summary, ANFIS algorithm gives a more accurate prediction result at the expense of the highest computation time and can provide a more meaningful relationship between inputs and output variables in terms of fuzzy if-then rules. On the contrary, the BPNN algorithm takes less computation effort. Due to the lack of tremendous actual historical data, we did not attempt to construct a whole cycle time estimator by this approach in this chapter. In the future, a cycle time estimator can be obtained by constructing all the tool models for both processing time and waiting time. The estimated cycle time of the IC product is the sum of the time estimated by all individual tool models. The relationship is given by:

$$CT = \sum_i^N CT_i \ , \qquad (14.7)$$

where CT_i : Cycle time of the i-th tool model; N : The total number of steps of a specific lot; CT : Overall cycle time. Using the system model in Figure 7 and making a comparison between estimated and scheduled cycle time, engineers or developed mechanisms can decide the urgency of each lot and then adjust the priority dynamically.

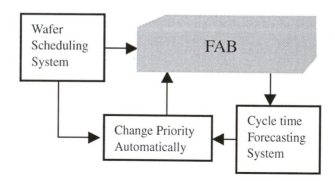

Figure 7 Architecture of the closed-loop cycle time forecasting system
based on tool models

References

1. T. Takagi and M. Sugeno, Fuzzy identification of systems and its applications to modeling and control, *IEEE Transactions on Systems, Man, and Cybernetics*, Vol. 15, pp. 116-132, 1985.

2. M. Sugeno and G. T. Kang, Structure identification of fuzzy model, *Fuzzy Sets and Systems,* Vol. 28, pp. 15-33, 1988.

3. J. S. Jang, ANFIS: Adaptive-network-based fuzzy inference system, *IEEE Transactions on Systems, Man, and Cybernetics*, Vol. 23, No. 3, pp. 665-684, 1993.

4. B. P. Graham and R. B. Newell, Fuzzy adaptive control of a first order process, *Fuzzy Sets and Systems*, Vol. 31, pp. 47-65, 1989.

5. Kazuo Ishii, Teruo Fujii, and Tamaki Ura, Neural network system for on-line controller adaptation and its application to underwater robot, *Proceedings of 1998 IEEE International Conference on Robotics and Automation,* Leuven, Belgium, May 16-20, pp. 756-761, 1998.

6. C. S. G. Lee, Neuro-fuzzy systems for robotics, in *Handbook of Industrial Robotics,* S. Nof, ed., John Wiley, New York, 1998.

7. C. T. Lin and C. S. G. Lee, *Neural Fuzzy Systems*, Prentice-Hall, Englewood Cliffs, 1996.

8. J. S. Jang, C. T. Sun, and E. Mizutani, *Neuro-Fuzzy and Soft Computing,* Prentice-Hall, Englewood Cliffs, 1997.

9. L. X. Wang and J. M. Mendel, Fuzzy basis functions, universal approximation, and orthogonal least-squares learning, *IEEE Transactions on Neural Networks,* Vol. 3, No. 5, pp. 807-814, 1992.

10. J. S. R. Jang and C. T. Sun, Functional equivalence between radial basis function networks and fuzzy inference systems, *IEEE Transactions on Neural Networks,* Vol. 4, No. 1, pp. 156-159, 1993.

11. C. T. Lin and C. S. G. Lee, Neural-network-based fuzzy logic control and decision system, *IEEE Transactions on Computers,* Vol. 40, No. 12, pp. 1320-1336, 1991.

12. D. Nauck, F. Klawonn, and R. Kruse *Foundations of Neuro-Fuzzy Systems* Wiley, New York, 1997.

13. H. R. Berenji and P. Khedkar, Learning and tuning fuzzy logic controllers through reinforcements, *IEEE Transactions on Neural Networks,* Vol. 3, No. 5, pp. 724-740, 1992.

14. M. K. Adl, et al., Hierarchical modeling and control of re-entrant semiconductor manufacturing facilities, *Proceedings of the 35th Conference on Decision and Control*, Kobe, Japan, 1996.

15. R. Alvarez-Vrgas, A study of the continuous flow model of production lines with unreliable machines and finite buffers, *Journal of Manufacturing Systems*, Vol. 13, No. 3, pp. 221-234, 1994.

16. R. W. Conway, et al., *Theory of Scheduling*, Addison-Wesley, Reading, MA, 1967.

17. B. Ehteshami, et al., Trade-off in cycle time management: Hot lots, *IEEE Transactions on Semiconductor Manufacturing*, Vol. 5, No. 2, pp. 101-106, 1992.

18. S. Gershwin, *Manufacturing Systems Engineering*, Prentice-Hall, Englewood Cliffs, 1994.

19. H. O. Hartley, The modified Gauss-Newton method for the fitting of nonlinear regression function by least squares, *Technometrics*, 3, pp. 269-280, 1961.

20. C. H. Hung, et al., Managing on time delivery during foundry fab production ramp-up, *National Conference on Semiconductor*, pp. 18-34, 1996.

21. A. R. Kumar, Re-entrant lines, *Technical report*, Coordinated Systems Laboratory, University of Illinois, Urbana, 1994.

22. C. H. Lu, et al., Efficient scheduling policies to reduce mean and variance of cycle-time in semiconductor manufacturing plants, *IEEE Transactions on Semiconductor Manufacturing*, Vol. 7, No. 3, pp. 374-388, 1994.

23. D. W. Marquardt, An algorithm for least squares estimation of nonlinear parameters, *Journal of the Society of Industrial and Applied Mathematics*, 11: pp. 431-441, 1963.

24. A. Raddon and B. Grigsby, Throughput time forecasting model, *IEEE/SEMI Advanced Semiconductor Manufacturing Conference*, pp. 430-433, 1997.

25. A. A. Rodriguez and M. Kawski, *Modeling and Robust Control of Re-entrant Semiconductor Fabrication Facilities: Design of Low-Level Decision policies*, Proposal to Intel Research Council, 1994.

26. L. Sattler, Using queueing curve approximation in a fab to determine productivity improvements, *IEEE/SEMI Advanced Semiconductor Manufacturing Conference*. Texas Instruments, Dallas, 1996.

27. K. S. Tsakalis, et al., Hierarchical modeling and control for re-entrant semiconductor fabrication lines: A mini-fab benchmark, *IEEE/SEMI Advanced Semiconductor Manufacturing Conference*, pp. 508-513, 1997.

Chapter 15

Data Preprocessing

Data preprocessing converts raw data and signals into data representation suitable for application through a sequence of operations. The objectives of data preprocessing include size reduction of the input space, smoother relationships, data normalization, noise reduction, and feature extraction. Several data preprocessing algorithms, such as data values averaging, input space reduction, and data normalization, will be briefly discussed in this chapter. Computer programs for data preprocessing are also provided.

15.1   Introduction

A pattern is an entity to represent an abstract concept or a physical object. It may contain several attributes (features) to characterize an object. Data preprocessing is to remove the irrelevant information and extract key features of the data to simplify a pattern recognition problem without throwing away any important information. It is crucial to the success of fuzzy modeling and neural network processing and when the quantity of available data is a limiting factor. In fact, data preprocessing converts raw data and signals into data representation suitable for application through a sequence of operations. It can simplify the relationship inferred by a model. Though preprocessing is an important role, the development of an effective preprocessing algorithm usually involves a combination of problem-specific knowledge and iterative experiments. In particular, the process is very time consuming and the quality of the preprocessing may vary from case to case.

The objectives of data preprocessing have five folds [3]: size reduction of the input space, smoother relationships, data normalization, noise reduction, and feature extraction.

Size Reduction of the Input Space: Reducing the number of input variables or the size of the input space are a common goal of the preprocessing. The objective is to get a reasonable generalization with a lower dimensionality of the data set without losing the most significant relationship of the data. If the input space is large, one may identify the most important input variables and eliminate the insignificant or independent variables by combining several variables as a single variable. This approach can reduce the number of inputs and the input variances, and therefore improve results if there are only limited data.

Smoother Relationships: Another commonly used type of preprocessing is problem transformation. The original problem is transformed into a simpler problem. It means that the associated mappings become smoother. The transformations can be obtained from intuition about the problem.

Normalization: For many practical problems, the units used to measure each of the input variables can skew the data and make the range of values along some axes much larger than others. This results in unnecessarily complex relationships by making the nature of the mapping along some dimensions much different from others. This difficulty can be circumvented by normalizing (or scaling) each of the input variables so that the variance of each variable is equal. A large value input can dominate the input effect and influence the model accuracy of the fuzzy learning system or the neural network system. Data scaling depends on the data distribution.

Noise Reduction: A sequence of data may involve useful data, noisy data, and inconsistent data. Preprocessing may reduce the noisy and inconsistent data. The data corrupted with noise can be recovered with preprocessing techniques.

Feature Extraction: The input data is a pattern in per se. If the key attributes or features characterizing the data can be extracted, the problem encountered can be easily solved. However, feature extractions are usually dependent upon the domain-specific knowledge.

15.2   Data Preprocessing Algorithms

For data preprocessing, the original raw data used by the preprocessor is denoted as a *raw input vector*. The transformed data output produced by the preprocessor is termed a *preprocessed input vector or feature vector* . The block diagram of the data preprocessing is shown in Figure 1.

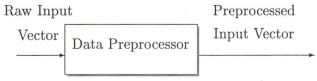

Figure 1   Block diagram of data preprocessing

In general, problem-specific knowledge and generic dimensionality reduction techniques will be used to construct the preprocessor. A better preprocessing algorithm can be arrived at by exercising several different forms of preprocessing techniques. Several data preprocessing algorithms are discussed below [1-6].

15.2.1   Data Values Averaging

Since the averaging effect can reduce the data sensitivity with respect to fluctuation, a noisy data set can be enhanced by taking average of the data. In a time series analysis, the moving average method can be adopted for filtering the small data fluctuation. Note that root-mean-square error between an average and the true mean will decrease at a rate of $\frac{1}{\sqrt{N}}$ for noisy data with a standard deviation of σ. The averaging result of noisy data is given in Figure 2.

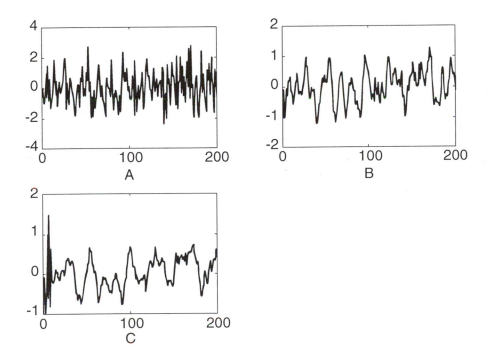

Figure 2   Noise reduction averaging (see MATLAB program in Appendix) A. Original noisy time-series data (the mean is 0; the std is 1), B. Time-series data averaged over a window of 5 data points, C. Time-series data averaged over a window of 10 data points.

15.2.2   Input Space Reduction

If the available data is not rich enough compared to the size of the input variables, the input space reduction should be employed. Several algorithms can be applied.

Principal Component Analysis (PCA)

PCA is used to determine a m-dimensional "most significant" subspace from the n-dimensional input space. Then the data is projected onto this m-dimensional subspace. Therefore, the number of input variables can be reduced from n to m. We shall discuss the PCA Theorem [2] next.

For a given data set (a training set) $T_{tra} = \mathbf{x} = \{\mathbf{x}^1, \mathbf{x}^2, \cdots, \mathbf{x}^N\}$, containing N n-dimensional zero-mean randomly generated patterns (i.e., $\mathbf{x}^i \in \mathcal{R}^n, i = 1, \cdots, N$) with real-valued elements, let $\mathbf{R}_x \in \mathcal{R}^{n \times n}$ be a symmetric, real-valued $n \times n$ co-variance matrix. Let the eigenvalues of the covariance matrix $\mathbf{R}_x$ be arranged in the decreasing order $\lambda_1 \geq \lambda_2 \geq \cdots \lambda_n \geq 0$ (with $\lambda_1 = \lambda_{max}$). Assume that the corresponding orthonormal eigenvectors (orthogonal with unit length $\|e\| = 1$) $\mathbf{e}^1, \mathbf{e}^2, \cdots, \mathbf{e}^n$ consist of the $n \times n$ orthonormal matrix

$$\mathbf{E} = [\mathbf{e}^1, \mathbf{e}^2, \cdots, \mathbf{e}^n] , \tag{15.1}$$

with columns being orthonormal eigenvectors. Then the optimal linear transformation

$$\mathbf{y} = \overset{\wedge}{\mathbf{W}} \mathbf{x} \tag{15.2}$$

transforms the original n-dimensional patterns $\mathbf{x}$ into m-dimensional ($m \leq n$) feature patterns by minimizing the mean least square reconstruction error,

$$J(F, T_{traz}) = \frac{1}{2} \sum_i^N \|\mathbf{x}^i - \overset{\wedge}{\mathbf{x}}^i \| = \frac{1}{2} \sum_i^N \sum_j^n (x_j^i - \hat{x}_j^i)^2.$$

The $m \times n$ optimal transformation matrix $\overset{\wedge}{\mathbf{W}}$, (under the constraints $\overset{\wedge}{\mathbf{W}} \overset{\wedge}{\mathbf{W}}^T = \mathbf{I}$), is given by

$$\overset{\wedge}{\mathbf{W}} = \begin{bmatrix} \mathbf{e}^1 \\ \mathbf{e}^2 \\ \cdots \\ \mathbf{e}^m \end{bmatrix} , \tag{15.3}$$

where the m rows are composed of the first m orthonormal eigenvectors of the original data covariance matrix $\mathbf{R}_x$.

Remarks

(1) The n-dimension mean vector is

$$\mu = E[\mathbf{x}] = [E[\mathbf{x}^1], E[\mathbf{x}^2], \dots, E[\mathbf{x}^n]]^T \tag{15.4}$$

and the square $n \times n$ dimensional covariance matrix is

$$\mathbf{R}_x = \sum = E[(\mathbf{x} - \mu)(\mathbf{x} - \mu)^T] \tag{15.5}$$

where $E[.]$ is the expectation operator, and μ is the mean vector of patterns $\mathbf{x}$.

(2) The square, semipositive definite, symmetric, real-valued covariance matrix $\mathbf{R}_x$ describes the correlations between the elements of pattern vectors (treated as random variables).

(3) The original data patterns are assumed to be zero-mean random vectors

$$\mu = E[\mathbf{x}] = \mathbf{0} \ . \tag{15.6}$$

If this condition is not satisfied, the original pattern $\mathbf{x}$ can be converted into the zero mean representation by the operation $\mathbf{x} - \mu$. For zero mean patterns, the covariance (autocorrelation) matrix is defined as

$$\mathbf{R}_x = \sum = E[\mathbf{x}\mathbf{x}^T] \ . \tag{15.7}$$

(4) Since the exact probability distribution of the patterns is not known, the true values μ and R_x (the mean vectors and the covariance matrix) are usually not available. The given data set T_{tra} contains a finite number of N patterns $\{\mathbf{x}^1, \mathbf{x}^2, \cdots, \mathbf{x}^N\}$. Therefore, the mean can be estimated by

$$\overset{\wedge}{\mu} = \frac{1}{N} \sum_{i=1}^{N} \mathbf{x}^i \tag{15.8}$$

and the covariance matrix (unbiased estimate) by

$$\overset{\wedge}{\mathbf{R}}_x = \frac{1}{N-1} \sum_{i=1}^{N} (\mathbf{x}^i - \overset{\wedge}{\mu})(\mathbf{x}^i - \mu)^T \tag{15.9}$$

based on a given limited sample.

For zero-mean data, the covariance estimate becomes

$$\overset{\wedge}{\mathbf{R}}_x = \frac{1}{N-1} \sum_{i=1}^{N} \mathbf{x}^i (\mathbf{x}^i)^T = \frac{1}{N-1} \mathbf{x}^T \mathbf{x} \ , \tag{15.10}$$

where $\mathbf{x}$ is a whole $N \times n$ original data pattern matrix.

The n eigenvalues λ_i and the corresponding eigenvectors $\mathbf{e}^i$ can be solved by

$$\mathbf{R}_x \mathbf{e}^i = \lambda_i \mathbf{e}^i, i = 1, 2, \cdots, n \ , \tag{15.11}$$

The orthonormal eigenvectors are considered since the covariance matrix $\mathbf{R}_x$ is symmetric and real-valued. In other words, the eigenvectors are orthogonal $(\mathbf{e}^i)^T \mathbf{e}^j = 0$ $(i, j = 1, 2, \cdots, n, i \neq j)$ with unit length.

(5) The eigenvalues and corresponding eigenvectors must be in descending order since only the first m dominant eigenvalues will be considered as performing the dimensionality reduction.

Eliminating Correlated Input Variables

The input space reduction can also be achieved by removing highly correlated input variables. The correlation among input variables can be examined through statistical correlation tests (e.g., s-test) and visual inspection. All highly correlated variables should be eliminated except one. Unimportant input variables can also be eliminated.

Combining Noncorrelated Input Variables

Several dependent input variables can be combined to form a single input variable. Therefore, the input space and the complexity of the system modeling can be reduced.

15.2.3   Data Normalization (Data Scaling)

Data normalization can provide a better modeling and avoid numerical problems. Several algorithms can be used to normalize the data.

Min-Max Normalization

Min-max normalization is a linear scaling algorithm. It transforms the original input range into a new data range (typically 0-1). It is given as

$$y_{new} = \left(\frac{y_{old} - min1}{max1 - min1} \right)(max2 - min2) + min2 \ , \qquad (15.12)$$

where y_{old} is the old value, y_{new} is the new value, $min1$ and $max1$ are the minimum and maximum of the original data range, and $min2$ and $max2$ are the minimum and maximum of the new data range.

Since the min-max normalization is a linear transformation, it can preserve all relationships of the data values exactly, as shown in Figure 3. The two diagrams in Figure 3 resemble to each other except the scaling on y-axis.

Zscore Normalization

In Zscore normalization, the input variable data is converted into zero mean and unit variance. The mean and standard deviation of the input data should be calculated first. The algorithm is shown below

$$y_{new} = \frac{y_{old} - mean}{std} \ , \qquad (15.13)$$

where y_{old} is the original value, y_{new} is the new value, and $mean$ and std are the mean and standard deviation of the original data range, respectively.

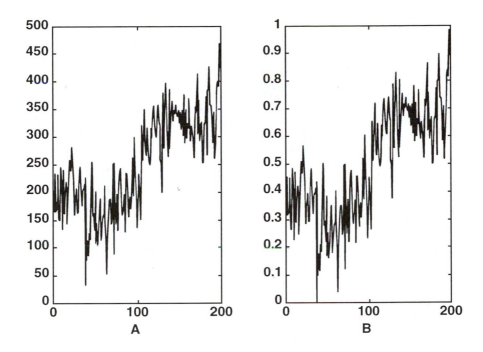

Figure 3 Min-max normalization (see MATLAB program in Appendix) A. Original
unnormalized data B. Data normalized using min-max normalization

For the case that the actual minimums and maximums of the input variables are
unknown, the Zscore normalization can be used. The algorithm is based on the
normalization of the standard deviation of the sample population. The example is
shown in Figure 4.

Sigmoidal Normalization

Sigmoidal normalization is a nonlinear transformation. It transforms the input
data into the range -1 to 1, using a sigmoid function. Again, the mean and standard
deviation of the input data should be calculated first. The linear (or quasi-linear)
region of the sigmoid function corresponds to those data points within a standard
deviation of the mean, while those outlier points are compressed along the tails of
the sigmoid function.

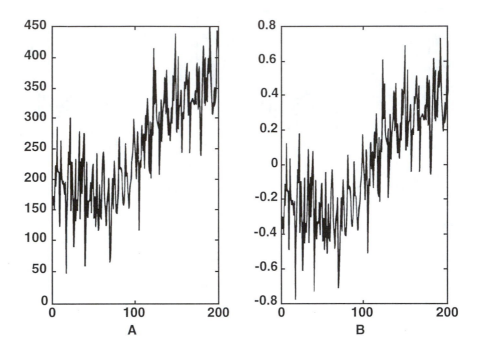

Figure 4 Zscore Normalization (see MATLAB program in Appendix) A. Original
unnormalized data B. Data normalized using Zscore normalization

The algorithm is given below:

$$y_{new} = \frac{1 - e^{-\alpha}}{1 + e^{-\alpha}} \; , \tag{15.14}$$

where

$$\alpha = \frac{y_{old} - mean}{std} \; . \tag{15.15}$$

The outliers of the data points usually have large values. In order to represent
those large outlier data, the sigmoidal normalization is an appropriate approach.
The data shown in Figure 5 have two large outlier points. Clearly, the sigmoidal
normalization can still capture the very large outlier values while mapping the input
data to the range -1 to +1.

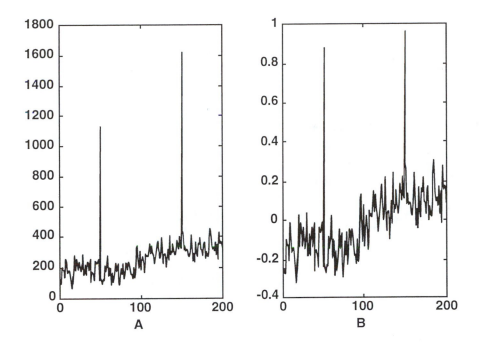

Figure 5 Sigmoidal normalization (see MATLAB program in Appendix) A.
Original unnormalized data B. Data normalized using sigmoidal normalization

15.3   Conclusions

Data preprocessing is very important and useful for data modeling. The key issue is
to maintain the most significant data while throwing away unimportant data. The
overprocessing of the data may result in catastrophe. One should always start with
minimal preprocessing and incrementally add more preprocessing while evaluating
the result. In particular, there exist many other algorithms for data preprocessing.

15.4   Appendix: Matlab Programs

15.4.1   Example of Noise Reduction Averaging

```
clear;
y1=randn(1,200);% the mean is 0; the stand deviation is 1
y2=y1;
y3=y2;
```

```
for i=1:195
    y2(i+4)=(y1(i)+y1(i+1)+y1(i+2)+y1(i+3)+y1(i+4))/5;
end
for i=1:190
    y3(i+9)=(y1(i)+y1(i+1)+y1(i+2)+y1(i+3) +y1(i+4)+...
        y1(i+5)+y1(i+6)+y1(i+7)+y1(i+8)+y1(i+9))/10;
end
figure(4)
%title('Noise Reduction Averaging')
subplot(221),plot(y1),xlabel('A')
subplot(222),plot(y2),xlabel('B')
subplot(223),plot(y3),xlabel('C')
```

15.4.2   Example of Min-Max Normalization

```
clear;
x=5:0.005:6;
y1=exp(x)+25*sin(100*x)+43*randn(1,201)+36*cos(10*x);
min1=min(y1); max1=max(y1);
min2=0; max2=1;
y2=((y1-min1)/(max1-min1))*(max2-min2)+min2;
subplot(121),plot(y1(1:200)),xlabel('A')
subplot(122),plot(y2(1:200)),xlabel('B')
```

15.4.3   Example of Zscore Normalization

```
clear;
x=5:0.005:6;
y1=exp(x)+25*sin(100*x)+43*randn(1,201)+36*cos(10*x);
mean1=sum(y1)/length(y1);
std1=sqrt((norm(y1)^2)/length(y1)-mean1);
y3=(y1-mean1)/std1;
subplot(121),plot(y1(1:200)),xlabel('A')
subplot(122),plot(y3(1:200)),xlabel('B')
```

15.4.4   Example of Sigmoidal Normalization

```
clear;
x=5:0.005:6;
y1=exp(x)+25*sin(100*x)+43*randn(1,201)+36*cos(10*x);
y1(50)=y1(50)+986;y1(150)=y1(150)+1286;
mean1=sum(y1)/length(y1);
std1=sqrt((norm(y1)^2)/length(y1)-mean1);
alpha=(y1-mean1)/std1;
y4=(1-exp(-alpha))./(1+exp(-alpha));
```

subplot(121),plot(y1(1:200)),xlabel('A')
subplot(122),plot(y4(1:200)),xlabel('B')

15.4.5 The Definitions of Mean and Standard Deviation

mean:
let n values be $x_1, x_2, \cdots, x_n$, then the mean

$$\overline{X} = \frac{1}{n}(x_1 + x_2 + \cdots + x_n)$$

$$= \frac{1}{n}\sum_{i=1}^{n} x_i$$

standard deviation:

$$std = \sqrt{\frac{1}{n}\sum_{i=1}^{n}(x_i - \overline{X})^2}$$

$$= \sqrt{\frac{1}{n}\sum_{i=1}^{n}x_i^2 - \overline{X}^2}$$

References

1. J. A. Anderson, Data representation in neural networks, *AI Expert,* pp. 30-37, June 1990.

2. K. Cios, W. Pedrycz, and R. Swiniarski, *Data Mining Methods for Knowledge Discovery,* Kluwer Academic Publishers, Boston, 1998.

3. R. L. Kennedy, Y. Lee, B. V. Roy, C. D. Reed, and R. P. Lippman, *Solving Data Mining Problems Through Pattern Recognition,* Prentice-Hall, Englewood Cliffs, 1998.

4. J. Lawrence, Data preparation for a neural network, *AI Expert,* pp. 34-41, Nov. 1991.

5. R. Stein, Selecting data for neural networks, *AI Expert,* pp. 42-47, Feb. 1993.

6. R. Stein, Preprocessing data for neural networks, *AI Expert,* pp. 32-37, March 1993.

Chapter 16

Control of a Flexible Robot Arm using a Simplified Fuzzy Controller

A flexible robot arm is a distributed system per se. Its dynamics are very complicated and coupled with the non-minimum phase nature due to the non-collocated construction of the sensor and actuator. This gives rise to difficulty in the control of a flexible arm. In particular, the control of a flexible arm usually suffers from control spillover and observation spillover due to the use of a linear and approximate model. The robustness and reliability of the fuzzy control have been demonstrated in many applications, particularly, it is perfect for a nonlinear system without knowing the exact system model. However, a fuzzy control usually needs a lot of computation time. In order to alleviate this restraint, a simplified fuzzy controller is developed for real-time control of a flexible robot arm. Furthermore, the self-organizing control based on the simplified fuzzy controller is also developed. The simulation results show that the simplified fuzzy control can achieve the desired performance and the computation time is less than 10 ms so that the real-time control is possible.

16.1 Introduction

Modeling and control of flexible robot arms have been actively investigated for several years. In the past, the Euler-Bernoulli beam model was frequently used to model a one-link flexible arm and various control strategies were proposed to compensate for beam vibration [2,3,5,6,7,20,21,28]. Even the finite element model was applied to model the flexible arms [4,18]. The major difficulty in the control of a flexible arm arises from its flexible nature. Basically, a flexible arm is an infinite dimension and nonlinear system per se, while most existing modeling and control techniques are based on the finite dimension and linear model. Hence, those techniques should compensate for control spillover and observation spillover. In addition, a flexible arm with end-point feedback is a non-minimum phase system. Namely, the system has unstable zeros due to the non-collocated sensors and actuators [7,11]. Hence, the feedforward control based on the inverse dynamics is not directly applicable. In contrast with the control of the human arm, the control of the flexible robot arm seems awkward. Since the control of the human arm is based on the sensory feedback and knowledge base, this reminds us the knowledge base control may be useful for the control of the flexible robot arm. The fuzzy control is one kind of expert controls [1]. The structure of fuzzy control is rather simple in comparison with other knowledge base control systems. Thus, it is perfect for a nonlinear system without knowing the exact system model. The robustness and reliability of the fuzzy control have been demonstrated in many applications [14,15,17]. However, fuzzy control usually takes up a lot of computation time. Although quantization and look-up table can reduce the computation time, they give rise to other problems. The quantization leads to worse precision, while the look-up table is only adequate to a special controlled process and can not be changed unless the table is set up again. In particular, when the system is large (too many system states and rules) and complex, the memory requirement becomes excessively large. Hence, the research in fuzzy hardware systems [26] and fuzzy memory devices [25] was developed. Alternatively, the simplification in the fuzzy control algorithm may lead to less computation time as well as less cost. Therefore, a simplified fuzzy controller is developed for real-time control purpose. The basic properties of the simplified fuzzy control and its relation to PID control will be addressed. Furthermore, the self-organizing control based on the simplified fuzzy controller is also developed so that the system dynamics and performance can be continuously and automatically learned and improved. Finally, the developed controllers are applied to the flexible robot arm. The results show that the simplified fuzzy control can achieve the desired performance and the computation time is less than 10 ms so that the real-time control is possible. The organization of the chapter is as follows. Section 1 describes the model of an one-link flexible arm followed by the simplified fuzzy control and the self-organizing fuzzy control. Then, the simulation results of the fuzzy control for a flexible arm are presented and followed by conclusion.

16.2   Modeling of the Flexible Arm

The one-link flexible arm is considered and shown in Figure 1. It is a thin uniform beam of stiffness EI and mass per unit length ρ. The total length of the arm is l and the torque actuation point is at $x = l$. Let I_B and I_H denote the moment of inertia of the beam and the hub, respectively. I_T is the sum of I_B and I_H. The external applied torque at the joint is τ. Using the Euler-Bernoulli beam model and neglecting structural damping, the dynamic equations can be derived from Hamilton principle.

The kinetic energy of the beam and the hub is given by

$$T = \frac{1}{2}\left[I_H\dot{\theta}^2 + \int_0^l \rho(\frac{\partial y}{\partial t})^2 dx\right] = \frac{1}{2}\sum_{i=0}^{\infty} I_T\,\dot{q}_i^2 \ , \tag{16.1}$$

where θ is the joint angle; $q_i(t)$ are time-dependent modal coordinates; and $y(x,t)$ is the displacement of the arm at the distance x from the joint. From Figure 1, $y(x,t)$

is given by

$$y(x,t) = w(x,t) + x\theta , \qquad (16.2)$$

where w is the deflection of the beam at the distance x. The potential energy stored in the beam is given by

$$V = \frac{1}{2}\left[\int_0^l EI(\frac{\partial^2 y}{\partial x^2})^2 dx - \tau\theta \right] = \frac{1}{2}\sum_{i=0}^{\infty} I_T \omega_i^2 q_i^2 + \tau \sum_{i=0}^{\infty} \phi_i^{(1)}(0)q_i , \qquad (16.3)$$

where $\phi_i(x)$ are the normalized mode shape functions, $\phi_i^{(1)}$ is the first derivative of ϕ_i, τ is the applied torque, and ω_i is the angular frequency. The complete solution of $y(x,t)$ is characterized by

$$y(x,t) = \sum_{i=0}^{\infty} \phi_i(x)q_i(t) . \qquad (16.4)$$

The partial differential equations, the boundary conditions and the detailed derivation can be found in [16]. For simplicity, only first $n+1$ modes in the dynamic equations are considered and the higher modes are assumed negligible. In fact, the fuzzy controller need not use an exact model of the system. Then the dynamic equations are derived with respect to the mode shape functions and modal coordinates q_i as

$$\ddot{q}_i + \omega_i^2 q_i = \frac{\tau}{I_T}\phi_i^{(1)}(0), \quad i = 0, 1, \cdots, n . \qquad (16.5)$$

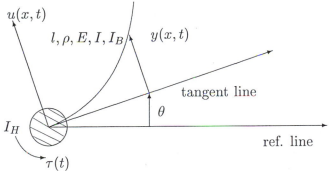

where
$u(x,t)$: displacement measured from tangent line
ρ: mass per unit length
EI : rigidity of arm
I_B : moment of inertia about the root
l : length of arm
I_H : inertia of hub
$\tau(t)$: external applied torque
$y(x,t)$: total displacement measured from reference line

Figure 1   Schematic diagram of a one-link flexible arm

In a physical system, there exists damping forces, such as transverse velocity damping forces and strain velocity damping moments. These damping forces are considered in Rayleigh model [19]. In fact, the damping coefficients in Rayleigh model are characterized by the corresponding natural frequency and the coefficients of damping force and damping stress. Therefore, a damping term should be added to Equation (16.5). The resultant equation is:

$$\ddot{q}_i + 2\zeta_i\omega_i\,\dot{q}_i + \omega_i{}^2 q_i = \frac{\tau}{I_T}\phi_i{}^{(1)}(0), i = 0, 1, \cdots, n \,, \tag{16.6}$$

where the damping coefficient ζ_i can be determined by experiments. In addition, the output of the system we wish to control is the displacement at the end-point, which is given by

$$y(l, t) = \sum_{i=0}^{n} \phi_i(l) q_i(t) \,. \tag{16.7}$$

From Equations (16.6) and (16.7), the linear state space model of the flexible arm can be denoted as:

$$\begin{cases} \dot{\mathbf{Q}} = \mathbf{A}\mathbf{Q} + \mathbf{B}\tau \\ y = \mathbf{D}\mathbf{Q} \end{cases}, \tag{16.8}$$

where

$$\mathbf{Q} = [q_0 \quad \dot{q}_0 \quad q_1 \dot{q}_1 \cdots q_n \quad \dot{q}_n]^T$$
$$\mathbf{D} = [l \quad 0 \quad \phi_1(l) \quad \cdots \phi_n(l) \quad 0]$$
$$\mathbf{B} = \frac{1}{I_T}[0 \quad 1 \quad 0 \quad \phi_1{}^{(1)}(1) \cdots 0 \quad \phi_n{}^{(1)}(0)]^T$$

$$\mathbf{A} = \begin{pmatrix} 0 & 1 & 0 & 0 & \cdots & \cdots & \cdots & \cdots \\ 0 & 0 & 0 & 0 & \cdots & \cdots & \cdots & \cdots \\ 0 & 0 & 0 & 1 & \cdots & \cdots & \cdots & \cdots \\ 0 & 0 & -\omega_1{}^2 & -2\zeta_1\omega_1 & \cdots & \cdots & \cdots & \cdots \\ \cdots & \cdots & \cdots & \cdots & \cdots & \cdots & \cdots & \cdots \\ \cdots & \cdots & \cdots & \cdots & \cdots & \cdots & 0 & 1 \\ \cdots & \cdots & \cdots & \cdots & \cdots & \cdots & \omega_n{}^2 & -2\zeta_n\omega_n \end{pmatrix}.$$

Note that if the viscosity and deformation in the motor are considered, then A(2,1) and A(2,2) elements should not be equal to zero.

16.3   Simplified Fuzzy Controller

A typical fuzzy control system is shown in Figure 2.

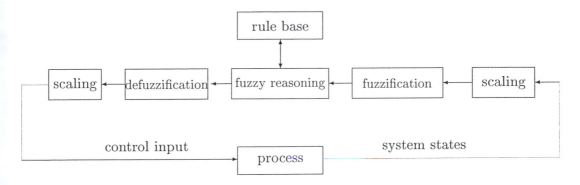

Figure 2   Structure of a typical fuzzy control

The fuzzy controller is constituted of five main parts: scaling, fuzzification, inference engine, defuzzification, and rule base. From the aspect of controller design, three fundamental parts should be decided first. They are system states selections, rule base constitution, and shapes of membership functions. As to the constitution of rule base, some static properties, such as completeness, interaction, and consistency of control rules, must be considered [8,9,10,12,13,22,27]. Given the membership functions, what kinds of shapes (basically a fuzzy number) are adequate? There is still no systematic procedure to make the optimal decisions due to the heuristic factors among them. The most common used functions are trapezoid and triangular shapes because of the simplicity and linearity. Normal distribution function and rational polynomial function [10] are also adopted frequently because the shapes can be tuned easily and meaningfully by some parameters in the functions. Since nonlinearities exist mainly in reasoning and defuzzification, the highly coupled variables cause the difficulties in analysis. In order to have a convenient way to look into fuzzy control, some simplified procedures are proposed. At the beginning, the specifications and descriptions about the controller are stated:

(1) System States:

The controller input and output can be defined as follows. Suppose that there are n rules for a SISO control system and the connections between rules are the linguistic word 'or'. One of the rules is expressed as below

$$\text{rule } i : \text{if } E \text{ is } E_i \text{ and } C \text{ is } C_i \text{ then } U \text{ is } U_i , \ i = 1, \cdots, n , \qquad (16.9)$$

where E stands for the error between the set point and the plant output; C stands for the change of error; U stands for the control input; E_i , C_i , and U_i are linguistic values (fuzzy sets) and belong to collections of reference fuzzy sets $RE = \{E_i, \ i = 1 \cdots p\}$, $RC = \{C_i, \ i = 1 \cdots q\}$, and $RU = \{U_i, \ i = 1 \cdots r\}$, respectively.

(2) Reference Fuzzy Sets And Membership Functions:

The triangular type membership function is chosen because of its linearity. The collections of the reference fuzzy sets for the error, the change of error, and the control input are the same. The linguistic meaning of the reference fuzzy sets and the corresponding labeled numbers are listed below:

linguistic term	meaning	antecedents	consequence	
NB	Negative Big	0	−3	
NM	Negative Medium	1	−2	
NS	Negative Small	2	−1	
ZO	Zero	3	0	(16.10)
PS	Positive Small	4	1	
PM	Positive Medium	5	2	
PB	Positive Big	6	3	

All the definitions are shown in Figure 3, where x axis is either "error", "change of error", or "control input".

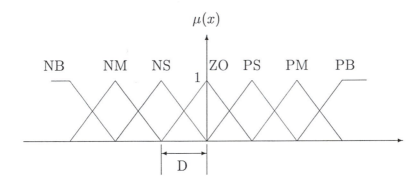

error : $x = e, D = D_e$

change of error : $x = c, D = D_c$

control input : $x = u, D = D_u$

Figure 3   Definitions of the reference fuzzy sets, membership functions, and universes of discou

(3) Rule Base:

The type of rules used here have the following form :

$$\text{if } E \text{ is } i \text{ and } C \text{ is } j \text{ then } U \text{ is } f(i,j) , \qquad (16.11)$$

where i, j, and $f(i,j)$ are numeric numbers in Equation (16.10). The total rule numbers are 49(7×7). It is clear that the rule setup turns to find a proper mapping f. We call these kind of rules parameterized rules. Note that the mapping f is also a function of another parameters; hence, we can adjust rules by tuning these parameters. For example,

$$f(i,j) = k_1 \times (i - 3) + k_2 \times (j - 3) . \qquad (16.12)$$

(4) Fuzzy Implication (t-norm, and t-conorm)

Fuzzy implication is selected as t-norm, and t-norm is chosen as an algebraic product operator; i.e.,

$$xTy = x \times y \ . \tag{16.13}$$

Although the corresponding t-conorm can be derived by De Morgan's law, under the consideration of defuzzification we use an approximate method instead of t-conorm to connect rules. It will be discussed in detail in the next section.

16.3.1 Derivation of Simplified Fuzzy Control Law

Let e and c be the values sensed from the plant output and scaled into the universe of discourses, respectively. As e and c are exact values, the following equation can be applied for reasoning:

$$
\begin{aligned}
\mu_{U_i'}(w) &= sup\{\mu_{E_i} \ T \ \mu_{C_i}(c_0), 0\} \ T \ \mu_{U_i}(w) \\
&= \mu_{E_i}(e_0) \ T \ \mu_{C_i}(c_0) \ T \ \mu_{U_i}(w)
\end{aligned}
\tag{16.14}
$$

where U' is the resultant fuzzy set rather than a value. A transformation called defuzzification which is opposite to the fuzzification procedure is needed to transform U' into a real value. Now, consider Figure 4. There are four rules fired:

$$\text{if } E \text{ is } i \text{ and } C \text{ is } j \text{ then } U \text{ is } f(i,j) \ ; \tag{16.15}$$

$$\text{if } E \text{ is } i \text{ and } C \text{ is } j+1 \text{ then } U \text{ is } f(i,j+1) \ ; \tag{16.16}$$

$$\text{if } E \text{ is } i+1 \text{ and } C \text{ is } j \text{ then } U \text{ is } f(i+1,j) \ ; \tag{16.17}$$

$$\text{if } E \text{ is } i+1 \text{ and } C \text{ is } j+1 \text{ then } U \text{ is } f(i+1,j+1) \ . \tag{16.18}$$

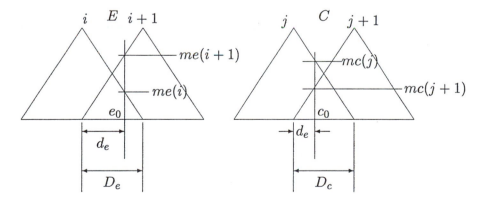

Figure 4   Computation of membership values

From the geometric relation in Figure 4, the membership values can be easily obtained as:

$$me(i) = \frac{D_e - d_e}{D_e} \tag{16.19}$$

$$me(i+1) = \frac{d_e}{D_e} \tag{16.20}$$

$$mc(j) = \frac{D_c - d_c}{D_c} \tag{16.21}$$

$$mc(j+1) = \frac{d_c}{D_c} \tag{16.22}$$

From Equation (16.14), the contribution of the rule (16.15), for example, will have the shape as shown in Figure 5.

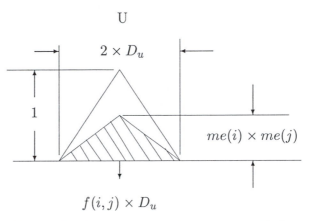

Figure 5   Control inferred by the rule (i,j)

The next step is to connect the fired rules and then proceed the defuzzification procedure so that a real control signal is produced. In order to obtain a simple result but not to violate "fuzzy spirit," we use the method called weighted area procedure to obtain the real control input. The control can be computed by the following formula

$$u = \frac{\sum(\text{area}) \times (\text{center of gravity})}{\sum(\text{area})} = \frac{D_u \times \sum_{i,j} me(i) \times me(j) \times f(i,j)}{\sum_{i,j} me(i) \times mc(j)} . \tag{16.23}$$

Note that

$$\sum_{i,j} me(i) \times mc(j) = 1 .$$

Thus, for the firing rules (16.15) to (16.18) the control (not scaled) is

$$u = D_u \times [me(i) \times mc(j) \times f(i,j) + me(i) \times mc(j+1) \times f(i,j+1) \\ + me(i+1) \times mc(j) \times f(i+1,j) + me(i+1) \times mc(j+1) \times f(i+1,j+1)] . \tag{16.24}$$

Equation (16.24) is called the simplified fuzzy control law.

16.3.2 Analysis of Simplified Fuzzy Control Law

The simplified fuzzy control law shown in Equation (16.24) is, in fact, a nonlinear function of the error and the change of error. The expression provides a way to reduce the efforts of fuzzy computation. If the controller has n inputs, then the total summation terms in the right-hand side of Equation (16.24) are 2^n. When $n > 3$, we can see that it is not practical to apply this control law because too many terms in Equation (16.24) cost a lot of computation time. On the other hand, if n is large, then the product of the membership values is small and the resultant value can not reflect the real situation. In this section, we discuss the controller with three inputs and one output. Let e, c, s, denote the error, the change of error, and the sum of error (scaled), respectively. There are three controller inputs. The definition is similar to that in Figure 3. The rules are still parameterized rules, i.e.,

$$\text{if } E \text{ is } i \text{ and } C \text{ is } j \text{ and } S \text{ is } k \text{ then } U \text{ is } f(i,j,k) \,. \tag{16.25}$$

The control law now is

$$u = D_u \times \sum_{i,j,k} me(i) \times mc(j) \times ms(k) \times f(i,j,k) \,, \tag{16.26}$$

where

$$ms(k) = \frac{D_s - d_s}{D_s}$$
$$ms(k+1) = \frac{d_s}{D_s} \,.$$

The e, c, s can be expressed as

$$e = (i - z) \times D_e + d_e$$
$$c = (j - z) \times D_c + d_c$$
$$s = (k - z) \times D_s + d_s$$

Then

$$d_e = e + (z - i) \times D_e \tag{16.27}$$

$$d_c = c + (z - j) \times D_c \tag{16.28}$$

$$d_s = s + (z - k) \times D_s \tag{16.29}$$

where

$$i = \sum_{m=1-z}^{z-1} u_s(e + m \times D_e)$$

$$j = \sum_{m=1-z}^{z-1} u_s(c + m \times D_c)$$

$$k = \sum_{m=1-z}^{z-1} u_s(k + m \times D_s)$$

and $u_s(\cdot)$ is a unit step function. Substituting Equation (16.27) to (16.29) into Equation (16.26), we obtain

$$u = D_u(c_1 e + c_2 c + c_3 s + c_4 ec + c_5 es + c_6 cs + c_7 ecs + c_8) , \qquad (16.30)$$

where

$$\begin{aligned}
c_1 =& [(j - z + 1)(k - z + 1)(f(i+1, j, k) - f(i, j, k)) \\
&+ (j - z + 1)(z - k)(f(i+1, j, k+1) - f(i, j, k+1)) \\
&+ (z - j)(k - z + 1)(f(i+1, j+1, k) - f(i, j+1, k)) \\
&+ (z - j)(z - k)(f(i+1, j+1, k+1) - f(i, j+1, k+1))]/D_e, \quad (16.31)
\end{aligned}$$

$$\begin{aligned}
c_2 =& [(i - z + 1)(k - z + 1)(f(i, j+1, k) - f(i, j, k)) \\
&+ (i - z + 1)(z - k)(f(i, j+1, k+1) - f(i, j, k+1)) \\
&+ (z - i)(k - z + 1)(f(i+1, j+1, k) - f(i+1, j, k)) \\
&+ (z - i)(z - k)(f(i+1, j+1, k+1) - f(i+1, j, k+1))]/D_c, \quad (16.32)
\end{aligned}$$

$$\begin{aligned}
c_3 =& [(i - z + 1)(j - z + 1)(f(i, j+1, k) - f(i, j, k)) \\
&+ (i - z + 1)(z - j)(f(i, j+1, k+1) - f(i, j+1, k)) \\
&+ (z - i)(j - z + 1)(f(i+1, j, k+1) - f(i+1, j, k)) \\
&+ (z - i)(z - j)(f(i+1, j+1, k+1) - f(i+1, j+1, k))]/D_s, \quad (16.33)
\end{aligned}$$

$$\begin{aligned}
c_4 =& \{(z - k)[f(i+1, j+1, k+1) - f(i+1, j, k+1)) \\
&- f(i, j+1, k+1) + f(i, j, k+1)] \\
&+ (k - z + 1)[f(i+1, j+1, k) - f(i+1, j, k)) \\
&- f(i, j+1, k) + f(i, j, k)]\}/[D_e \times D_c], \quad (16.34)
\end{aligned}$$

$$c_5 = \{(z-j)[f(i+1,j+1,k+1) - f(i+1,j+1,k)) - f(i,j+1,k+1)$$
$$+ f(i,j+1,k)] + (j-z+1)[f(i+1,j,k+1) - f(i+1,j,k))$$
$$- f(i,j,k+1) + f(i,j,k)]\}/[D_e \times D_c], \tag{16.35}$$

$$c_6 = \{(z-i)[f(i+1,j+1,k+1) - f(i+1,j+1,k)) - f(i+1,j,k+1)$$
$$+ f(i+1,j,k)] + (i-z+1)[f(i,j+1,k+1) - f(i,j+1,k))$$
$$- f(i,j,k+1) + f(i,j,k)]\}/[D_c \times D_s], \tag{16.36}$$

$$c_7 = [f(i+1,j+1,k+1) - f(i+1,j+1,k) - f(i+1,j,k+1)$$
$$+ f(i+1,j,k) - f(i,j+1,k+1) + f(i,j+1,k)$$
$$+ f(i,j,k+1) - f(i,j,k)]/[D_e \times D_c \times D_s], \tag{16.37}$$

$$c_8 = (i-z+1)(j-z+1)(k-z+1)f(i,j,k)$$
$$+ (i-z+1)(j-z+1)(z-k)f(i,j,k+1)$$
$$+ (i-z+1)(z-j)(k-z+1)f(i,j+1,k)$$
$$+ (i-z+1)(z-j)(z-k)f(i,j+1,k+1)$$
$$+ (z-i)(j-z+1)(k-z+1)f(i+1,j,k)$$
$$+ (z-i)(j-z+1)(z-k)f(i+1,j,k+1)$$
$$+ (z-i)(z-j)(k-z+1)f(i+1,j+1,k)$$
$$+ (z-i)(z-j)(z-k)f(i+1,j+1,k+1) \tag{16.38}$$

Note that the control u in Equation (16.30) is a nonlinear function of states e, c, s. Is it possible to be linear under certain conditions? Equation (16.30) is a linear controller if the nonlinear terms vanish; i.e., c_4, c_5, c_6, c_7 are equal to zero. Note that what we can manipulate is the mapping f; hence, f must satisfy some restrictions so as to achieve the above intention. Consider coefficients in Equations (16.31) to (16.37). In Equations (16.31) to (16.33), if we let

$$f(i+1,j,k) - f(i,j,k) = \alpha$$
$$f(i+1,j,k+1) - f(i,j,k+1) = \alpha$$
$$f(i+1,j+1,k) - f(i,j+1,k) = \alpha$$
$$f(i+1,j+1,k+1) - f(i,j+1,k+1) = \alpha \tag{16.39}$$

and

$$f(i,j+1,k) - f(i,j,k) = \beta$$
$$f(i,j+1,k+1) - f(i,j,k+1) = \beta$$
$$f(i+1,j+1,k) - f(i+1,j,k) = \beta$$
$$f(i+1,j+1,k+1) - f(i+1,j,k+1) = \beta \tag{16.40}$$

and

$$f(i,j,k+1) - f(i,j,k) = \gamma$$
$$f(i,j+1,k+1) - f(i,j+1,k) = \gamma$$
$$f(i+1,j,k+1) - f(i+1,j,k) = \gamma$$
$$f(i+1,j+1,k+1) - f(i+1,j+1,k) = \gamma$$

(16.41)

then Equations (16.39) to (16.41) guarantee that c_4, c_5, c_6, and c_7 vanish. The remaining coefficients become:

$$c_1 = \frac{\alpha}{D_e}$$

(16.42)

$$c_2 = \frac{\beta}{D_c}$$

(16.43)

$$c_3 = \frac{\gamma}{D_s}$$

(16.44)

$$c_8 = \alpha(z-i) + \beta(z-j) + \gamma(z-k) + f(i,j,k)$$

(16.45)

All the equations from Equations (16.39) to (16.41) can be understood from Figure 6.

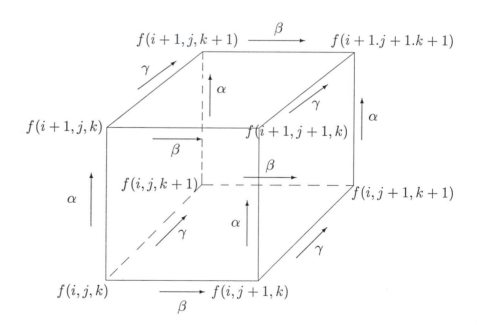

Figure 6   Rule cube to show the relations of Equations (16.39), (16.40), and (

The final result becomes

$$u = D_u(c_1 e + c_2 c + c_3 s + c_8) \ .$$

(16.46)

It is intuitive and reasonable that there always exists a rule with the following form:

if E is ZO and C is ZO and S is ZO then U is ZO .

This means

$$f(z, z, z) = 0 \tag{16.47}$$

and

$$c_8 = 0 . \tag{16.48}$$

Hence,

$$f(i, j, k) = \alpha(i - z) + \beta(j - z) + \gamma(k - z) . \tag{16.49}$$

It is interesting that Equation (16.49) can be regarded as the mapping required. In fact, from the simulation result, the rule base which is specified by the mapping (16.49) is effective and reasonable, especially for linear systems. The control law now is exact a PID controller

$$u = D_u(c_1 e + c_2 c + c_3 s) . \tag{16.50}$$

The above results are summarized in the following theorem.

Theorem 1

The simplified fuzzy control (SFC) law Equation (16.30) behaves like a PID controller if the increment of the mapping f in each direction is constant. In addition, if e, c, s fall into intervals $(-De, De)$, $(-Dc, Dc)$, $(-Ds, Ds)$, respectively, then the fuzzy controller is a pure PID controller.

16.3.3 Neglected Effect in Simplified Fuzzy Control

In the last section, we have mentioned that the weighted area method is only a convenient means in order to get the result Equation (16.24). There, t conorm is a maximum operator and center of gravity procedure is used. But, what are the side effects of a simplified fuzzy controller? Owing to the regular order of the reference fuzzy sets, there are only two possible overlap situations, as shown in Figure 7.

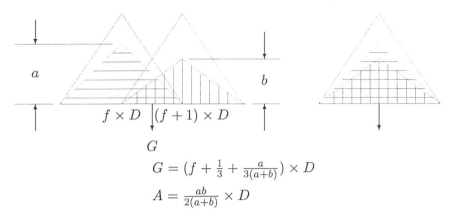

$$G = (f + \tfrac{1}{3} + \tfrac{a}{3(a+b)}) \times D$$

$$A = \tfrac{ab}{2(a+b)} \times D$$

Figure 7   Possible overlap conditions

Clearly, the shaded areas are the neglected parts in the simplified fuzzy control. The area and gravity of the shaded area can be obtained from the simple geometric computations. Therefore, the control law has the following expression

$$u_g = \frac{u - \frac{1}{D_u}\sum(A \times G)}{1 - \frac{1}{D_u}\sum A} \,, \tag{16.51}$$

where u is the simplified fuzzy control law. The additional minus terms in Eq. (16.51) depends on membership values and rules fired. These terms are not easy to regulate when we try to analyze the controller. However, if the overlap among the membership functions are not significant, the simplified fuzzy control is almost the same as the original fuzzy control but with much less computation effort.

16.4   Self-Organizing Fuzzy Control

The kernel of fuzzy control is the rule base. In other words, a fuzzy system is characterized by a set of linguistic rules. The rule base is termed as the fuzzy model of the controlled process. From the point of view of traditional control, the determination of the rule base is equivalent to system identification [23]. The rules can be acquired in many ways. In the last section, parameterized rules are used. Although parameterized rules are easy to use, the tuning of parameters are trial and error and tedious. Procyk and Mamdani [24] proposed a structure of a self-organizing fuzzy controller (SOC) and simulated it by fuzzy relation (quantization approach). Basically, we will follow the idea of Mamdani with slight modification.

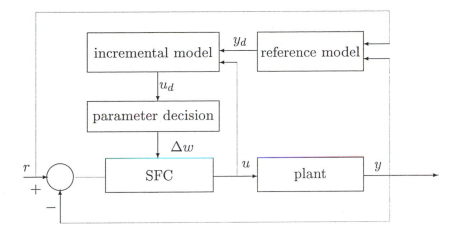

Figure 8   Structure of the self-organizing control system

The block diagram of SOC proposed here is shown in Figure 8. Three additional blocks other than the fuzzy controller in Figure 8 are the reference model, the incremental model, and the parameter decision. They constitute the rules modification procedures. Under normal condition, the modification operates at every sample instant. The idea of modification procedures is stated below.

(1) The reference model senses the plant output and reference input, then compares them with the desired responses (or trajectory). The model output is the performance measure which shows how good the controller works. For simplicity, the performance measure is the change in the desired output.

(2) The incremental model is a simplified linear model of the plant. It can be derived by linearization techniques or experiments. The model relates the input change of the plant to the output change of the plant; hence, it can be used to evaluate the required control input. The incremental model does not have to be accurate because its principal function is to reflect the approximate behavior of the controlled plant.

(3) The parameter decision unit receives the information from the incremental model and proceeds with the decision making or parameter estimation. It maybe involve heuristic or mathematical approaches, or a combination of both. As to the mathematical approach, it uses the estimation technique described in adaptive control. The heuristic approach needs more a priori knowledge and depends on the designers.

For the case of two controller inputs, we have had the control law as

$$u = G_u[me(i)mc(j)w(i,j) + me(i)mc(j+1)w(i,j+1) \\ + me(i+1)mc(j)w(i+1,j) + me(i+1)mc(j+1)w(i+1,j+1)] \tag{16.52}$$

or

$$u = G_u(c_1 e + c_2 c + c_3 ec + c_4) . \tag{16.53}$$

Because of the parameterized rule base, the undetermined factors except scaling factors in the control law are $w(i,j)$ in Equation (16.52) or c_i in Equation (16.53). Thus, the first work is to estimate coefficients $w(i,j)$ or c_i via some mathematical or heuristic processes, called rules modification procedure, at every one (or two more) sampling instants. The reference model, incremental model, and parameter decision are further described below.

16.4.1 Reference Model

In model reference adaptive control, the reference model is a prescribed mathematical model and the objective is to design the control law so that the closed-loop transfer function is close to the reference model. In fact, it is a pole placement procedure. In fuzzy control, the reference model proposed here is somewhat different from MRAC (model reference adaptive control). It is a fuzzy model and not only specifies the desired response but also gives the required quantity for adjustment. The fuzzy model is all due to control objective and has nothing to do with the controlled plant.

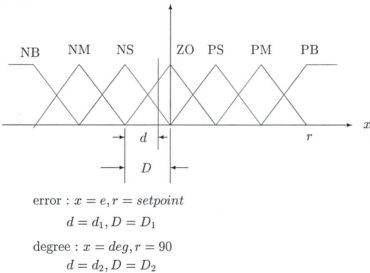

error : $x = e, r = setpoint$
$$d = d_1, D = D_1$$

degree : $x = deg, r = 90$
$$d = d_2, D = D_2$$

Figure 9   Meta rules specification

The reference model is created as meta rules. These meta rules are used to supervise the response of plant output and give the amount of modification. The meta rule has the form as

if error is PS and slope is NM then change of plant output is PM .

Note that the error is defined as

$$e(k) = setpoint - y(k) \qquad (16.54)$$

and the slope is defined as

$$m(k) = \frac{y(k) - y(k-1)}{T} ,$$ (16.55)

where T is the sampling period. The dimension in Equation (16.55) should be transformed into degree

$$deg(k) = \frac{180 \times tan^{-1} m(k)}{\pi} .$$ (16.56)

The meta rule base is listed in Table 1. The corresponding notations are defined in Equation (16.10). Note that zero elements in the rule table are the desired response regions, while the others need to be modified. The meta rules bear no relation to the controlled plant. They are based on the control objective. The universe of discourses and reference fuzzy sets are shown in Figure 9.

The required change of the output can be obtained again through the simplified fuzzy computation as:

$$y_d = G_y \Big[\frac{(D_1 - d_1)(D_2 - d_2)R(i,j) + (D_1 - d_1)d_2 R(i,j+1)}{D_1 D_2}$$
$$+ \frac{d_1(D_2 - d_2)R(i+1,j) + d_1 d_2 R(i+1,j+1)}{D_1 D_2} \Big] .$$ (16.57)

16.4.2 Incremental Model

Consider the following SISO system

$$\dot{\mathbf{x}} = f(\mathbf{x}, u), \ \mathbf{x} \in \mathbf{R}^n$$ (16.58)

$$y = g(\mathbf{x}) .$$ (16.59)

Variations in $\mathbf{x}$ and u cause variation in $\dot{\mathbf{x}}$

$$\delta \dot{\mathbf{x}} = \delta f(\mathbf{x}, u) = \frac{\partial f}{\partial \mathbf{x}} \delta \mathbf{x} + \frac{\partial f}{\partial u} \delta u .$$ (16.60)

Note that

$$\delta \dot{\mathbf{x}} = \frac{d}{dt}(\delta \mathbf{x}) .$$ (16.61)

Let T be the sampling period. Then, Equation (16.60) becomes

$$\Delta \mathbf{x} \approx T \delta \dot{\mathbf{x}} = T \frac{\partial f}{\partial \mathbf{x}} \Delta \mathbf{x} + T \frac{\partial f}{\partial u} \Delta u$$

$$\Delta \mathbf{x} = (\mathbf{I} - T \frac{\partial f}{\partial \mathbf{x}})^{-1} T \frac{\partial f}{\partial u} \Delta u .$$ (16.62)

From Equation (16.59), we have

$$\delta y = \frac{\partial g}{\partial \mathbf{x}} \delta \mathbf{x} \ .$$

(16.63)

Therefore,

$$\Delta y = \frac{\partial g}{\partial \mathbf{x}} \Delta \mathbf{x} = M \Delta u$$

$$M = \frac{\partial g}{\partial \mathbf{x}} (\mathbf{I} - T \frac{\partial f}{\partial \mathbf{x}})^{-1} T \frac{\partial f}{\partial \mathbf{x}}$$

(16.64)

and

$$\Delta u = N \Delta y \ ,$$

(16.65)

where $N = M^{-1}$ is called the incremental value (or a matrix for MIMO case). Some remarks are made for the incremental model.

(i) For a linear system

$$\dot{\mathbf{x}} = \mathbf{A}\mathbf{x} + \mathbf{B}u, \ \mathbf{x} \in \mathbf{R}^n \ .$$

(16.66)

$$y = \mathbf{C}\mathbf{x} \ .$$

(16.67)

Then

$$N = [\mathbf{C}(\mathbf{I} - T\mathbf{A})^{-1}T\mathbf{B}]^{-1} \ is \ a \ constant \ .$$

(16.68)

(ii) For an unknown plant, the Jacobian matrices $\frac{\partial f}{\partial \mathbf{x}}, \cdots$ etc., can be approximately obtained from experiments. Indeed, the incremental model is a linearization of a nonlinear system. Various operating points result in various incremental value N.

(iii) There is a convenient way to compute the incremental value, i.e.,

$$N = \frac{u(k) - u(k-1)}{y(k) - y(k-1)}$$

(16.69)

and take the average value from experiments or simulations.

(iv) The existence of the inverse of $\mathbf{I} - T\frac{\partial f}{\partial \mathbf{x}}$ is trouble. The possible way to avoid the singular condition is that we can adjust sampling period T under the stability and the implementation of digital systems.

Table 1 Meta Rule Base

$e(k)$

	NB	NM	NS	ZO	PS	PM	PB
NB	ZO	PS	PS	PM	PM	PB	PB
NM	NS	ZO	PS	PS	PM	PB	PB
NS	NS	NS	ZO	PS	PS	PM	PM
ZO	NM	NS	NS	ZO	PS	PS	PM
PS	NM	NM	NS	NS	ZO	PS	PS
PM	NB	NB	NM	NS	NS	ZO	PS
PB	NB	NB	NM	NM	NS	NS	ZO

$deg(k)$ labels the rows.

(a) linguistic definition

$e(k)$

	0	1	2	3	4	5	6
0	0	1	1	2	2	3	3
1	-1	0	1	1	2	3	3
2	-1	-1	0	1	1	2	2
3	-2	-1	-1	0	1	1	2
4	-2	-2	-1	-1	0	1	1
5	-3	-3	-2	-1	-1	0	1
6	-3	-3	-2	-2	-1	-1	0

$deg(k)$ labels the rows.

(b) corresponding labeled numbers in (a)

16.4.3 Parameter Decision

From Equations (16.57) to (16.65), we have the required change of the control input Δu_d . Hence, from Equation (16.52), we have

$$u_d(k) = u(k-1) + \Delta u_d \ . \tag{16.70}$$

However, only the knowledge of the value Δu_d can not decide four weights $w(\cdot, \cdot)$ in Equation (16.52) or four coefficients c_i in Equation (16.53). Since we have already known that

$$\sum_{i,j} me(i)mc(j) = 1 \tag{16.71}$$

we obtain

$$\Delta w(i,j) = \frac{\Delta u_d}{G_u} \ . \tag{16.72}$$

Namely,

$$w_d(i,j) = w(i,j) + \frac{\Delta u_d}{G_u} \ . \tag{16.73}$$

Equation (16.73) is used to adjust the weights.

16.5   Simulation Results

The simulation results of the fuzzy control of the one-link flexible arm are presented in this section. The parameters of the flexible arm are listed in Table 2. Four modes are selected (including rigid mode); thus, the state space model is eight orders. The four vibration modes are listed in Table 3. The motor is modeled as a second order system, the parameters are also shown in Table 2. The simulations are performed for cases with and without the motor dynamics. In the simulation, the desired end-point position is 3 meters.

Table 2 Flexible Arm and dc Motor Parameters
Euler-Bernoulli beam:

$$l = 1 \ M$$
$$E = 2 \times 10^{11} \ N/M^2$$
$$\rho = 0.8 \ kg/M$$
$$I = 2.5 \times 10^{-11} \ M^4$$
$$A = 10^{-2} \ M^2$$
$$I_H = 0.1 \ kg \cdot M^2$$
$$I_B = \frac{\rho l^3}{3} \ kg \cdot M^2$$
$$I_T = I_B + I_H$$

DC Motor:

torque constant $K_i = 1.1\ N \cdot M/Amp$

back emf constant $K_b = 1\ V/rad/s$

rotor inertia of motor $J = 0.09\ kg \cdot M$

armature resistance $R_a = 3.8\ \Omega$

armature inductance $L_a = 10\ mH$

viscous frictional coefficient $B = 0.02\ N \cdot M/rad/s$

Table 3 Four Vibration Modes of Euler Beam

vibration mode	0	1	2	3
natural freq. (Hz)	0	2.3	8.11	22.05
$\phi_i(l)$	1	-1.5597	1.8562	-1.9675
$\phi_i^{(1)}(0)$	2.7273	4.7437	1.9151	0.7046
damping factor	0.01	0.015	0.02	0.02

First, the simulation results of the simplified fuzzy controller vs. conventional fuzzy controller are presented in Figure 10, Figure 11, and Figure 12. In Figure 10, the motor dynamics are not considered. The solid line denotes the simplified fuzzy controller with parameterized rule base. The dash line denotes the conventional fuzzy controller with min-max principle and center of gravity. The shapes of membership functions are adjustable (trapezoid). In contrast, the motor dynamics are considered in Figure 11. From Figure 10 and Figure 11, we can see that negative position takes place in the transient response. That is the nature of non-minimum phase system. Physically, because the beam is flexible, when the external torque is applied to the root of the beam in the beginning, the beam bends back relative to the tangent line (or rigid mode) then the tip position becomes negative. In order to reduce the negative position, the control rules in the beginning of the response should not be the original rule base. The other rule base is used for the situation that tip position is negative. The result is given in Figure 12. Clearly, the negative position of the tip has been reduced but the settling time becomes a little bit longer. Although the second rule base is applied, the phenomenon of the negative position can not be completely eliminated. The positive displacement part of the tip is acted only by the original rule base. Therefore, its response is similar to the previous figures.

The comparisons of these controllers are listed in Table 4. Obviously, the simplified fuzzy control is 40 times faster than the conventional fuzzy control. Since the computation times of the simplified fuzzy control are all within 10 ms (sampling period), the real time control can be achieved. Note that the computation time includes the computation of the mathematical model of the flexible arm; thus, these values are conservative. In addition, the use of two rule bases can reduce the negative position at the expense of longer settling time, more computation time, and larger steady state error, although they are not seriously reflected to the response.

Table 4 Comparison of Simplified and Conventional Fuzzy Controllers

	Steady state error(m)	number of rules	computation time(ms)	settling time (s)
Figure 10 (Simplified)	0.0055	49	8.5	1.67
Figure 10 (Conventional)	0.0168	49	350	1.66
Figure 11 (Simplified)	0.0055	49	9.1	1.76
Figure 11 (Conventional)	0.0055	49	400	1.69
Figure 12	0.016	I:25 II:49	9.3	1.98

Next, let us consider the case of self-organizing fuzzy control (SOC). The learning results for the flexible arm are shown in Figure 13. Figure 13(a) is the SOC with two learning sampling periods at the first run, and the remaining runs are given in Figure 13(b). The response at the first run is bad because we give the rules arbitrarily. Although the learning results are not perfect, the convergence is verified. In order to improve the overshoots and negative positions, a more reasonable and effcient updating law is needed.

From the simulation results of the flexible arm, we can see that the coefficients of function which tune the rules are the same, $k_1 = 0.6$ and $k_2 = 0.6$; i.e.,

$$f(i, j) = k_1 \times (i - 3) + k_2 \times (j - 3) . \tag{16.74}$$

In fact, for linear plants the rules are 'linear' and have similar form. Because the flexible arm is modeled as a linear system, the linear function Equation (16.74) can create a reasonable rule base for the linear plant. Table 5 lists two rule tables. Table 5(a) is obtained by inspecting step response, and Table 5(b) is created from Equation (16.74). It is clear that the two rule bases are basically the same. Table 5(b) has more rules than Table 5(a). Too many rules ($7 \times 7 = 49$) are the disadvantage of the parameterized rule base because some conflicts may exist among rules. Note that from Table 5, the simplified fuzzy controller behaves like a PD controller at the regions that satisfy the following relation:

$$f(i + 1, j + 1) - f(i + 1, j) - f(i, j + 1) + f(i, j) = 0 . \tag{16.75}$$

At the neighborhood of origin in the phase plane, the four fired rules satisfy Equation (16.75), and therefore the controller is a pure PD controller.

16.6   Conclusions

This chapter presented a simplified fuzzy controller for the one-link flexible arm. The control law is due to the simplification on reasoning and defuzzification. It offers a convenient way to compute the control input and reduces the reasoning time.

In fact, a fuzzy controller is a highly nonlinear PD controller (if only the error and the change of error are fedback). From the simplified control law, we can see that the nonlinear term remains in the product of the error and the change of error, and the simplified fuzzy controller behaves exactly like a linear PD controller if the rules satisfy certain conditions. If the overlap between membership functions (depending on rule base) is not serious, the complete control law can be approximated by the simplified control law. This means that the fuzzy controller behaves like a variable coefficients PD controller with slight nonlinearity. The overlap of membership functions associated with the rule base constitutes the variable coefficients in the control law, and this results in robustness.

The flexible arm is a non-minimum phase system. Morris and Vidyasagar [19] showed that the Euler beam model can not be stabilized by a finite-dimensional controller (rational function) from the viewpoint of controller design. But the fuzzy controller is a rule base system, it does not care about which mathematical model is chosen. In addition, it is a nonlinear controller; hence, the resultant closed-loop system is nonlinear. The facts shown in [19] are not adequate in these conditions. From the simulation results by the simplified fuzzy control, the tip position can be controlled well.

A self-organizing fuzzy controller using the simplified fuzzy control law was also presented in this chapter. Although the simulation results show that the justification is simple and the response converges gradually after many times learning; however, the rules' justification seems to be coarse and unreasonable because the four fired rules are adjusted in the same weight at parameter decision procedure. There should be some heuristics to adjust the fired rules individually.

For a complex process, we may use the simplified fuzzy control as a 'pre-controller' so as to obtain the coarse structure of the fuzzy controller. According to the coarse controller, the work on rules' acquirement and membership function shapes adjustment may not be so tedious.

Table 5 Comparison of Rule Tables

Change of error

error	0	1	2	3	4	5	6
0				-3	-2		
1				-2			
2				-1	0		
3	-3	-2	-1	0	1	2	3
4	-2		-1	1			
5				2			
6			2	3			

(a) lists the rules derived from the investigation of step response

Change of error

error	0	1	2	3	4	5	6
0	-3	-3	-2	-2	-1	-1	0
1	-3	-2	-2	-1	-1	0	1
2	-2	-2	-1	-1	0	1	1
3	-2	-1	-1	0	1	1	2
4	-1	-1	0	1	1	2	2
5	-1	0	1	1	2	2	3
6	0	1	1	2	2	3	3

(b) lists the rules created from Equation (16.74)

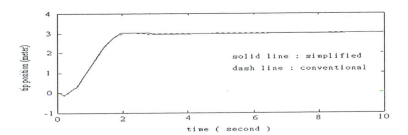

Figure 10   Simplified vs. conventional fuzzy controllers without motor dynamics

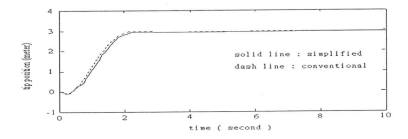

Figure 11   Simplified vs. conventional fuzzy controllers without motor dynamics

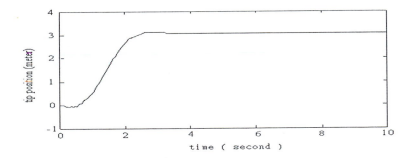

Figure 12   Simplified vs. conventional fuzzy controller with two rule bases

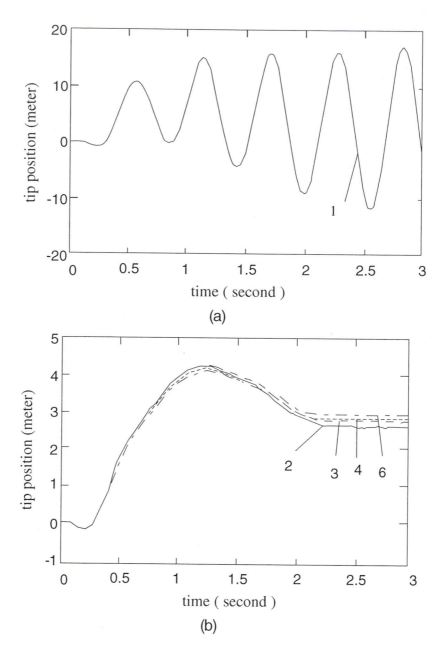

Figure 13   SOC for the 2nd system with two learning sampling periods,
(a) first run, (b)remaining runs

References

1. K. J. Strom, Auto-tuning, adaptation and expert control, *American Control Conference,* pp. 1514-1519, 1987.

2. M. Balas and D. Meldrum, Application of model reference control to a flexible remote manipulator arm, *American Control Conference*, pp. 825-832, 1985.

3. E. Bayo, Computed torque for the position control of open chain flexible robots, *IEEE International Conference on Robotics and Automation*, pp. 316-321, 1988.

4. E. Bayo, A finite-element approach to control the end-point motion of a single-link flexible robot, *Journal Robotic Systems*, Vol. 4, No. 1, pp. 63-75, 1987.

5. W. J. Book and G. G. Hastings, Experiments in optimal control of a flexible arm, *American Control Conference*, pp. 728-729, 1985.

6. W. J. Book and G. G. Hastings, A linear dynamic model for flexible robotic manipulators, *IEEE Control Systems Magazine*, Vol. 7, No. 1, pp. 61-64, 1987.

7. R. H. Jr. Cannon and E. Schmitz, Initial experiments on the end-point control of a flexible one link robot, *International Journal of Robotics Research*, Vol. 3, pp. 62-75, 1984.

8. Y. Y. Chen and T. C. Tsao, A description of the dynamical behavior of fuzzy systems, *IEEE Transactions on System, Man, and Cybernetics*, Vol. 19, No. 4, pp. 745-755, 1989.

9. S. Daley, et al., Comparison of a fuzzy logic controller with a P+D control law, *ASME Journal of Dynamics, Measurement, and Control*, 1989.

10. J. Domi, Membership function as an evaluation, *Fuzzy Sets and Systems*, North-Holland, pp. 1-21, 1990.

11. S. D. Eppinger and W. P. Seering, Modeling robot flexibility for endpoint force control, *IEEE International Conference on Robotics and Automation*, pp. 165-170, 1988.

12. C. C. Lee, Fuzzy logic in control systems: Fuzzy logic controller-Part one, *IEEE Transactions on Systems, Man, and Cybernetics*, pp. 404-418, 1990.

13. C. C. Lee, Fuzzy logic in control systems: Fuzzy logic controller-Part two, *IEEE Transactions on Systems, Man, and Cybernetics*, pp. 419-435, 1990.

14. Y. F. Li and C. C. Lau, Development of fuzzy algorithms for servo systems, *IEEE Control Systems Magazine*, pp. 65-72, April 1989.

15. M. H. Lim and Y. Takefuji, Implementing fuzzy rule-based systems on silicon chips, *IEEE Expert,* pp. 31-45, February 1990.

16. H. P. Huang and W. M. Lin, Control of a flexible robot arm using simplified fuzzy controllers, *International Journal of Fuzzy Systems,* Vol. 2, No.1, pp. 78-88, March 2000.

17. E. H. Mamdani, Application of fuzzy algorithms for control of simple dynamic plant, *Proceedings of IEE* Vol. 121, pp. 1585-1588, 1974.

18. C. H. Menq and J. S. Chen, Dynamic modeling and payload-adaptive control of a flexible manipulator, *IEEE International Conference on Robotics and Automation,* pp. 488-493, 1988.

19. K. A. Morris and M. Vidyasagar, A comparison of different models for beam vibrations from the standpoint of control design, *ASME Journal of Dynamic Systems, Measurement, and Control,* pp. 349-356, 1990.

20. W. L. Nelson and D. Mitra, Load estimation and load adaptive optimal control for a flexible robot arm, *IEEE International Conference on Robotics and Automation,* pp. 206, 1986.

21. J. H. Park and H. Asada, Design and control of minimum-phase flexible arms with torque transmission mechanisms, *IEEE International Conference on Robotics and Automation,* pp. 1790-1795, 1990.

22. W. Pedrycz, *Fuzzy Control and Fuzzy Systems,* Research Studies Press, John Wiley and Sons, New York, 1989.

23. X. T. Peng, Generating rules for fuzzy logic controllers by functions, *Fuzzy Sets and Systems,* North-Holland, pp. 83-89, 1990.

24. T. J. Procyk and E. H. Mamdani, A linguistic self-organizing process controller, *Automatica,* pp. 15-30, 1979.

25. M. Togai and H. Watanabe, Expert system on a chip: An engine for real-time approximate reasoning, *IEEE Expert Systems Magazine,* pp. 55-62, 1986.

26. T. Yamakawa, High-speed fuzzy controller hardware system: The mega FIPS machine, *Information Sciences,* pp. 113-128, 1988.

27. H. Ying, W. Siler, and J. J. Buckley, Fuzzy control theory: A nonlinear case, *Automatica,* Vol. 26, pp. 513-520, 1990.

28. K. Yuan and L. C. Lin, Motor-based control of manipulators with flexible joins and links, *IEEE International Conference on Robotics and Automation,* pp. 1809-1814, 1990.

Chapter 17

Application of Neuro-Fuzzy Systems: Development of a Fuzzy Learning Decision Tree and Application to Tactile Recognition

Since a decision tree has a simple, apparent, and fast reasoning process, it has been applied to many fields such as pattern recognition and classification. This chapter discusses an algorithm to generate a fuzzy learning decision tree by determining its structure and parameters. This algorithm first collects enough training data for generating a practical decision tree. It then uses fuzzy statistics to calculate fuzzy sets for representing the training data in order to save computation memory and increase generation speed. A suboptimal criterion is used to determine a decision tree from the resultant fuzzy sets. The generated fuzzy decision tree was then mapped into a rule-based system and a multi-layer perceptron. The performance among fuzzy decision trees, rule-based systems, and neural networks was then compared. A general-purpose tactile sensing and recognition system was constructed and the proposed algorithm was applied to that system. As a cost-effective method, this system uses fuzzy logic to interpolate the measured tactile force data in the software. Finally, the proposed algorithm was used to generate the desired decision tree from the tactile force data. Based on the decision tree, objects can be precisely recognized online.

17.1 Introduction

The decision tree is a well-known decision structure [17]. Decision trees can be applied to many fields [1,15,18] such as pattern recognition, classification, decision support systems, expert systems, robot control, and dynamics identification. A decision tree has several nodes arranged in a hierarchical structure. Each of these nodes has a different function and meaning. Figure 1 is an example of a decision tree applied to a classification problem in a two-dimensional feature space, where each object to be classified is represented by a two-dimensional feature. In Figure 1, the condition $g_i(X) \geq 0$, $i = 1 \cdots 4$, is the decision function or splitting rule that splits the corresponding region into two parts and decides to which part the object should belong. The label $C_j, j = I \cdots V$, is one of the names of the predefined classes. The ellipse containing $g_0(X) \geq 0$ is the root node where the classification begins. The

ellipse containing $g_i(X) \geq 0, i = 1 \cdots 3$, is the internal or non-terminal node where
the classification continues. The square containing $C_i, i = I \cdots V$, is the terminal
node where the classification stops.

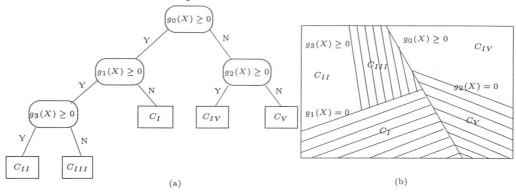

(a)   (b)

Figure 1 An example decision tree. (a) The tree structure, (b) The partition of the feature space

From the example in Figure 1, it is easy to see that the decision tree implements
decisions or classifications in a simple, apparent, multistage manner. Furthermore,
since each node of a decision tree uses only a simple splitting rule and a small subset of all of the features, the entire decision process is fast and efficient [15]. As
for the performance of decision trees, it is interesting to understand the superiority
of artificial neural networks and observe the relationship between them. Artificial
neural networks are powerful in many fields such as object localization [10], optimization, recognition, classification, prediction, diagnosis, decision making, association,
approximation, and generalization. Decision trees can be related to artificial neural networks by mapping them into a class of artificial neural networks or entropy
nets with far fewer connections [19]. In some prediction and recognition applications, decision trees have comparable performance with artificial neural networks
[1,2]. Decision trees and artificial neural networks can also be combined into a new
system with better performance [7,20].

To generate a practical decision tree, numerous training data are necessary. However, if the number of training data is enormous, a long processing time and large
memory space are required to generate a useful decision tree. To overcome this problem, this chapter proposes a method using fuzzy statistics to generate several fuzzy
sets from the training data. This method effectively compresses the training data
and meaningfully represents its statistical distribution. This has the advantages of
increasing computation speed, saving storage memory, and maintaining satisfactory
performance.

Given a set of training data, an optimal decision tree is inherently difficult to
obtain. This is primarily because of the following two reasons [15]: (a) optimizing a
node's performance does not necessarily optimize the performance of the entire tree,
and (b) the possible number of tree topologies is extremely large. To cope with this

difficulty, this study adopted a sub-optimal splitting rule at each node to generate a decision tree. This is a cost-effective method that quickly obtains a decision tree with satisfactory performance.

The experimental results suggest that a human tactile perception system can quickly and accurately recognize real-world objects [21]. Hence, it is reasonable to infer that tactile sensors mimicking human tactile perception should be useful in object recognition and classification applications. With the rapid advances in tactile sensor technology, there have been more and more tactile sensors applied to numerous fields, such as biomedicine [5,9], computer peripherals, and industry [26,27]. This study constructed a fuzzy logic based tactile sensing and recognition system. As tactile perception is the human inherent capacity and fuzzy logic is the nature of human reasoning, it is straightforward to apply fuzzy logic [6,16,24,25] to increase the resolution of a tactile sensor and improve the performance of decision trees [4]. This is a cost-effective method which can integrate human expertise as well as fulfill nonlinear interpolation. To examine the performance of the proposed fuzzy learning decision tree and complete the functions of the system, the proposed algorithm was applied to a tactile sensing and recognition system and then used to classify several geometric objects and the human hand.

The organization of this chapter is as follows. Section 2 introduces tactile sensors and the construction of a tactile sensing, and recognition system. The proposed fuzzy interpolation is also described in this section. A fuzzy learning decision tree is developed in Section 3. The developed tree is then mapped into and compared with a rule-based system and a multi-layer perceptron. In Section 4 we study a tactile system that classifies geometric objects and human hand using the proposed fuzzy learning decision tree. Experimental results are given and discussed. Finally, several concluding remarks are drawn in Section 5.

17.2 Tactile Sensors and a Tactile Sensing and Recognition System

The tactile sensor used in this chapter is one type of force sensing resistor (FS-RTM). The FSR is a polymer thick film device that exhibits a decreasing resistance with increasing force normally applied to the device surface [26]. Compared to other types of tactile sensors, FSRs are cheap, have less hysteresis than conductive rubber, and are less sensitive to vibration and heat than piezo film [27]. The FSR is suitable for human-touch control of electronic or mechanical devices, but not for exact force measurement, as in a load cell or strain gauge [26]. The following subsection will briefly introduce several types of FSRs.

17.2.1 Types of FSRs

Currently, four types of FSRs are available [26,27]:

(i.) **Basic FSR and simple arrays:** A basic FSR is a sensing element consisting of two polymer sheets laminated together. The applied force (or pressure) vs.

resistance curve for FSRs is approximately linear in a log plot.

(ii.) **Linear potentiometer:** The FSR linear potentiometer can sense the amount and position of the applied force. It can detect only one point of contact at a time.

(iii.) **X-Y-Z digitizer pad:** The X-Y-Z digitizer pad is a three-layer structure: two layers of FSR linear potentiometers for detecting positions in the x and y directions, and another layer for the amount of applied force in the z direction. It can detect only one point of contact at a time.

(iv.) **Matrix array:** A FSR matrix array is a two-dimensional tactile sensing pad comprised of basic FSRs adjacently arranged in one plane. It can measure multiple-point contact. Each basic FSR in a matrix array is a sensor pixel. The measured data of these sensor pixels can be addressed serially or in parallel by a personal computer through an interface device. This chapter discusses the application of passive tactile sensing and recognition, namely, the tactile sensors are fixed and objects come to touch the tactile sensors. Therefore, the type of FSR adopted in this study was the FSR matrix array. The adopted FSR matrix array is comprised of 16 × 16 basic FSRs.

17.2.2 A Tactile Sensing System

The tactile sensing and recognition system built in this study is shown in Figure 2.

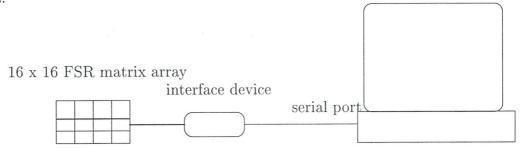

Figure 2 The tactile sensing and recognition systems built in this study

This system consists of three parts: hardware devices, software kernel, and man-machine interface. The software kernel controls the hardware devices, receives tactile data from them, outputs information to the man-machine interface, and receives command input. Those three parts are described in the following subsections.

17.2.2.1 Hardware Devices

The important hardware devices in Figure 2 are described below briefly:

• Personal Computer: A personal computer shown in Figure 2 can receive tactile data through a RS 232 serial port and display the relative tactile data magnitude in terms of different colors on the screen.

- Tactile Sensing Device: A two-dimensional 16×16 tactile sensing matrix (FSR matrix array) manufactured by Interlink Electronics [27] is connected to the PC through an interface device. The changes in resistance are the raw tactile data. The size of each basic FSR is 0.2" $\times$ 0.2".
- Interface Device: The interface device reads tactile data from a tactile sensing matrix and sends them online to the RS 232 port of the personal computer through an RS 232 line.

17.2.2.2 Software Kernel

The functions of the software kernel are outlined in the following:
- Receive the raw tactile data from the interface device, preprocess them to form the tactile image, and output the image to the man-machine interface.
- Interpolate the tactile image to increase its resolution and calculate its features for recognition and classification purposes.
- Receive commands from the keyboard and control the man-machine interface.

17.2.2.3 Man-machine Interface

As shown in Figure 2, the main components of the man-machine interface of the tactile sensing and recognition system are two output windows, one result window, and one color table. The system can be equipped with up to two FSR array matrices, each corresponding to an output window. Each output window can display online the pressure distribution of the corresponding FSR matrix array. The mapping relationship between the pressure values and the 16 colors is stored in the color table and shown on the computer screen between the two output windows. According to the color table, the pressure values at each sensor pixel can be represented by different colors. The result window, just below the output windows, is set up for showing the object-recognition results.

By pressing predefined hotkeys, users can control some functions of the man-machine interface, including depicting the interpolated or original pressure distribution, freezing the online display in the output windows, and changing the range of pressure values in the color table. In different applications, the weight-level of the objects to be classified varies and the function for changing the range of pressure values in the color table is designed accordingly.

17.2.3 Interpolation to Increase Resolution

Software interpolation is a cost-effective method to increase the resolution of tactile sensors. This study applied this method to the application of tactile sensing and recognition. Using 3-times interpolation as an example, the interpolation operation is shown in Figure 3. In Figure 3, $P_{ij}, i = 1, 2, \cdots, j = 1, 2, \cdots$, represents the grid (sensor pixel) pressure value at row i and column j, and i_{ij} represents the grid pressure value at row i and column j in P_{22} after interpolation, where P_{ij} are given and i_{ij} are the interpolated pressure values to be calculated. Therefore, interpolation can be defined as determining i_{ij} for each P_{ij}. The following subsections present several methods for implementing software interpolation, including the proposed fuzzy interpolation and fractal interpolation.

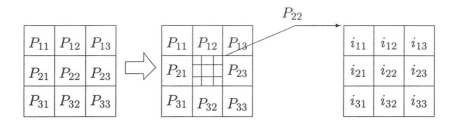

Figure 3 The operation of interpolating P_{22}

17.2.3.1 Linear Interpolation

The most simple and intuitive interpolation is linear interpolation. It is suitable in applications where the data distribution is smooth enough. This method uses a straight line to connect adjacent discrete data points in order to interpolate the data between the given data points. Using Figure 3 as an example, after applying line equations and setting the pressure value i_{22} of the central grid to P_{22}, the interpolated pressure values i_{ij} can then be obtained using the following line equation:

$$i_{ij} = P_{22} + (1/3)(P_{ij} - P_{22}) \, . \tag{17.1}$$

17.2.3.2 Polynomial Interpolation

Replacing the linear equation with a polynomial equation, linear interpolation can be expanded to polynomial interpolation. If the nth order polynomial, $1 + x + x^2 + \cdots + x^n$, is applied to connect the adjacent discrete data points, the i_{ij} in Figure 3 can be calculated using the following equation:

$$i_{ij} = P_{22} + n^{-1}[(1/3) + (1/3)^2 + \cdots + (1/3)^n](P_{ij} - P_{22}) \, , \tag{17.2}$$

where the first n is a normalization scaling constant for satisfying the boundary condition, and the second n is the order of the polynomial. In the case of m-times interpolation, the $1/3$ in the above equation should be replaced with x/m, where x is an integer representing the distance from the current grid to P_{22} in terms of the grid. According to the above equation, it is apparent that polynomial interpolation is more time-consuming than linear interpolation. Therefore, polynomial interpolation is applied only when the pressure distribution is known to be some type of polynomial.

17.2.3.3 Fractal Interpolation

The proposed fractal interpolation assumes that the pressure distribution satisfies the fractal geometry assumption: geometric shapes in the macro view are similar to those in the micro view. To satisfy this assumption, using Figure 3 as an example, the proposed fractal interpolation interpolates P_{22} into i_{ij}:

$$i_{ij} = \frac{9 P_{22} P_{ij}}{\sum_{i,j=1}^{3} P_{ij}} \, , \qquad (17.3)$$

where '9' is a scaling constant for converting the mean value of i_{ij} to P_{22}. In the above equation, since '9', 'P_{22}', and '$\sum P_{ij}$' are constants, i_{ij} is proportional to P_{ij} and hence satisfies the fractal geometry assumption. Since fractal interpolation interpolates a data point by symmetrically making use of the surrounding data to satisfy the fractal geometry assumption, it can only implement odd-numbered interpolations. For n-times interpolations, where n is an odd number, the proposed fractal interpolation interpolates P_{22} into i_{ij}:

$$i_{ij} = \frac{n^2 P_{\frac{n+1}{2} \frac{n+1}{2}} P_{ij}}{\sum_{i,j=1}^{n} P_{ij}} \, . \qquad (17.4)$$

From the above equation, it can be seen that fractal interpolation has the advantage of fast computation, even faster than linear interpolation. By its nature, fractal interpolation can not be applied to boundary data points. Furthermore, when the number of fractal interpolations increases, more surrounding data is required. These are the disadvantages of fractal interpolation.

17.2.3.4 Fuzzy Interpolation

This study used a two-dimensional fuzzy number [6] to depict the pressure information carried by the tactile data in each grid of a tactile sensor, as shown in Figure 4. The fuzzy number in Figure 4 is a regular pyramid with its top vertex located at the center of the gird, and its four bottom vertices at the centers of diagonally adjacent grids. It reflects the fact that the closer a point is to the center of the grid, the closer are their pressure values. The fuzzy number also makes an assumption about the pressure distribution in terms of its geometric shape. Different geometric shapes may be suitable for different applications and should be designed according to that parameter.

Using the fuzzy number in Figure 4, every point in a tactile sensor will have four fuzzy numbers covering it. To interpolate the pressure value of a point using the fuzzy interpolation, we should calculate its membership grades with respect to its four fuzzy numbers and treat the resultant membership values as weights. The pressure value of this point can then be computed as the weighting sum for the pressure values of its four adjacent grids. The computation steps are formulated as follows:

(1) For a given point, read the pressure values of the four adjacent grids and assign them as $\nu_i, i = 1, 2, \cdots, 4$.

(2) Calculate the membership values of the point with respect to the fuzzy numbers of the four adjacent grids and assign them as $\nu_i, i = 1, 2, \cdots, 4$.

(3) The pressure value of point ν_i is calculated as

$$\nu_i = \frac{\sum_{i=1}^{4} m_i \nu_i}{\sum_{i=1}^{4} m_i} \ . \tag{17.5}$$

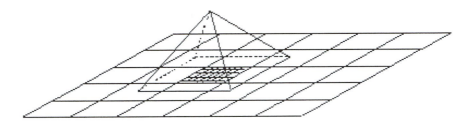

Figure 4 A fuzzy number depicting the pressure information

17.3 Development of a Fuzzy Learning Decision Tree

This study used the following steps to obtain a practical decision tree for classifying objects: (1) select a set of features that can well describe the objects and then massively measure every feature of each object to acquire sufficient training data; (2) apply fuzzy statistics to find fuzzy sets for compressing and representing the training data; and (3) determine a decision tree from the fuzzy sets. The above-mentioned three steps are described in the subsequent context. The performance of the developed fuzzy decision tree is then examined qualitatively by comparing it with a rule-based system and a multi-layer perceptron.

17.3.1 Architecture of the Fuzzy Learning Decision Tree

The complete procedure for determining the proposed fuzzy decision tree can be detailed in a flowchart, as shown in Figure 5. It can be seen from Figure 5 that determining a fuzzy decision tree from the representative fuzzy sets is a recursive process. The components and tree-generation process in Figure 5 are explained below.

17.3.2 Features Selection

Several types of features have been applied to tactile image recognition, including geometric features [17], tactile features [3], and tactile moments [11,12], etc. The above-mentioned features are presented and compared in the following paragraphs. This study chose complex moment invariants, one type of tactile moment, as the feature to describe the objects in the application.

The geometric features include area, perimeter, diameter, radius, convex perimeter, convex area, elongation, thinness ratio, and smoothness, etc. The convex perimeter and convex area are, respectively, the perimeter and area of the convex hull of the contacted point set, where the convex hull of a point set is the intersection of all convex sets containing the point set. One of the advantages of the geometric features is that they are simple and, hence, can easily be computed. These geometric features, however, can not effectively capture the characteristics of objects. For example, a rectangle may have the same area as a circle. Therefore, this chapter did not adopt geometric features. The tactile features include hardness, surface smoothness, surface curvature, surface roughness, oriented edges, etc. These tactile features are useful in applications where the surfaces of objects are much larger than that of the tactile sensor. In the application for this study, however, the surface of the tactile sensor is larger than that of the objects. Hence, this study did not adopt tactile features. By mathematical transformation, a lot of useful features can be extracted from the tactile image [8]. This study selected complex moment invariants as the describing features. The advantages of complex moment invariants are that they are (1) invariant to translation and rotation; (2) simpler and more powerful than other moment features in object recognition [11]; and (3) effective in capturing the characteristics of objects. The complex moment invariants used in this study were derived as shown below.

The definition of a continuous complex moment is:

$$C_{pq} = \int \int (x + iy)^p (x - iy)^q g(x, y) dx dy , \qquad (17.6)$$

where p and q are nonnegative integers; $g(x, y)$ is the gray level function of the tactile image. Because the resolution of the tactile sensor is limited, the above formula is modified into discrete complex moments as:

$$C_{pq}^d = \sum_{j=1}^{n} \sum_{k=1}^{m} (j\Delta x + ik\Delta y)^p (j\Delta x - k\Delta y)^q g(j\Delta X, k\Delta y)\Delta x \Delta y . \qquad (17.7)$$

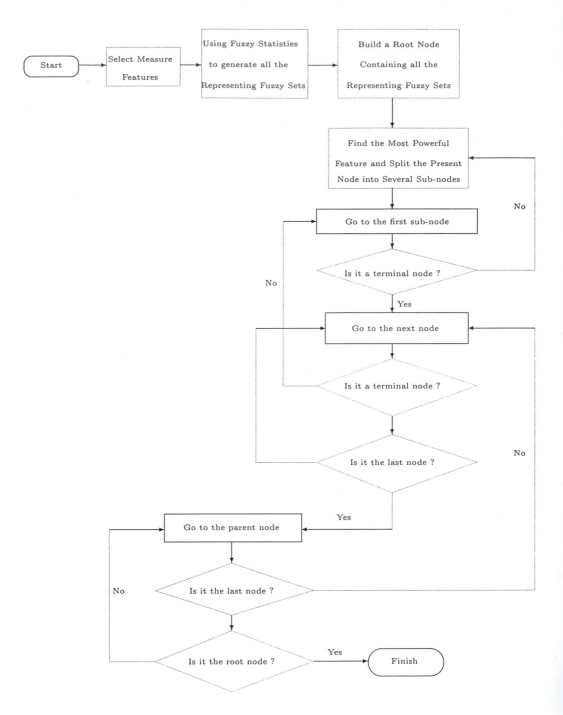

Figure 5   The flowchart of the algorithm of determining a fuzzy decision tree

The equation can be simplified by setting $\Delta x = \Delta y = 1$. The simplified equation is:

$$C_{pq}^{ds} = \sum_{j=1}^{n} \sum_{k=1}^{m} (j + ik)^p (j - ik)^q g(j, k) , \qquad (17.8)$$

where (j, k) represents the coordinates, relative to the corresponding center of pressure, of the grids (or pixels) of the tactile image. A set of complex moment invariants then can be obtained as [11]:

$$
\begin{aligned}
S_1 &= C_{11}^{dsc} & \qquad S_7 &= |C_{40}^{dsc}|^2 \\
S_2 &= |C_{20}^{dsc}|^2 & \qquad S_8 &= |C_{31}^{dsc}|^2 \\
S_3 &= |C_{30}^{dsc}|^2 & \qquad S_9 &= C_{22}^{dsc} \\
S_4 &= |C_{21}^{dsc}|^2 & \qquad S_{10} &= (C_{40}^{dsc})^* (C_{31}^{dsc})^2 + C.C. \\
S_5 &= (C_{30}^{dsc})^2 (C_{21}^{dsc})^* + C.C. & \qquad S_{11} &= C_{31}^{dsc}(C_{20}^{dsc})^* + C.C. \\
S_6 &= (C_{21}^{dsc})^2 (C_{20}^{dsc})^* + C.C.
\end{aligned}
\qquad (17.9)
$$

where the $*$ represents a complex conjugate, and $C.C.$ denotes the complex conjugate of its previous term.

17.3.3 Fuzzy Sets for Compressing Training Data

This study adopted LR type fuzzy sets [6] for compressing and representing training data. Fuzzy statistics [22] were then used to determine several points of the membership functions of the fuzzy sets. Finally, the curve fitting method and the pre-determined points were used to identify the LR type fuzzy sets. The procedure for identifying the representative LR type fuzzy sets is described in the following subsections.

17.3.4 Determining Several Points on a Fuzzy Set

The training data is denoted as $T = \{T_1, T_2, \cdots, T_n\}$, where $T_i = \{T_{i1}, T_{i2}, \cdots, T_{im}\}$ is the training data for the object i, and T_{ij} is the training data for the feature j of the object i. The goal is to find fuzzy sets F_{ij} to represent T_{ij}, $i = 1, 2, \cdots, n, j = 1, 2, \cdots, m$. The algorithm for estimating several points on F_{ij} for identification is outlined as follows: (1) The standard deviation and the mean of T_{ij} are calculated as SD_{ij} and M_{ij}, respectively. (2) The number of points, $(2n + 1)$, to be identified is determined. (3) For the $2n + 1$ points at $M_{ij}, M_{ij} \pm (1/n)SD_{ij}$, $M_{ij} \pm (2/n)SD_{ij}, \cdots, M_{ij} \pm SD_{ij}$, determine their membership grades $\mu_{ij}(k)$ with respect to F_{ij}, as shown in Figure 6, using $\mu_{ij}(k) = S_{ij}(k)/Sij(0)$.

$S_{ij}(k)$ is the number of data located in the interval $[M_{ij} + (k - 1)SD_{ij}/n, M_{ij} + (k + 1)SD_{ij}/n]$, $k = 0, \pm 1, \pm 2, \cdots, \pm n$. From the above equation and Figure 6, it is apparent that $\mu_{ij}(k)$ is proportional to $S_{ij}(k)$, the number of data located around the identified point. Hence, the above equation consists of the physical meaning of fuzzy sets and statistics and $\mu_{ij}(k)$ is the desired point.

This algorithm is valid under the assumption that the training data probability distribution has the property $S_{ij}(0) > S_{ij}(k), \pm 1, \pm 2, \pm, \pm n$. Under this assumption, the identified fuzzy sets are normal fuzzy sets. If the symmetrical fuzzy set assumption is further assumed, only $n + 1$ points need to be estimated.

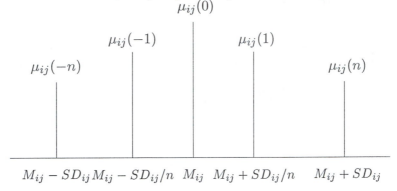

Figure 6 $2n + 1$ points on a fuzzy set F_{ij}

17.3.5 Identifying a LR Type Fuzzy Set

A fuzzy set M is of the LR type if there exists reference functions L (for left) and R (for right), and scalars $a > 0, b > 0$ with

$$
\mu_m = \begin{cases} L(\dfrac{m - x}{a}), & \text{for } x \leq m \\[2mm] R(\dfrac{x - m}{b}), & \text{for } x \geq m \end{cases}, \tag{17.10}
$$

where m is the mean value of M [6]. If the reference functions are approximated by polynomials of order n, L and R can be formulated as

$$
\begin{aligned}
L &= a_0 + a_1 x + a_2 x^2 + \cdots + a_n x^n \\
R &= b_0 + b_1 x + b_2 x^2 + \cdots + b_n x^n
\end{aligned}. \tag{17.11}
$$

Since we have determined in the previous subsection $n + 1$ points for L and R, respectively, the coefficients $a_0, a_1, \cdots, a_n$ and $b_0, b_1, \cdots, b_n$ can be computed by curve fitting method.

17.3.6 Learning Procedure of a Decision Tree

The learning procedure of the fuzzy decision tree is stated in the following:

Step 1. Select a set of features suitable for classification, as described in Section 17.3.2, and massively measure each feature to get sufficient training data.

Step 2. Find the fuzzy sets $F_{ij}, i = 1 \cdots n, j = 1 \cdots m$, to represent the training data, as depicted in Section 17.3.3. Use the fuzzy sets to form fuzzy pattern vectors $F_{pvi} = (F_{i1}, F_{i2}, \cdots, F_{im},)$ for representing object $i, i = 1 \cdots n$.

Step 3. Build the root node and start the learning process. The root node contains all objects represented by F_{pvi}, $i = 1 \cdots n$.

Step 4. Find the most powerful feature for the present node. A feature is the most powerful feature if it can split the node into the greatest number of sub-nodes. If more than one feature has the same power, the feature requiring the least computation time is chosen.

Step 5. Split the present node into several sub-nodes using the feature selected in step 4 and the splitting rule described below.

According to a heuristic threshold T and the distances between the objects, the splitting rule clusters the objects in the present node into several groups. These groups are the content of the desired sub-nodes. The heuristic threshold T is specific to the selected feature and the current tree level (root node is level 0). It is defined by the following equation:

$$T = M_{SD} \times \frac{L_{MAX} - L_{current}}{2} , \qquad (17.12)$$

where M_{SD} is the mean of the standard deviation of the selected feature; L_{MAX} and $L_{current}$ are the maximum allowable level and current level of the fuzzy decision tree under design, respectively. It is easy to see that the above equation is reasonable by considering the following facts:

• The splitting process with a larger threshold is more rigorous, accurate, and slower than that with a smaller threshold.

• When the standard deviation of the selected feature of any object increases, M_{SD} also increases and the threshold T should increase in order to guarantee a satisfactory and correct classification rate.

• When the splitting rule applies to the deeper node, $L_{current}$ increases and the threshold T should become smaller in order to reach terminal nodes more easily with an insignificant decrease in the correct classification rate. This insignificant decrease is due to fewer objects left in the deeper node.

The distance between objects i and j can be viewed from any feature and is defined by the following equation:

$$
\begin{aligned}
d_k(i,j) &= \alpha|M_{ik} - M_{jk}| + \beta|SD_{ik} - SD_{jk}| \\
\alpha + \beta &= 1 \quad \alpha \text{ and } \beta \text{ are real}
\end{aligned} \qquad (17.13)
$$

where $d_k(i,j)$ is the distance between objects i and j, viewed from feature k. M_{ij} and SD_{ij} are defined in Section 17.3.3. α and β are heuristically selected weighting factors for indicating the relative importance between M_{ij} and SD_{ij}. The above definition has the advantage of considering both M_{ij} and SD_{ij}. The experiment presented in Section 17.4 sets α and β as 0.9 and 0.1, respectively. Because M_{ij} are dominant, it is reasonable that α should be larger than β. Several values for α and β may be tried in order to get a good result. A node is a terminal-node if it satisfies a predefined stopping criterion, otherwise it is a non-terminal node. A node satisfies the predefined stopping criterion if it contains only one fuzzy pattern vector, or contains more than one fuzzy pattern vector but are all too similar to split.

Step 6. For each non-terminal node, repeat Steps 4 and 5 until there are no non-terminal nodes.

Step 7. For every terminal node containing more than one fuzzy pattern vectors, the next most powerful feature is linearly combined to split it. This step is repeated until each terminal node contains only one fuzzy pattern vector.

Based on the above steps, a practical fuzzy decision tree can be obtained. The resultant decision tree generates a piecewise smooth decision surface. Note that the proposed fuzzy decision tree is determined in a batch manner from the training data. Namely, the proposed algorithm processes all of the training data and then generates a decision tree by determining its structure and parameters. The structure of a decision tree means the topology of the tree. The parameters of a decision tree include the objects and the threshold in each non-terminal node. If it is desirable to adapt to any changes in the training data, with even a small change, such as adding one data point, the proposed algorithm must re-process all of the training data to obtain a new decision tree. This is equivalent to forgetting all the old data and then re-learning the old data plus new data. Although this learning manner is not efficient, the proposed algorithm is so efficient that it can overcome this drawback.

17.3.7 Comparing to Rule Based Systems

In a decision tree, a path from the root node to any one of the terminal nodes contains at least two nodes and hence at least one splitting rule. The splitting rules in a path must be simultaneously satisfied so that the terminal node of the path can be reached from the root node. Using 'and' and 'not' operators, the splitting rules of a path can be combined to form a new rule to represent the path. Repeating this procedure for every path in a decision tree, the decision tree can then be mapped into a rule-based system with each path represented by a rule. The fuzzy decision tree developed in this study can be mapped into a fuzzy rule-based system. In this mapping method, the number of the paths in the decision tree is equal to the number of rules in the corresponding rule-based system. In symmetrical binary trees, the relationship among the numbers of paths, tree levels, and rules is

$$\text{the number of paths} = \text{the number of rules} = 2^{(\text{the number of tree levels}-1)},$$

$$(17.14)$$

where the number of tree levels includes the levels containing the root node and the terminal nodes. A symmetrical tree is defined as a tree in which every sub-tree on one side has the same topology as its counterpart on the other side. An example of the mapping relationship is shown in Figure 7.

In Figure 7, $SR_{ij}, i = 0, 1, \cdots, j = 1, 2, \cdots$, is the splitting rule in the jth node of the ith level. The root node is in level 0 and has the splitting rule SR_{00}. TN_k, $k = 1, 2, \cdots$, is the kth terminal node. R_m, $m = 1, 2, \cdots$, is the rule representing the path from the root node to the mth terminal node. In Figure 7(a), the root node receives the data input, the internal or non-terminal nodes then make decisions, and finally the terminal nodes output the results. In Figure 7(b), inference engine receives the data input, then infers and outputs results according to the rules in the rule base and the data in the database. The rules in Figure 7(b) can be obtained from the splitting rules in Figure 7(a) using the following equations:

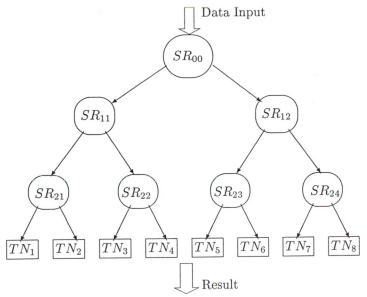

(a) An example biary tree

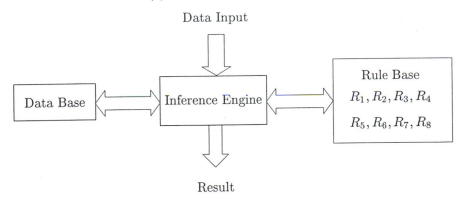

Result

(b) The corresponding rule-based system

Figure 7   An example binary tree and its corresponding rule-based system

$$R_1 = SR_0 \wedge SR_{11} \wedge SR_{21} \quad R_5 = (\sim SR_0) \wedge SR_{12} \wedge SR_{23}$$
$$R_2 = SR_0 \wedge SR_{11} \wedge (\sim SR_{21}) \quad R_6 = (\sim SR_0) \wedge SR_{12} \wedge (\sim SR_{23})$$
$$R_3 = SR_0 \wedge (\sim SR_{11}) \wedge SR_{22} \quad R_7 = (\sim SR_0) \wedge (\sim SR_{12}) \wedge SR24$$
$$R_4 = SR_0 \wedge (\sim SR_{11}) \wedge (\sim SR_{22}) \quad R_8 = (\sim SR_0) \wedge (\sim SR_{12}) \wedge (\sim SR_{24}) \,,$$
$$(17.15)$$

where '$\wedge$' is 'and' operator and '$\sim$' is 'not' operator. The 'and' and 'not' operators have their well-developed definitions in traditional two-value logic and fuzzy logic. Which definitions will be applied depend upon the requirements of the applications.

In every decision process of a decision tree, only the splitting rules in one path will be checked. On the other hand, in the corresponding rule-based system, every rule will be checked in every decision process, where each rule is composed of all of the splitting rules in the corresponding path. Therefore, the inference speed of a decision tree is approximately n-times faster than that of its corresponding rule-based system, where 'n' is the number of rules (or paths). For the same reason, however, a decision tree is less reliable than its corresponding rule-based system. Namely, by considering every rule, rule-based systems increase the reliability at the expense of decreasing its inference speed. If a decision tree is the proposed fuzzy decision tree, there is another reason for less reliability: there is fuzziness in each non-terminal node and the fuzziness will accumulate from the root node through the terminal nodes. Fortunately, by selecting suitable threshold values, limiting the depth of the trees, and collecting good training data, the problem of fuzziness can be diminished and reliable results can be obtained.

The storage spaces required by the decision trees and rule-based system are different and worthy of comparison. This chapter uses n-ary trees as examples to make the comparison. An n-ary tree refers to a decision tree in which every non-terminal node contains n sub-nodes. Assume that every splitting rule occupies the same amount of storage space. Setting this amount to unity, the storage space Sd required by an n-ary decision tree of the m-level is

$$S_d = 1 + n + n^2 + \cdots + n^{m-2} = (n^{m-1} - 1)/(n - 1) . \qquad (17.16)$$

The storage space S_r required by the corresponding rule-based system is

$$
\begin{aligned}
S_r &= \text{(number of rules or paths)} \times \text{(number of splitting rules in one path)} \\
&= \text{(number of terminal nodes)} \times \text{(number of splitting rules in one path)} . \\
&= n^{m-1} \times (m - 1) = (m - 1)n^{m-1}
\end{aligned}
$$

$$(17.17)$$

The difference for storage space D_s is

$$
\begin{aligned}
D_s = S_r - S_d &= (m - 1)n^{m-1} - (n^{m-1} - 1)/(n - 1) \\
&= \{[(m - 1)(n - 1) - 1]n^{m-1} + 1\}/(n - 1)
\end{aligned} \qquad (17.18)
$$

Because both m and n are greater than or equal to 2, D_s in the above equation is always greater than zero. Furthermore, D_s increases with m linearly and with n in the $(m - 1)^{th}$ power. Therefore, the storage space required by a decision tree is smaller than that required by its corresponding rule-based system, and the difference increases with the depth and width of the decision tree. The differences between a decision tree and its corresponding rule-based system are summarized in Table 1.

Table 1 The Differences between Decision Trees and Their
Corresponding Rule-based Systems

	Decision Trees	Rule-based Systems
Speed	Fast	Moderate
Reliability	Moderate	Good
Storage Space	Small	Large

17.3.8 Comparison with Artificial Neural Networks

As the decision surfaces of decision trees are piecewise smooth, it can be assumed that the decision surfaces of some kinds of decision trees are piecewise linear. Under the piecewise linear assumption, the regions split by the decision surfaces in the feature space can be obtained by 'and' and 'or' operations of the linear inequalities in the nodes of the decision tree. These operations can be performed not only by decision trees but also multi-layer perceptrons (MLPs). Therefore, it is straightforward to infer that a decision tree can be mapped into an equivalent neural network. Entropy nets were proposed in [19] as the equivalent neural networks mapped from decision trees. In fact, entropy nets are not new types of neural networks but one type of MLPs. An example for explaining the mapping from decision trees to MLPs is shown in Figure 8.

In Figure 8, x_1 and x_2 are the input features. SR_{ij} has the same meaning as that in Figure 7. C_i, $i = 1, 2, 3$, are the classes to be classified. The neuron in hidden layer 1 performs the function of SR_{ij}. The neurons 'AND' in hidden layer 2 perform the logical 'and' operations, and the neurons 'OR' in the output layer perform the logical 'or' operations. In the corresponding MLP, the neurons in the input layer receive the input data and then distribute it to each neuron in the hidden layer 1. The neurons in hidden layer 1 linearly partition the feature space to implement the functions of the non-terminal nodes in the decision tree. The neurons in hidden layer 2 produce the hierarchical structure of the decision tree.

The neurons in the output layer manage the situation where at least one class contains more than one region. The realization of the neuron in hidden layer 1 is given in Figure 9.

Note that S_{ij} is the shift constant of SR_{ij}, and net is the polynomial part of SR_{ij}. The realization of the neuron 'AND' is given in Figure 10, while the realization of the neuron 'OR' is given in Figure 11.

The mapping procedure from a decision tree to its corresponding MLP is outlined in the following:

i. Construct the input layer so that every feature corresponds and connects to one neuron in the layer.

ii. Construct hidden layer 1 so that every neuron in the layer receives all of the feature output of the input layer. The function of every node of the decision tree is performed by one neuron in this layer. Through this construction, the number of neurons in the layer should be equal to the number of non-terminal nodes in the decision tree.

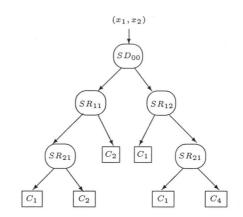

(a) A symmetrical non-full binary tree

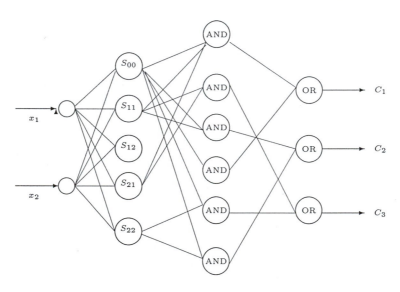

(b) The corresponding MLP of (a)

Figure 8   A Symmetrical non-full binary tree and its corresponding MLP as an example

iii. Construct hidden layer 2 with each neuron in the layer performing the function of logical 'and'. The purpose of the neurons in the layer is to represent the terminal nodes of the decision tree and hence the path to them. Therefore, each neuron should perform the 'and' function for combining the non-terminal nodes in the path and hence represent the path. It should be noted that the number of neurons in the layer should be equal to the number of terminal nodes in the decision tree.

iv. Let each neuron in hidden layer 2 receive the output of those neurons in hidden layer 1. Those neurons in hidden layer 1 represent the nodes in the corresponding path.

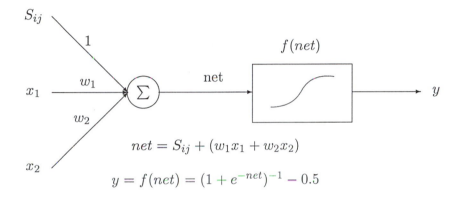

$$net = S_{ij} + (w_1 x_1 + w_2 x_2)$$

$$y = f(net) = (1 + e^{-net})^{-1} - 0.5$$

Figure 9 The neuron performing the splitting function SR_{ij}

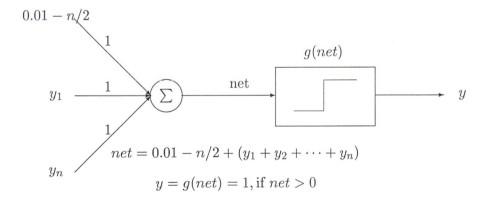

$$net = 0.01 - n/2 + (y_1 + y_2 + \cdots + y_n)$$

$$y = g(net) = 1, \text{ if } net > 0$$

Figure 10 The neuron performing the logical 'and' function

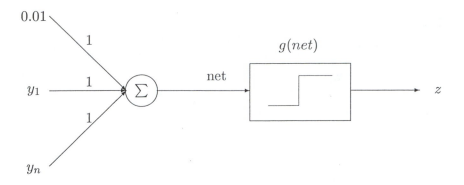

Figure 11 The neuron performing the logical 'or' function

v. Construct the output layer so that every class is represented by one neuron in the layer. This neuron performs the logical 'or' function to combine all of the regions composed by the corresponding class. Through this construction, the number of output layers should be equal to the number of classes to be classified. If each class contains only one region, the logical 'or' function output layer is not necessary and hidden layer 2 becomes the output layer.

vi. Let each neuron in the output layer receive the output of those neurons belonging to the corresponding class in hidden layer 2.

It can be proved that if a decision tree is a binary tree, symmetrical or non-symmetrical, the number of terminal nodes is always greater than the number of non-terminal nodes by 1. Therefore, in the corresponding MLP, the number of neurons in hidden layer 2 is always greater than the number of neurons in hidden layer 1 by 1. The above facts can be formalized by the following theorem:

Theorem 1: If a decision tree is a binary tree, symmetrical or non-symmetrical, the number of terminal nodes is always greater than the number of non-terminal nodes by 1.

Proof: A decision tree is full, if and only if, the two sub-nodes of every non-terminal node are of the same type (non-terminal or terminal). If the binary decision tree is symmetrical and full, the number of non-terminal nodes N_n is

$$N_n = 1 + 2 + 2^2 + \cdots + 2^{d-2} = 2^{d-1} - 1 \, ,$$

where d is the depth of the decision tree. The number of terminal nodes N_t is

$$N_t = 2^{d-1}$$

Hence, N_t is greater than N_n by 1 and the theorem can be applied to binary decision trees that are symmetrical and full.

In this proof, the **basic sub-tree** is defined as a tree consisting of a non-terminal node and two terminal nodes. A **basic removal** is defined as the action of replacing a basic subtree with a terminal node. It can easily be understood that any non-symmetrical or non-full tree can be determined by performing several basic removals. Because both the numbers of terminal and non-terminal nodes decrease by 1 after one basic removal, the difference between the two numbers will not change after several basic removals. Consequently, the theorem can be applied to any type of binary decision tree, symmetrical or non-symmetrical, full or non-full, and the proof is valid.   **Q.E.D.**

It can be observed from Figure 8 that the mapped MLP is not fully connected and has fewer connections than an ordinary MLP. This feature has the advantage of saving storage space and increasing learning speed. Almost all artificial neural networks, including MLPs, are black box in structure. The structure of the mapped MLP, however, is as clear as the decision tree. In the mapped MLP, the function of each neuron and the meanings of the weights are clear. Furthermore, the partition of the feature space into regions, the 'and' and 'or' operations of regions, can be easily observed in hidden layer 1, hidden layer 2, and the output layer, respectively. Applying the sigmoid function to the neurons in hidden layer 1 of the mapped MLP, the generalization capacity can be better than that of the decision tree. Comparing the batch learning characteristics of the decision tree, another advantage of the mapped MLP is its incremental learning capacity. For new unlearned data, this capacity makes the mapped MLP have a faster learning speed over the decision tree. For initial training data, however, the decision tree learns much faster than the mapped MLP. The mapped MLP has an initial satisfactory result and can be further trained to gain more precise results. Although the mapped MLP has the advantages mentioned above, it can be seen from Figure 8 that their computational load and computation time are greater than those of the decision trees. Fortunately, with parallel computation, the computation time of the mapped MLP can be less than that of the decision tree.

17.4 Experiments

This section describes the application of the proposed fuzzy decision tree to a tactile image classification problem to examine the performance and functions of the tactile sensing and recognition system.

Table 2 Object Names and Their Corresponding Class Names

Class Name	C1	C2	C3	C4	C5	C6	C7
Object Name	Rectangle in 0°	Rectangle in 27°	Rectangle in 45°	Circle	Square	Hand in 0°	Hand in 45°

17.4.1 Experimental Procedures

The experimental procedures are described below:

(1) Select the objects to be classified. The objects selected in this experiment were circle, square, rectangle, and human hand. For testing the orientation and recognition capacity, the same hand and rectangle in different orientations were regarded as different objects. Therefore, the objects used in this experiment were circle, square, rectangle in $0°$, rectangle in $27°$, rectangle in $45°$, hand in $0°$, and hand in $45°$. Table 2 lists the object names and their corresponding class names.

(2) Choose the complex moment invariants, $S_1, S_2, \cdots, S_{11}$, described in Section 17.3.2 as the features for describing the objects in this experiment.

(3) Thoroughly measure the features of the objects in Table 2 to get sufficient training data. In this experiment, every feature of each object is measured 50 times in different positions. The mean values of the measured training data are summarized in Table 3.

Table 3 Mean Values of the Training Data

Features\ Objects	C1	C2	C3	C4	C5	C6	C7
Feature 1: S1	27.55	22.50	32.18	6.72	23.49	972.12	804.93
Feature 2: S2	512.47	237.06	625.92	1.94	1.05	1.92e5	1.59e5
Feature 3: S3	2.49	21.05	42.49	1.90	3.11	1.19e6	1.71e6
Feature 4: S4	1.87	2.37	9.96	0.12	1.40	5.81e6	5.19e6
Feature 5: S5	2.49	41.21	212.25	0.37	4.68	4.62e9	6.06e9
Feature 6: S6	58.61	55.49	412.59	0.08	2.37	5.03e9	4.04e9
Feature 7: S7	7088.06	1566.41	9512.83	2.56	358.88	7.70e7	1.07e8
Feature 8: S8	1.35e4	4497.79	1.91e4	10.81	18.83	4.12e8	4.01e8
Feature 9: S9	129.68	82.85	157.84	10.78	63.37	3.83e4	3.29e4
Feature 10: S10	2.28e6	3.65e5	3.78e6	32.46	577.80	7.06e12	8.03e12
Feature 11: S11	5260.11	2060.90	6918.19	9.10	8.83	1.77e7	1.58e7

(4) Identify the representative fuzzy sets for every feature of each object using the method proposed in Section 17.3.3. Considering speed and simplicity, the study assumed that the shapes of the fuzzy sets are all triangular. Hence, the L and R reference functions are line equations and there are only two parameters which need to be estimated for each reference function. Using the representative fuzzy set F_{ij} as an example, the membership at M_{ij} is 1 and the memberships at $M_{ij} \pm Sd_{ij}$ are 0. The two points on the L and R reference functions can then be obtained and used to estimate their parameters.

(5) Determine a decision tree from the training data using the algorithm proposed in Section 17.3.4. The resulting decision tree is shown in Figure 12.

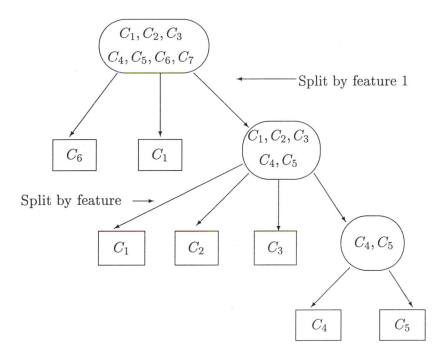

Figure 12 The resultant decision tree

(6) Install the generated decision tree in Step (5) into the tactile sensing and recognition system depicted in Section 17.2.2.

(7) Place an object on the tactile sensing device, measure its tactile image, and on-line interpolate the tactile image.

(8) On-line classify the object given in Step (7).

17.4.2 Experiment Results and Discussions

We will discuss the tactile images of a test object after the fuzzy interpolation in this section.

Observing the boundaries and pressure distribution of the tactile images, it is easy to see that fuzzy interpolation can produce smooth and reasonable results. On the other hand, the accuracy of fuzzy interpolation depends upon the accuracy of the fuzzy numbers assigned to the sensor grids. Theoretically, if the assigned fuzzy numbers are exact, the tactile image can be interpolated to infinite resolution with exact accuracy. Physically, the resolution of the fuzzy interpolation, however, depends on available computer memory. The correct classification rates in Table 3 are satisfactory but not perfect. The reasons for the imperfections are the limited resolution of the tactile sensor and the variable contact conditions between the objects and the tactile sensor. The limited resolution causes the selected features of an object not to remain the same, even for slight changes in position and orientation. The variable contact condition makes the selected features of an object not to remain unchanged, even for the same position and orientation.

It should be noted that n-time interpolation means that the resolution in each dimension of the tactile image increases n times. The correct classification rates of the objects are listed in Table 4.

Table 4 Correct Classification Rate

Objects Name	Correct Number	Fail Number	Correct Classification Rate
Circle	47	3	94 %
Square	40	10	80 %
Rectangle in 0°	46	4	92 %
Rectangle in 27°	48	2	96 %
Rectangle in 45°	44	6	88 %
Hand in 0°	45	5	90 %
Hand in 45°	44	6	88 %

17.5 Conclusions

This study used fuzzy sets to represent and compress training data in order to save computation time and storage space. By learning the fuzzy sets, the proposed tree generation algorithm can rapidly generate a satisfactory decision tree. To examine the performance of the algorithm, the generated tree was mapped into a rule-based system and a multi-layer perceptron and several comparisons were made. Furthermore, the algorithm was installed in a tactile sensing and recognition system and experiments were conducted on tactile image classification. The generated decision tree can perform classifications at a satisfactory correct rate.

As the resolution of a tactile sensor is limited, the complex moment invariants can not be exactly invariant under translation and rotation. This is a drawback that may make tactile images more difficult to classify. This study applied the proposed fuzzy interpolation method to increase the resolution and use it to recognize the orientation of an object. The proposed fuzzy interpolation is cost-effective and proven to be useful and satisfactory. A heuristic threshold was used in the proposed tree generation algorithm. It reflects the probability distribution of the training data and hence increases the correct classification rate. As the tree grows deeper and deeper, the heuristic threshold becomes smaller and smaller and hence reduces the tree size because terminal nodes are easier to obtain with smaller thresholds. Therefore, the heuristic threshold makes the generated tree close to optimal in the sense of correct classification rate and tree size.

References

1. L. Atlas, R. Cole, Y. Muthusamy, A. Lippman, J. Connor, D. Park, M. E. Sharkawi, and R. J. Marks II, A performance comparison of trained multilayer perceptrons and trained classification trees, *Proceedings of the IEEE,* Vol. 78, No. 10, pp. 1614-1619, Oct. 1990.

2. D. E. Brown, V. Corruble, and C. L. Pittard, A comparison of decision tree classifiers with backpropagation neural networks for multimodal classification problems, *Pattern Recognition,* Vol. 26, No. 6, pp. 953-961, 1993.

3. R. A. Browse, Feature-based tactile object recognition, *IEEE Transactions on Pattern Analysis and Machine Intelligence,* Vol. PAMI-9, No. 6, pp. 779-786, Nov. 1987.

4. R. L. P. Chang and T. Pavlidis, Fuzzy decision tree algorithms, *IEEE Transactions on Systems, Man, and Cybernetics,* Vol. SMC-7, No. 1, pp. 28-35, January 1977.

5. P. Dario, M. Bergamasco, D. Femi, A. Fiorillo, and A. Vaccarelli, Tactile perception in unstructured environments: A case study for rehabilitative robotics applications, *International Conference on Robotics and Automation,* pp. 2047-2054, 1987.

6. D. Duboid and H. Prade, *Fuzzy Sets and Systems: Theory and Applications,* Academic Press, New York, 1980.

7. H. Guo and S. B. Gelfand, Classification trees with neural network feature extraction, *IEEE Transactions on Neural Networks,* Vol. 3, No. 6, Nov. 1992.

8. J. Jurczyk and K. A. Loparo, Mathematical transforms and correlation techniques for object recognition using tactile data, *IEEE Transactions on Robotics and Automation,* Vol. 5, No. 3, pp. 359-362, June 1989.

9. P. Kazanzides, J. Zuhars, B. Mittelstadt, and R.H. Taylor, Force sensing and control for a surgical robot, *International Conference on Robotics and Automation,* pp. 612-617, 1992.

10. A. Leung and S. Payandeh, Application of adaptive neural network to localization of objects using pressure array transducer, *Robotica,* Vol. 14, pp. 407-414, 1996.

11. R. C. Luo and H. H. Loh, Tactile array sensor for object identification using com-lex moments, *International Conference on Robotics and Automation,* pp. 1935-1940, 1987.

12. R. C. Luo and W. H. Tsai, Object recognition using tactile image array sensors, *International Conference on Robotics and Automation,* pp. 1248-1253, 1986.

13. G. J. Monkman and P. K. Taylor, Thermal tactile sensing, *IEEE Transactions on Robotics and Automation,* Vol. 9, No. 3, pp. 313-318, June 1993.

14. C. Muthukrishnan, D. Smith, D. Myers, J. Rebman, and A. Koivo, Edge detection in tactile images, *International Conference on Robotics and Automation,* pp. 1500-1505, 1987.

15. Y. Park and J. Sklansky, Automated design of linear tree classifiers, *Pattern Recognition,* Vol. 23, No. 12, pp. 1393-1412, 1990.

16. W. Pedrycz, Fuzzy sets in pattern recognition: Methodology and methods, *Pattern Recognition,* Vol. 23, No. 1-2, pp. 121-146, 1990.

17. J. W. Roach, P. K. Paripati, and M. Wade, Model-based object recognition using a large-field passive tactile sensor, *IEEE Transactions on Systems, Man, and Cybernectics,* Vol. 19, No. 4, pp. 846-853, July/August 1989.

18. T. D. Sanger, A tree-structured adaptive network for function approximation in high-dimensional spaces, *IEEE Transactions on Neural Networks,* Vol. 2, No. 2, March 1991.

19. I. K. Sethi, Entropy nets: From decision trees to neural networks, *Proceedings of the IEEE,* Vol. 78, No. 10, pp. 1605-1613, Oct. 1990.

20. I. K. Sethi and J. H Yoo, Design of multicategory multifeature split decision trees using perceptron learning, *Pattern Recognition,* Vol. 27, No. 7, pp. 939-947, 1994.

21. S. A. Stansfield, Primitives, features, and exploratory procedures: Building a robot tactile perception system, *International Conference on Robotics and Automation,* pp. 1274-1279, 1986.

22. P. Z. Wang, From the fuzzy statistics to the falling random subsets, *Advances on Fuzzy Sets Theory and Applications,* P. P. Wang, ed., Pergamon Press, pp. 81-95, 1983.

23. S. Yaniger and J. P. Rivers, Force and position sensing resistors: An emerging technology, *Interlink Electronics,* Feb. 1992.

24. L. A. Zadeh, Outline of a new approach to the analysis of complex systems and decision processes, *IEEE Transactions on Systems, Man, and Cybernetics,* Vol. SMC-1, pp. 28-44, 1973.

25. H. P. Huang and C. C. Liang, A learning fuzzy decision tree and its application to tactile image, *IEEE/RSJ International Conference On Intelligent Robots and Systems,* Oct. 13-17, 1998, Victoria, B.C., Canada.

26. FSRTM Technical Specifications, *Interlink Electronics.*

27. What is a Force Sensing Register, *Interlink Electronics.*

Chapter 18

Fuzzy Assessment Systems of Rehabilitative Process for CVA Patients

In recent years, cerebrovascular accidents have become a very serious disease in our society. How to assess the states of cerebrovascular accident (CVA) patients and rehabilitate them is very important. Therapists train CVA patients according to the functional activities they need in their daily lives. During the rehabilitative therapeutic activities, the assessment of motor control ability for CVA patients is very important. In this chapter, a fuzzy diagnostic system is developed to evaluate the motor control ability of CVA patients. The CVA patients will be analyzed according to the motor control abilities defined by kinetic signals. The kinetic signals are fed into the proposed fuzzy diagnostic system to assess the global control ability and compare with the FIM (Functional Independent Measurement) score, which is a clinical index for assessing the states of CVA patients in hospitals. It is shown that the proposed fuzzy diagnostic system can precisely assess the motor control ability of CVA patients.

18.1 Introduction

In recent years, cerebrovascular accidents have become a very serious disease all over the world. Since cerebrovascular accidents always make part of a person's brain cells lack oxygen, his/her brain usually cannot control certain muscles volitionally. When the cerebrovascular accidents occur, they affect the abilities of a person's daily life and become very burdensome to families and the whole society. Ways to assess the states of patients and help them to recover and take care of themselves, are very important for therapists to train cerebrovascular accident patients.

Traditionally, therapists used to train cerebrovascular accident (CVA) patients according to the functional activities in daily life, such as from lying to sitting, from sitting to standing, and from standing to walking. If a patient can not stand well, he/she is trained to stand upright. If a patient can not walk well, he/she is trained to walk again until the skill is regained. It is a direct and effective method. But in some aspects, the rehabilitative process will accustom patients to using the sound side and ignoring the affected side. Though the experienced therapists adjust these problems in some ways during the rehabilitative process, it may not be the same case for young therapists. How to rehabilitate CVA patients more objectively and efficiently is very important.

Medical doctors usually assess the CVA patients based on FIM (Functional Independent Measurement) scores and Brunnstrom Stage. FIM scores are used to assess the independent ability of the subjects, and Brunnstrom Stage is used to assess the muscle control ability of the affected side. Kinetic signals and kinematic signals are used for evaluating the motor control ability.

The trajectory of center of gravity (COG) and the trajectory of the center of pressure (COP) are two typical kinetic signals. They are often used to appraise the postural stability of subjects. The kinetic signals can be analyzed in terms of some typical features, such as the sway path, the sway area, and the sway frequency.

Many researchers used the COP trajectory of the test period to realize the normal sway area of the subject. Some researchers analyzed the maximum sway position in A/P or M/L direction [5,6,9,12,19], while others used the recuperated time, from perturbed to stable, to define the ability of balance [12,18]. The COP movement speed was used to classify the subjects' ability [5,8,12]. The weight difference between two feet was also used to define the degrees of balance [6,8,10,14]. Shue et al. analyzed the entire period of the trajectory to identify the several states of a functional activity [18]. Dieners and coworkers [3] analyzed the response in frequency domain in the process of functional activities.

The trajectory of every main segment of a man is a typical kinematic signal. A functional activity usually has its basic patterns. The most obvious way to distinguish abnormal people from normal ones is to find out the patterns of a functional activity. Based on the basic patterns of certain activities, the subjects can be diagnosed. Some people used a camera to record the trajectory of every main segment to identify the normal pattern [15]. Some used inclinometers fixed on the angles to identify the angular variance in every phase of certain functional activities [4,10,13,16,18].

The kinetic and kinematic signals are used to explore the states of patients and help the rehabilitation. Artificial Neural Network (ANN) and fuzzy logic are useful tools to identify the rehabilitative strategies. Patterson [17] used a neural network approach with EMG signals to determine the status of anterior cruciate ligament (ACL) rehabilitation. Abel [1] used neural networks for analyzing and classifying healthy subjects and patients with myopathic and neuropathic disorders. Graupe [7] used ANN controlling functional electric stimulation (FES) to facilitate patient-responsive ambulation. Loslever [11] used the fuzzy logic to draw gait patterns at which the kinetic signals and kinematic signals are involved. Barreto [2] used ANN and fuzzy logic as associative memories to build an expert system for aiding medical diagnosis. Therefore, the medical decision support system can be constructed using ANN and fuzzy logic. The ANN and fuzzy logic evaluate the states of patients and learn the experience of experts to make the decision.

In order to rehabilitate CVA patients, the state of the subjects must be defined first. In this chapter, the concept of motor control ability is used to assess the state of CVA patients. Motor control is an interface between the neuroscience, kinesiology, and biomechanics. The neuro-physiological, kinematic, and kinetic signals can realize the motor control ability of subjects. In this chapter, the kinetic response and the neuro-physiological response are used to define the state of motor control ability. If the states of kinetic response and the neuro-physiological response are the two axes of motor control ability, the concept of motor control ability plane can be formed, as shown in Figure 1.

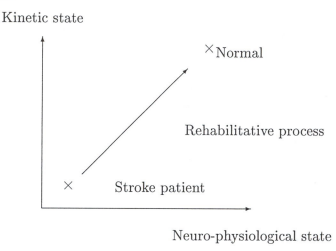

Figure 1 The concept of motor control ability plane

In this chapter, a fuzzy kinetic state assessment system will be constructed so that the kinetic signals, i.e., COP signals, can be characterized. The system is used to evaluate the kinetic states of the subjects. In the kinetic state assessment system, the rule base, database and its inference will be developed.

18.2   COP Signals Feature Extraction

The force distribution on the foot during a test period can be used to analyze the postural control ability of a subject. In this chapter, the postural control ability is defined as the ability that can maintain the center of mass near the center of support base and can result in less body sway when the posture is disturbed by external or internal forces. In order to achieve these goals, the central neural system (CNS) must integrate several systems. The visual system, vestibular system and proprioceptor system are used to detect the posture of the body. The musculoskeletal system is used to arrange every segment of the body to keep the COP within the support base.

It is difficult to identify the postural control ability using the neural- physiological signals because many kinds of actions may be performed. The easiest way to identify the postural control ability is to observe the trajectory of the COP when an external force or an internal force disturbs the posture. For a human system, the response of the system characterizes the features of the system, as shown in Figure 2.

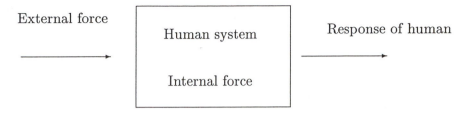

Figure 2 The response of human system when an external

force or internal force is input

The response signals of subjects in this chapter will be discussed in four domains: space domain, time domain, frequency domain, and force domain. By calculating the trajectory of the COP during the test period and fitting into the events that disturb the balance states, some postural control ability features can be found. Figure 3 shows the four domains used to analyze the COP trajectory.

18.2.1   Space Domain Analysis

The COP trajectory on the space domain is the most direct data about the postural control ability. A man who can keep his balance may keep his COP near the center of the support base when the postural is disturbed. The minimum circle, which can include the trajectory of COP during the test period, is an index to identify the postural control ability. Figure 4 shows the concept of the minimum circle of sway during the test period. In order to decrease the variance of each subject, the trajectory of the COP must be normalized by the support base as

$$X_{normalized} = X_{original}/W \qquad (18.1)$$

$$Y_{normal} = [Y_{original} - (Y_{upper} + Y_{lower})]/L . \qquad (18.2)$$

All symbols are defined in Figure 4. Hence, the features of COP trajectory on the space domain can be defined by two indices: A_{min} and R_{min} · A_{min} is the area of minimum circle of sway. The circle includes the trajectory of the COP during the test period. R_{min} is the radius of the minimum circle of sway.

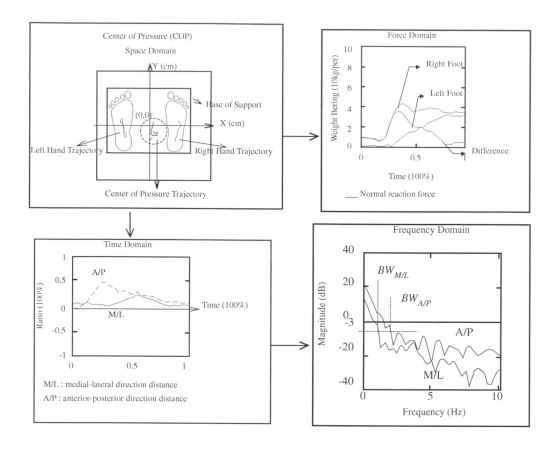

Figure 3 The process of analyzing COP data

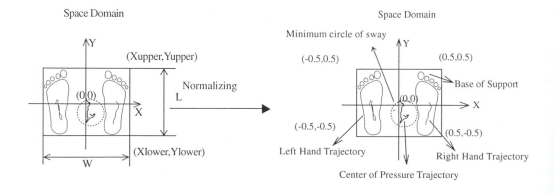

Figure 4 The concept of minimum circle of sway on the space domain

18.2.2   Time Domain Analysis

In order to realize the postural control ability of the subjects, the balance index, including $Maxd_{A/P}$ and $Maxd_{M/L}$, will be defined in the time domain. Before calculating the balance index, the detected signals should be preprocessed. The trajectories of the COP are separated into two different directions, anterior-posterior (A/P) direction and medial-lateral (M/L) direction. There are some particular features in every test action. For example, when the movement platform moves in the A/P direction, the most important feature is the sway amplitude of the COP in the A/P direction rather than the COP position in space. When the subject acts from sitting to standing, the sway amplitude in the M/L direction is an important feature due to the abnormal sway in the M/L direction. In order to decouple these features, the trajectory of the COP is divided into A/P direction and M/L direction. Figure 5 shows the process of data separation and normalization on time domain.

Different testing periods exist in certain actions, such as the STS (sit to stand) periods. Thus, the testing time should be normalized. The method is to filter the original data with a low-pass filter and cut these data into certain parts. The representative value of each part is the mean value of each part. All parts form a new time series data, which represents the normalized curve. In this chapter, the second order butterworth digital filter with cut-off frequency 5Hz is designed to achieve the goal. The transfer function of the filter is given as

$$H(z) = \frac{0.6389 + 1.2779z^{-1} + 0.6389z^{-1}}{1 - 1.143z^{-1} + 0.4128z^{-2}} \ . \tag{18.3}$$

This method is useful and reasonable because the sway of the COP is a continuous and low-frequency response. The low-pass filter can not only decay the noise generated by the recording machine but also rebuild the position that the COP really goes. Furthermore, the sway amplitude should be normalized due to the different base of support. The normalizing equations are shown in Equations (18.1) and (18.2).

The postural control ability indices in time domain include $Maxd_{A/P}$ and $Maxd_{M/L}$ is the maximum sway distance in the A/P direction. It is defined by $Maxd_{A/P} = AP_{upper} - AP_{lower} \cdot Maxd_{M/L}$ $Maxd_{A/P}$ is the maximum sway distance in the M/L direction. It is defined by $Maxd_{M/L} = ML_{upper} - ML_{lower}$. Their relations are given in Figure 6. All symbols are also defined in Figure 6.

18.2.3   Frequency Domain Analysis

If the time domain signals are transformed into frequency domain, other features of the signals will appear. During the test of the movement platform, the movement platform can be regarded as an impulse input to the human system.

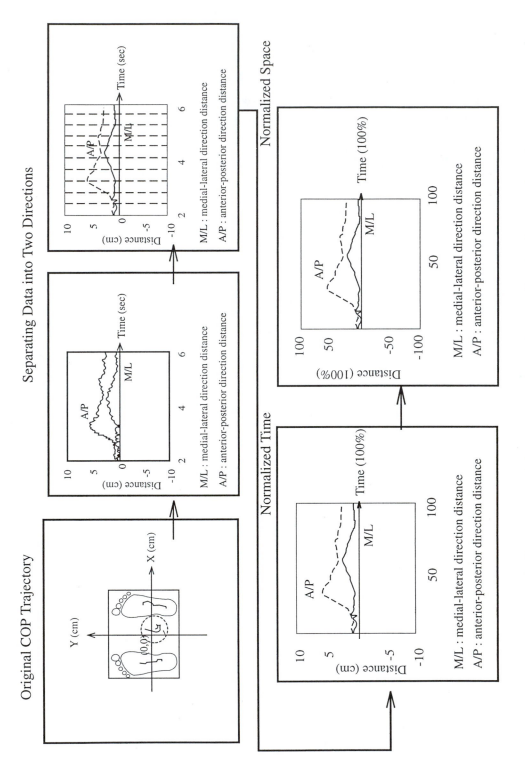

Figure 5 The process of data separation and normalization in time domain

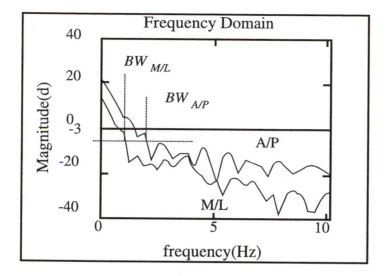

Figure 6 The balance indices in time domain

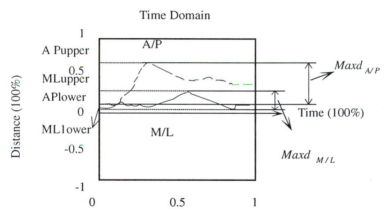

M/L : medial-lateral direction distance

A/P : anterior-posterior direction distance

Figure 7 The concept of the COP signals bandwidth on the Bode plot

The COP signals in time domain are transformed into frequency domain by Fast
Fourior Transformation (FFT). Then, the bandwidth of the COP signals is defined
as the frequency when the magnitude of the COP signals on the Bode plot is equal
to -3 dB. Figure 7 shows the concept of bandwidth on the Bode plot.

Two indices, $BW_{A/P}$ and $BW_{M/L}$, are used to define the features of the COP
trajectory in the frequency domain, as given in Figure 7. $BW_{A/P}$ is the bandwidth
in the A/P direction, while $BW_{M/L}$ is in the M/L direction.

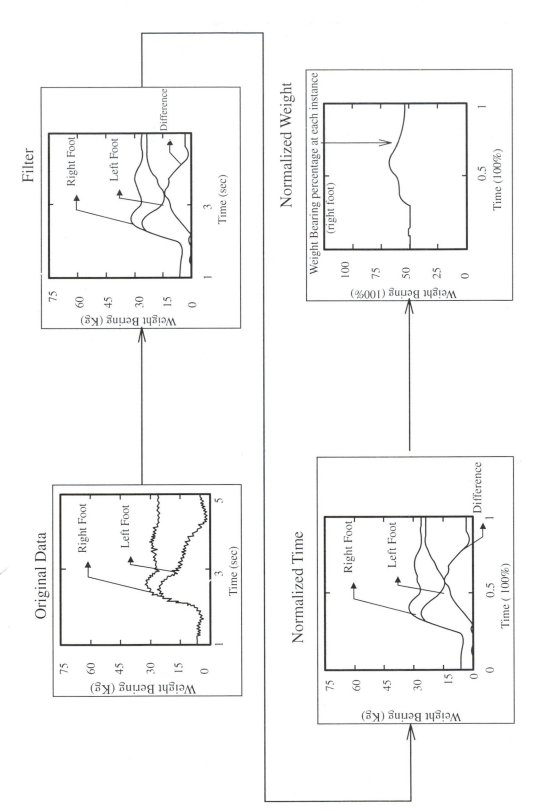

Figure 8 The process of data separation and normalization in force domain

18.2.4   Force Domain Analysis

In addition to the above three domains, the postural control ability indices in force domain are also discussed. Though most healthy elders can maintain their balance with a single foot, people usually distribute their body weight into two feet equally. Based on the biomechanical analysis, the potential of standing with two feet is more than that with a single foot. Therefore, standing with two feet during the test period has a good postural control ability. Figure 8 shows the data preprocessing in the force domain.

The first step to calculate the weight bearing indices is to filter the noise signals from hardware. A low-pass filter is used to eliminate the high frequency noise because the electric noises exist in the high frequency and the weight bearing signals fall into the low frequency. The second order butterworth digital filter with cut-off frequency 5Hz is designed for this purpose. Its transfer function is again given by

$$H(z) = \frac{0.6389 + 1.2779z^{-1} + 0.6389z^{-2}}{1 - 1.143z^{-1} + 0.4128z^{-2}} \cdot \tag{18.4}$$

Again, the test time is normalized. The entire test time is cut into 100 parts and the representative value of each part is its mean value. The weight bearing is also normalized. The average weight bearing percentage of the affected side or the lighter side of the normal subjects, $W_{Average}$, is given as

$$W_{Average} = \frac{\sum_{k=1}^{n} \frac{W_k}{F_k}}{n} , \tag{18.5}$$

where W_k is the weight bearing value of the affected feet or the right feet of the normal subject in the kth interval, F_k is the total weight bearing value in the kth interval, and n is the number of samples. Then $W_{Average}$ is used as the balance index to assess the balance states in the force domain.

18.3   Relationship between COP Signals and FIM Scores

FIM (Functional Independent Measurement) scores are traditionally used to assess the independent ability of CVA patients. There are six items in the FIM scores.
  1. Self Care: it includes (a) eating, (b) grooming, (c) bathing, (d) dressing-upper body, (e) dressing-lower body, (f) toileting.
  2. Sphincter Control: it includes (a) bladder management, (b) bowel management.
  3. Mobility: it includes (a) bed, chair, wheelchair, (b) toilet, (c) tub, shower.
  4. Locomotion: it includes (a) walk, wheelchair, (b) stairs.
  5. Communication: it includes (a) comprehension, (b) expression.
  6. Social Cognition: it includes (a) social interaction, (b) problem solving, (c) memory.
Among these items, the mobility and the locomotion are the items of most concern because they are the ability indices of the motor control. In order to understand the

relationship between FIM score and the stability of the subjects, the mobility scores and the locomotion scores are used to classify the subjects into three parts: normal subjects, high FIM score CVA patients, and low FIM score CVA patients. Here, the high FIM score is that the summing score of the mobility and the locomotion is greater than 25 (26–35), and the low FIM score is that less than or equal to 25 (15–25). If the subjects can practice the STS independently, the sum of the mobility scores and the locomotion scores are greater than 15.

Seven balance indices in four domains are defined below again.

$W_{Average}$:   The average weight bearing percentage of affected feet or the right feet of the normal subjects.

$Maxd_{A/P}$: The maximum sway distance in the A/P direction.

$Maxd_{M/L}$: The maximum sway distance in the M/L direction.

A_{min} :   The area of minimum circle of sway, which includes all the trajectories of the COP.

R_{min} :   The radius of the minimum circle of sway.

$BW_{A/P}$:   The bandwidth in the A/P direction.

$BW_{M/L}$:   The bandwidth in the M/L direction.

These seven balance indices are used to evaluate the normal subjects, high FIM and low FIM CVA patients during STS, movement platform moving in the A/P direction, and movement platform moving in the M/L direction. The relationship between the COP signal and the FIM score are described below.

Case 1 is the STS test. The results are given in Figure 9. The comparison results are summarized in Table 1. Clearly, only $W_{Average}$ and $Maxd_{M/L}$ are significant in STS test.

Table 1 Summary of Level of Significance (f-test) of the Differences Obtained from Various Parameters During STS (NS = not significant)

Comparison	Normal (27)	High FIM (16)	Low FIM (17)	Significance
$W_{Average}$	0.4880± 0.0104	0.4636± 0.0241	0.4126± 0.0517	$p < 0.05$
$Maxd_{M/L}$	0.3284± 0.0825	0.4248± 0.1050	0.5100± 0.1209	$p < 0.1$
$Maxd_{A/P}$	0.5531± 0.1230	0.4181± 0.1544	0.6497± 0.1441	NS
A_{min}	0.3219± 0.1956	0.2132± 0.1043	0.4093± 0.2363	NS
R_{min}	0.3174± 0.1732	0.2585± 0.1312	0.3520± 0.1731	NS
$BW_{A/P}$	0.9386± 0.7132	0.7125± 0.5893	0.8293± 0.6848	NS
$BW_{M/L}$	1.1538± 0.8452	1.2352± 0.8971	1.2113± 0.9437	NS

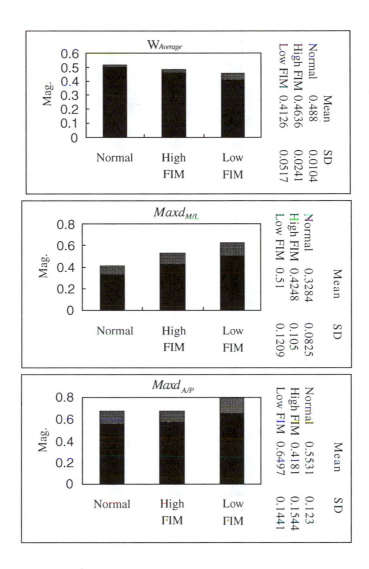

Figure 9 (a) The statistical comparison among normal subjects, high FIM, and low FIM
CVA patients during STS

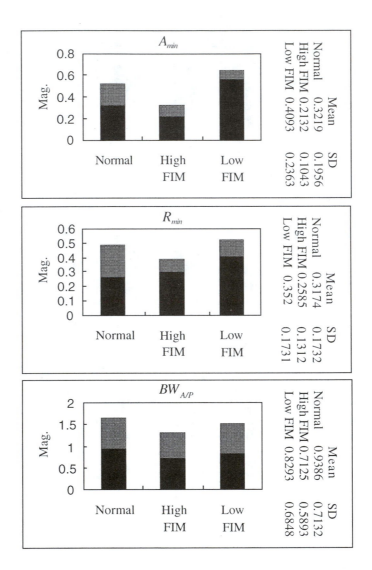

Figure 9 (b) The statistical comparison among normal subjects, high FIM, and low FIM
CVA patients during STS

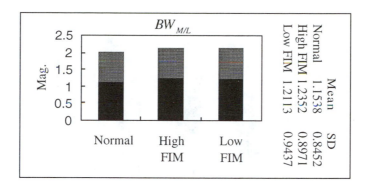

Figure 9 (c) The statistical comparison among normal subjects, high FIM, and low FIM
CVA patients during STS

Case 2 is the movement platform moving in the A/P direction. The results are
given in Figure 10. The comparison results are summarized in Table 2. Clearly,
$W_{Average}$, $Maxd_{A/P}$, A_{min}, and R_{min} are significant in the test of the movement
platform moving in the A/P direction.

Table 2 Summary of Level of Significance (f-test) of the Differences Obtained from Various
Parameters During the Movement Platform Moving in A/P Direction (NS = not
significant)

Comparison	Normal (27)	High FIM (16)	Low FIM (17)	Significance
$W_{Average}$	0.4826± 0.0102	0.4498± 0.0387	0.3815± 0.0647	<0.1
$Maxd_{M/L}$	0.1187± 0.0422	0.1459± 0.0457	0.2062± 0.0831	NS
$Maxd_{A/P}$	0.1752± 0.0663	0.2458± 0.1084	0.3826± 0.1403	<0.1
A_{min}	0.0534± 0.0382	0.0913± 0.0648	0.1233± 0.0810	<0.1
R_{min}	0.0913± 0.0691	0.1407± 0.0735	0.2068± 0.0518	<0.1
$BW_{A/P}$	1.2431± 0.8635	0.5312± 0.5123	0.5567± 0.3287	NS
$BW_{M/L}$	1.4498± 1.0263	1.2123± 0.8955	1.1431± 0.8812	NS

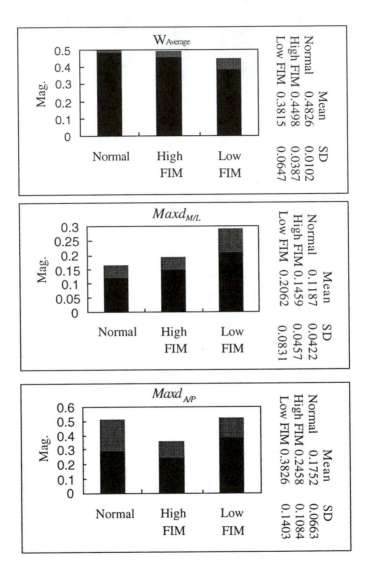

Figure 10 (a) The statistical comparison among normal subjects, high FIM, and low FIM CVA patients during the movement platform moving in the A/P direction

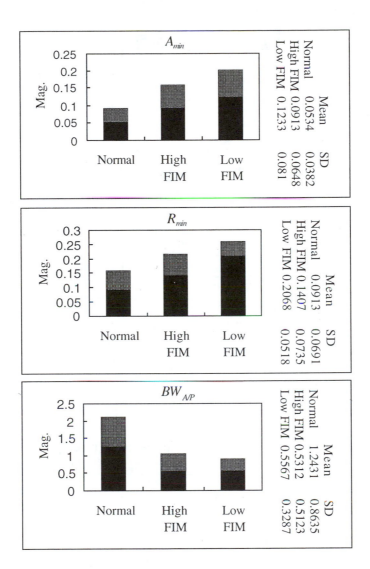

Figure 10 (b) The statistical comparison among normal subjects, high FIM, and low FIM
CVA patients during the movement platform moving in the A/P direction

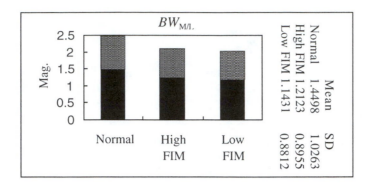

Figure 10. (c) Statistical comparison between normal subjects, high FIM and low FIM CVA patients during the movement platform moving in the A/P direction

Table 3 Summary of Level of Significance (f-test) of the Differences Obtained from Various Parameters During the Movement Platform Moving in M/L direction (NS = not significant)

Comparison	Normal (27)	High FIM (16)	Low FIM (17)	Significance
$W_{Average}$	0.4646± 0.0258	0.4007± 0.0745	0.3746± 0.0691	NS
$Maxd_{M/L}$	0.1257± 0.0436	0.1517± 0.0577	0.2813± 0.1025	<0.1
$Maxd_{A/P}$	0.1839± 0.0671	0.1610± 0.0831	0.2453± 0.1025	NS
A_{min}	0.0483± 0.0284	0.0515± 0.0320	0.0912± 0.0576	NS
R_{min}	0.1037± 0.0646	0.1074± 0.0493	0.1766± 0.0577	NS
$BW_{A/P}$	1.1084± 0.7998	0.8639± 0.5932	0.5637± 0.3389	NS
$BW_{M/P}$	1.7327± 1.2134	1.2482± 0.8963	1.1170± 0.7491	NS

Case 3 is the test of the movement platform moving in the M/L direction. The significances of the seven indices are calculated, as given in Figure 11. The comparison results are summarized in Table 3. Clearly, only index $Maxd_{M/L}$ is significant for the movement platform moving in the M/L direction.

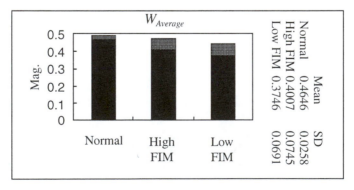

Figure 11 (a) The statistical comparison among normal subjects, high FIM, and low FIM CVA patients during the movement platform moving in the M/L direction

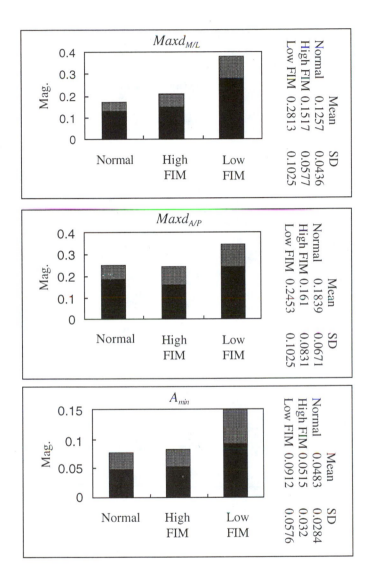

Figure 11 (b) The statistical comparison among normal subjects, high FIM, and low FIM
CVA patients during the movement platform moving in the M/L direction

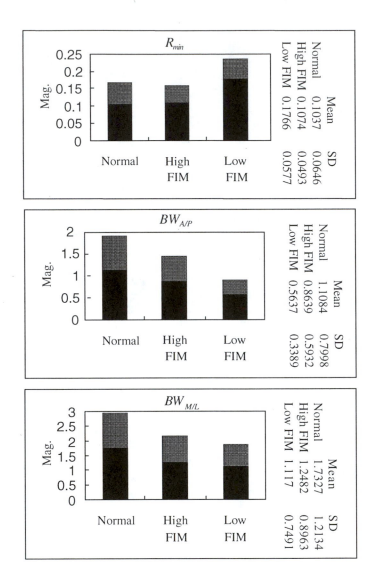

Figure 11 (c) The statistical comparison among normal subjects, high FIM, and low FIM CVA patients during the movement platform moving in the M/L direction

18.4   Construction of Kinetic State Assessment System

Motor control is an interface between the neuroscience kinesiology and kinetics. In order to assess the motor control ability of stroke patients, the kinetic state and neuro-physiological state are used to classify the CVA patients. The fuzzy theory is used to integrate several indices from the kinetic signal COP and the kinetic state of CVA patients.

A kinetic state assessment system is constructed using fuzzy logic to evaluate the kinetic states of the CVA patients. The fuzzy system receives the features of kinetic signals and sends out the kinetic state assessment to the rehabilitative therapeutic decision support system. The fuzzy rule base is constructed based on the experiences of experts. The fuzzy database is constructed in terms of the statistical results of the CVA patients and normal elders. The system is used to evaluate the kinetic states of the subjects. The kinetic state assessment system consists of rule base, database, and inference system, as shown in Figure 12.

The fuzzy model can be divided into four parts:

1. Kinetic Indices Input: The kinetic state indices are used as the inputs of the kinetic state assessment system.

2. Knowledge Base: The construction of the fuzzy model must have professional instructions, which come from experiential suggestions and statistical information.

3. Fuzzy Inference: It serves as the reasoning mechanism.

4. Defuzzification Output: The outputs of the fuzzy rules are some kinds of fuzzy messages. Therefore, the fuzzy messages must be transformed into crisp messages so that the real meaning of the fuzzy messages can be understood.

Each part of the kinetic state assessment system is elaborated below.

18.4.1   Balance Indices Input

Based on the analysis in Section 18.3, three balance indices are inputted to the kinetic state assessment system. They are:

1. $W_{Average}$:   the average weight bearing percentage of affected side during the STS.

2. $Maxd_{A/P}$:   the maximum sway distance in the A/P direction during the movement platform moving in the A/P direction.

3. $Maxd_{M/L}$:   the maximum sway distance in the M/L direction during the movement platform moving in the M/L direction.

The data preprocessing of these three indices are described before.

18.4.2   Knowledge Base

The knowledge base can be divided into two parts: database and rule base. The database is constructed by membership functions, which result from some statistical data. The rule base is established by some experienced doctors. The detailed description of the database and the rule base are described as follows.

Database: There exist large variances in the biomechanical experiences. It is unreasonable to use a crisp value to describe an appearance because the fundamental knowledge representation unit is a notion of linguistic variable.

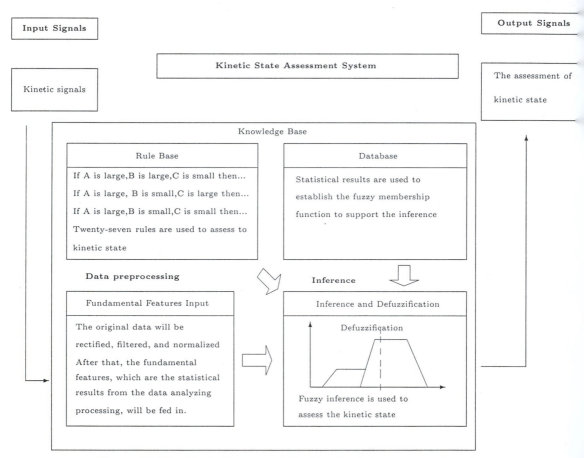

Figure 12 The block diagram of the kinetic state assessment system

The linguistic variables are represented by the term set, and the term set is composed of some subterms. For example, the "swaying magnitude" is a linguistic variable. It may be composed of large, medium, and small. It is illogical to separate these values from a crisp value. The efficient way to solve this problem is by using a membership function to describe a linguistic value. Figure 13 shows the membership function of the linguistic variable "swaying magnitude".

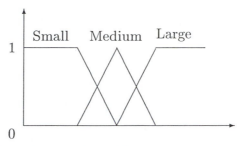

Figure 13 The membership function of linguistic variable "swaying magnitude"

The peak value is 1 and the break points can be decided by experiences or statistical results. In this chapter, the triangular membership function is used.

Rule base: Many intuitive assessments are used in our daily lives. By collecting these perceptive rules, most things can be distinguished easily. Based on the fuzzy propositions and linguistic description, such as "if then", "and", "or", "not", etc., a complex fuzzy proposition called a compound fuzzy proposition can be formed. For example, " if the sway magnitude is large, the balance ability is bad" is a typical compound fuzzy proposition, which can be treated as the mapping from the cause domain to the result domain. The rule base of a fuzzy model can be formed as follows.

$$
\begin{aligned}
&R_{1\times 1}: \quad \text{if } x = A_1 \text{ and } y = B_1 \text{ then } z = C_{1\times 1} \\
&R_{2\times 1}: \quad \text{if } x = A_2 \text{ and } y = B_1 \text{ then } z = C_{2\times 1} \\
&\qquad\quad \cdot \qquad \cdot \qquad \cdot \\
&\qquad\quad \cdot \qquad \cdot \qquad \cdot \\
&R_{i\times j}: \quad \text{if } x = A_i \text{ and } y = B_j \text{ then } z = C_{i\times j} \qquad\qquad (18.6) \\
&\qquad\quad \cdot \qquad \cdot \qquad \cdot \\
&\qquad\quad \cdot \qquad \cdot \qquad \cdot \\
&R_{m\times n}: \quad \text{if } x = A_m \text{ and } y = B_n \text{ then } z = C_{m\times n} \\
&R = R_{1\times 1} \cup R_{2\times 1} \cup \cdots \cup R_{m\times n}
\end{aligned}
$$

where x, y are input linguistic variables; z is an output linguistic variable; m, n are the numbers of linguistic subterms; A, B, C are the universes of x, y, z, respectively; $A_i, B_j, C_{i\times j}$ are the fuzzy subsets of A, B, C, respectively; $R_{i\times j}$ is a fuzzy relation matrix of A_j, B_j, and $C_{i\times j}$; and R is the fuzzy relation matrix of all rules.

18.4.3   Fuzzy Inference Engine

After the knowledge base is created, the fuzzy algorithm is used to calculate the results. Here, the Mamdani's method is chosen as the reasoning method with the balance indices found in the last section as the input. For the crisp inputs x_0 and

y_0 , the result of the $i \times j$ rule becomes

$$\alpha_{i \times j} = \mu_{A_i}(x_0) \wedge \mu_{B_i}(y_0)$$
$$\mu_{C'_{i \times j}} = \alpha_{i \times j} \wedge \mu_{C_{i \times j}}(z)$$
$$\mu_{C'}(z) = \mu_{C'_{1 \times 1}}(z) \vee \cdots \vee \mu_{C'_{i \times j}}(z) \vee \cdots \vee \mu_{C'_{m \times n}}(z) \tag{18.7}$$

where $\mu_{A_i}(x)$, $\mu_{B_j}(y)$, $\mu_{C_{i \times j}}(z)$ are the membership functions of $A_i, B_j, C_{i \times j}$, respectively; $\mu'_{i \times j}(z)$ is the membership function of $C'_{i \times j}$; $\mu_{C'}(z)$ is the membership function of the output fuzzy set C' .

18.4.4   Defuzzification

The reasoning result of the fuzzy inference is a fuzzy set C'. But the fuzzy set C' is difficult to understand. In order to understand the real meaning of the fuzzy set, it is important to defuzzify the fuzzy set C' . The defuzzification process transfers a fuzzy set into a crisp value, which represents the physical state. Here, the center of gravity method is chosen as the defuzzification method.

$$\text{Final Value} = \frac{\int \mu_{C'}(z)z\,dz}{\int \mu_{C'}(z)\,dz} \ . \tag{18.8}$$

18.4.5   Parameters and Rules Setup

According to the calculated statistical results, the input signals, membership functions, and the fuzzy rules of this fuzzy model can be defined.

Input Signal Setup
The following three input signals are chosen.

1. $W_{Average}$: The average weight bearing percentage of affected side or the lighter side of the normal subjects during STS.

2. $Maxd_{A/P}$: The maximum sway distance in the A/P direction during the movement platform swaying in the A/P direction.

3. $Maxd_{M/L}$: The maximum sway distance in the M/L direction during the movement platform swaying in the M/L direction.

Database Setup
According to the calculated statistical results, the membership functions of STS, A/P, M/L and the kinetic state are defined in Figures 14 to 17.

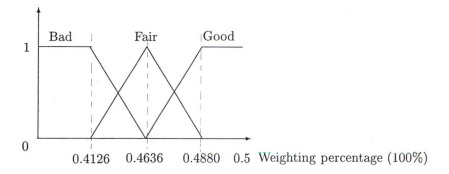

Figure 14 The membership function of the average weight bearing percentage of the affected side during STS

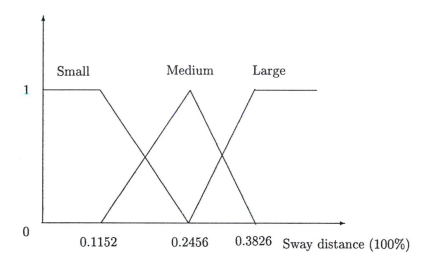

Figure 15 The membership function of the maximum sway distance in A/P direction during the movement platform swaying in A/P direction

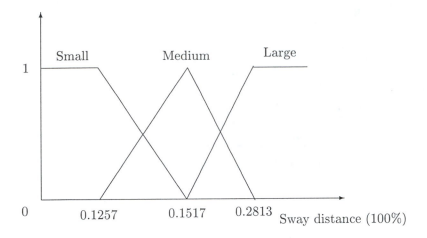

Figure 16 The membership function of the maximum sway distance in

M/L direction during the movement platform swaying in M/L direction

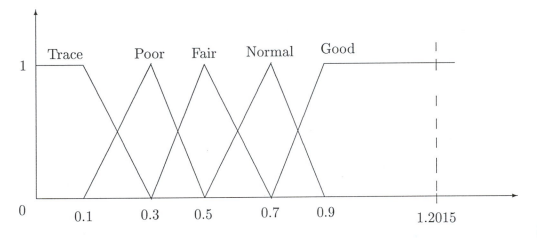

Figure 17 The membership function of the kinetic state

Table 4(a) The fuzzy rule of the kinetic state assessment system

Weighting: Good

A/P \ M/L	Small	Medium	Large
Small	Good	Good	Normal
Medium	Good	Normal	Fair
Large	Normal	Fair	Poor

Table 4(b) The fuzzy rule of the kinetic state assessment system
Weighting: Fair

A/P \ M/L	Small	Medium	Large
Small	Good	Normal	Fair
Medium	Normal	Fair	Poor
Large	Fair	Poor	Trace

Table 4(c) The fuzzy rule of the fuzzy kinetic state assessment system
Weighting: Bad

A/P \ M/L	Small	Medium	Large
Small	Normal	Fair	Poor
Medium	Fair	Poor	Trace
Large	Poor	Trace	Trace

Rule Base Setup

There are twenty-seven fuzzy rules for estimating the kinetic state. These rules are given in Tables 4(a) to 4(c). Note that "Weighting" is the average weight bearing percentage of the affected feet during STS, A/P is the maximum sway distance in the A/P direction during the movement platform swaying in the A/P direction, and M/L is the maximum sway distance in the M/L direction during the movement platform swaying in the M/L direction.

18.5   Results of Kinetic State Assessment System

The COP signals of thirty-three CVA patients are fed into the kinetic assessment system. The results are shown in Figure 18. The linear regression model with least square method [20] is used to analyze the output error. The linear regression model is given by

$$Out = -0.3493 + 0.0386 \times FIM , \qquad (18.9)$$

where Out is the output of the kinetic state assessment system; FIM is the FIM score. The relationship between the system output and the FIM scores is shown in Figure 18. The least square prediction equation is the characteristic function of the kinetic state assessment system. It is used to integrate the experiences of experts and the kinetic state assessment system. The sum of square error (SSE) is 0.4046, and the coefficient of correlation (CC) is 0.8438. Therefore, the kinetic state assessment system has high correlation with the FIM score. The kinetic state assessment can be evaluated from the COP signals. The result approaches the opinions of the experts.

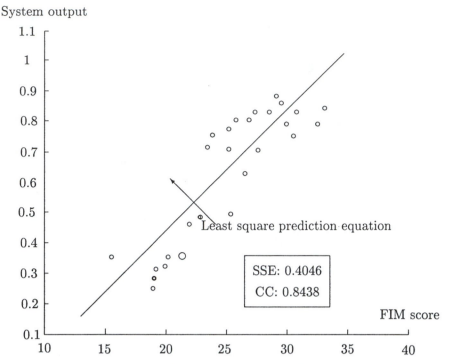

Figure 18 The relationship between the output of the kinetic state
assessment system and the FIM score

18.6   Conclusions

In this chapter, a fuzzy motor control ability assessment system was developed. Since the COP signals during the test period imply the global control abilities of the subjects, they are used to assess the kinetic states of the CVA patients. The fuzzy theory is used to integrate several indices and assess the global control ability of the CVA patients. The results of the kinetic state assessment system approach to the FIM scores. Traditionally, the states of the CVA patients are assessed subjectively by doctors. This chapter proposes an objective method to assess the kinetic state of the CVA patients. Furthermore, the kinetic state can be regarded as a risk of fall for the CVA patients due to its feature inputs.

References

1. E. W. Abel, P. C. Zacharia, A. Forster, and T. L. Farrow, Neural network analysis of the EMG interference pattern, *Medical Engineering & Physics,* Vol. 18, pp. 12-17, 1996.

2. J. M. Barreto and F. M. Azevdeo, Connectionist expert system as medical decision aid, *Artificial Intelligence in Medicine,* Vol. 5, No. 6, pp. 515-523, 1993.

3. H. C. Diener, J. Dichgans, W. Bruzek, and H. Selinka, Stability of human posture during induced oscillations of the body, *Experimental Brain Research,* Vol. 45, pp. 126-132, 1982.

4. C. A. Doorenbosch, J. Harlaar, M.E. Roebroeck, and J. Lankhorst, Two strategies of transfering from sit-to-stand: The activation of monoarticular and biarticular muscles, *Journal of Biomechanics,* Vol. 27, No. 11, pp. 1299-1307, 1994.

5. C. H. Geurts, T. W. Mulder, B. Nienhuis, and R. A. J. Rijken, Postural reorganization following lower limb amputation, *Scandinavian Journal Rehabilitation Medicine* Vol. 24, pp. 91-97, 1992.

6. P. A. Goldie, T. M. Bach, and O.M. Evans, Force platform measures for evaluating postural control: Reliability and validity, *Archives of Physical Medicine and Rehabilitation,* Vol. 70, pp. 510-517, 1989.

7. D. Grupe and H. Kordylewski, Artificial neural network control of FES in paraplegics for patient responsive ambulation, *IEEE Transactions on Biomedical Engineering,* Vol. 42, No. 7, pp. 699-707, 1995.

8. T. A. Hanke, Y. C. Pai, M. W. Rogers, Reliability of measurements of body center-of-mass momentum during sit-to-stand in healthy adults, *Physical Therapy,* Vol. 75, pp. 105-114, No. 2, Feb. 1995.

9. B. R. Hasselhus and G. M. Shambes, Aging and postural sway in woman, *Journal of Gerontology,* Vol. 30, No. 6, pp. 661-667, 1975.

10. E. Kwatny, D. H. Thomas, and H. G. Kwntay, An application of signal processing techniques to the study of myoelectric signals, *IEEE Transactions on Biomedical Engineering,* Vol. BME-17, No. 4, pp. 303-313, Oct. 1970.

11. B. E. Maki, P. J. Holliday, and G.R. Fernic, Aging and posture control: A comparison of spontaneous- and induced-sway balance test, *American Geriatrics Society Magazines,* Vol. 38, No. 1, pp. 1-8, Jan. 1990.

12. J. Massion, Postural changes accompanying voluntary movements: normal and pathological aspects, *Human Neurobiology,* Vol. 2, pp. 261-276, 1987.

13. J. Mizrahi, P. Solzi, H. Ring, and R. Nisell, Postural stability in stroke patients: Vectorial activity and relative sequence forces, *Medical & Biological Engineering & Computing*, Vol. 27, pp. 181-190, 1989.

14. M. P. Murray, A. Seirge, and R.C. Scholz, Center of gravity, center of pressure, and supportive forces during human activities, *Journal of Applied Physics*, Vol. 23, No. 6, pp. 831-838, Dec. 1967.

15. A. Nardones, A. Giordano, T. Corra, and M. Schieppati, Responses of leg muscles in humans displaced while standing, *Brain Research*, Vol. 113, pp. 65-84, 1990.

16. L. M. Nashner, *Motor Control Mechanisms in Health and Disease*, Raven Press, New York, pp. 607-619, 1983.

17. P. Z. Peebles, *Probability, Random Variables, and Random Signal Principles*, McGraw Hill, New York, 1993.

18. J. Z. Shue, Feature extraction and discrimination of electromyography (EMG) signal, *Master's Thesis*, Department of Electrical Engineering, National Taiwan University, 1993.

19. J. A. Waterston, M. B. Hawken, S. Tanyeri, P. Jantti, and C. Kennard, Influence of sensory manipulation on postural control in parkinson's disease, *Journal of Neurology, Neurosurgery, and Psychiatry*, Vol. 56, pp. 1276-1281, 1993.

20. A. M. Wing, S. Goodrich, N. V. Babul, J. R. Jenner, and S. Clapp, Balance evaluation in hemiparetic stroke patient using lateral force applied to the hip, *Archives of Physical Medicine and Rehabilitation*, Vol. 74, pp. 292-299, March 1993.

Chapter 19

A DSP-based Neural Controller for a Multi-degree Prosthetic Hand

The electromyographic (EMG) signal is used to discriminate eight hand motions: power grasp, hook grasp, wrist flexion, lateral pinch, flattened hand, centralized grip, three-jaw chuck, and cylindrical grasp. From the analysis of the PC-based control system, a three-channel EMG signal is used to distinguish eight hand motions for the short below elbow amputee. Pattern recognition is used in this discriminative system. Three surface electrodes are placed on palmaris longus, entensor digitorum, and flexor carpi ulnaris. Due to the complexity of the EMG signal and the portable consideration of the controller, a neural controller based on digital signal processor (DSP) is designed and implemented in this discriminative system. The DSP integrates the signal preprocessing module, the digital filter module, and pattern recognition module into the neural controller. The on-line DSP neural controller can provide an 87.5% correct rate for the discrimination of eight hand motions.

19.1 Introduction

The myoelectrically controlled prostheses provide amputees extra prosthetic options. However, most commercial prosthetic hands (e.g., Steeper Electric Hand, Otto Bock System Electric Hand, Swedish Systemteknik Hand) [2] are one degree of freedom grippers that are controlled by one or two channels of electromyographic (EMG) signals. Their surface electrodes were placed on the antagonist muscles. When the muscle tension reaches a threshold value, the prosthetic hands generate a digital on/off switch to control a motor to direct the hand in one direction or another. An alternate way is simple proportional control in terms of EMG signal [16]. The proportional control provides the user less sensitive and faster control of the hand, depending on the strength of muscle contraction. Proportional myoelectric control has been used in the Utah artificial arm [9]. Those EMG controlled systems mentioned above limit the ability of manipulation. Recently, a number of advanced dexterous robot hands have been successfully developed. A few of them can be used in prosthetics [8,12,13,17]. A modular prosthetic hand was developed by [5,13]. They are multi-fingered prosthetic hands capable of performing a variety of prehensile postures, including different grasp modes (e.g., power grasp, hook grasp, cylindrical grasp, three-jaw chuck, etc.). Nevertheless, how to control a multi-functional prosthetic hand using an EMG signal is the most difficult problem.

The human hand is a complex and amazingly versatile system. For performing dexterous tasks, the hand depends on control inputs from the central nervous system (CNS) and numerous sensors that provide feedback. If a specific task is defined, the hand is preshaped into a posture suitable to grasp, and then enclose the object [7]. The dexterous hand control has complex interaction between the hand and the object. Task level planning simplifies the process of controlling the dexterous hand [13,14]. The grasp planning involves selecting a grasp posture for the dexterous hand and determining a trajectory planning so that the hand contacts and approaches the object. A grasp posture is in terms of a grasp mode. If the grasp mode is defined, the preshaping controller determines the preshaping posture for the prosthetic hand and the enclosure controller closes the hand until the desired grasp forces are achieved. Once an EMG signal can be recognized as one of the eight hand motions, the relevant preshaping motion of the prosthetic hand can be determined. The complex discriminative system will be discussed in this chapter. For consideration of real-time performance, the complex algorithms of the discriminative system will be implemented in a DSP (digital signal processor) based neural control system.

19.2   EMG Discriminative System

19.2.1   EMG Signal Processing

The upper extremity has two main functional parts: the terminal prehension device (hand/wrist) and a crane system (arm/shoulder). The basic function of the terminal device is to provide the proper grip for functional activities. Prehension is one of the primary functions. In this chapter, eight types of prehensile postures, power grasp, hook grasp, wrist flexion, lateral pinch, flattened hand, centralized grip, three-jaw chuck, and cylindrical grasp, are selected from [3] for study.

In order to choose meaningful EMG signals for eight kinds of prehensile postures, the location of electrodes is important. According to the relations between the muscle locations and the prehensile postures [10], three channel electrodes are placed on palmaris longus, entensor digitorum, and flexor carpi ulnaris. In other words, we focus on short below elbow disarticulation.

The surface electrodes used for EMG signals are manufactured by B & L Engineering in the U.S.A. The B & L active electrode is available with an integrated ground. Though the surface EMG signal has been magnified in the electrode system, the output EMG signal is only about -0.2 V to 0.2 V (peak-to-peak). In order to avoid the distortion of converting from the analog EMG signal to digital signal, the EMG signal is further amplified about twenty times using an operational amplifier. The signal-preprocessing module is composed of the amplifier and analog filter circuit in a PC-based system [3,6]. In the DSP-based controller, the module only needs the amplifier circuit. For reducing the weight of a myoelectric controller, the digital filter will be implemented in a TMS320C31 DSP to replace the analog filter circuit.

19.2.2 Pattern Recognition

19.2.2.1 Feature Extraction

Surface EMG signals are nonlinear and stochastic. They are contributed by the summation of triggered motor units with respect to the measuring electrode location. Thus, different motions create different myoelectric signals with different characteristics. In this chapter, the features of myoelectric signals are calculated from time series and spectral parameters. Several kinds of features are used to represent the myoelectric signal patterns. They are given below.

Integral of EMG (IEMG): This is an estimate of the summation of absolute value of the EMG signal. It is given by

$$IEMG = \sum_{k=1}^{N} |x_k| \, , \tag{19.1}$$

where x_k is the kth sample data which has N samples raw data.

Waveform Length (WL): This is a cumulative variation of the EMG signal that can indicate the degree of variation about the EMG signal. It is given by

$$WL = \sum_{k=1}^{N} |\triangle x_k| \, . \tag{19.2}$$

Variance (VAR): This is a measure of the power density of the EMG signal, and is given by

$$VAR = \frac{1}{N-1} \sum_{k=1}^{N} x_k^2 \, . \tag{19.3}$$

Zero Crossings (ZC): This parameter counts the number of times that the signal crosses zero. A threshold needs to be introduced to reduce the noise induced at zero crossings. Given two continuous data x_k and x_{k+1}, the zero crossing can be calculated as

$$ZC = \sum_{k=1}^{N} [sgn(-x_k \times x_{k+1}) \text{ and } |x_k - x_{k+1}| \geq 0.02],$$

$$sgn(x) = \begin{cases} 1, & \text{if } x > 0 \\ 0, & \text{otherwise} \end{cases} \, . \tag{19.4}$$

Slope Sign Changes: This parameter counts the number of times the slope of the signal changes sign. Similarly, it needs to include a threshold to reduce noise induced at slope sign changes. Given three continuous data x_{k-1}, x_k, and x_{k+1} , the number of slope sign change increases if

$$(x_k - x_{k-1}) \times (x_k - x_{k+1}) \geq 0.03, \quad k = 1, \ldots, N \ . \tag{19.5}$$

Willison Amplitude (WAMP): It is the number of counts for each change of the EMG signal amplitude that exceeds a pre-defined threshold. It can indicate the muscle contraction level, and is given by

$$WAMP = \sum_{i=1}^{N} f(|x_k - x_{k+1}|),$$

$$f(x) = \begin{cases} 1, & \text{if } x > 0.3 \\ 0, & \text{otherwise.} \end{cases} \tag{19.6}$$

The above six parameters are extracted according to Equations (19.1) to (19.6). The first three parameters are time-domain calculation and all floating-point type. The last three parameters are rough frequency measure and all integer type. Hence, six parameters are divided into two groups: three floating point parameters and three integer parameters.

Autoregressive (AR) Model: It is difficult to analyze the EMG signal for its non-linear and nonstationary nature. But in a short time interval the EMG signal can be regarded as a stationary Gaussian process. Graupe et al. [4] addressed a linear model for a Gaussian process. They used a pure autoregressive (AR) model to identify the EMG time-series as

$$y_k = -\sum_{i=1}^{N} a_i y_{k-1} + w_k \ , \tag{19.7}$$

where y_k is the EMG time series, and k is the interval. N is the order of AR model, a_i are the estimate of the AR parameters, and w_k is the white noise. A least squares method that minimizes the sum of squared error is used to obtain the parameters of the signal model. Furthermore, the AR parameters can be calculated using the adaptive least mean square (LMS) algorithm by DSP programming. The adaptive AR method is similar to the adaptive predictor. It predicts the input EMG sample y_n according to previous input samples by minimizing the output error. The adaptive predictor structure is shown in Figure 1.

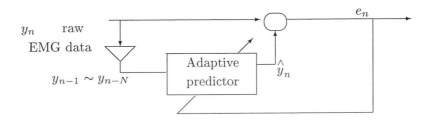

Figure 1   Adaptive linear predictor

Cepstrum Analysis (c(t)): Cepstrum analysis is concerned with the deconvolution of two signal types that are the basic wavelet and the impulse signal. It can be defined as the inverse Fourier transform of the logarithm of the magnitude of the power spectrum of the signal data [1]. It can be represented as

$$c(t) = \frac{1}{2\pi} \int_{-\pi}^{\pi} ln[X(e^{j\omega})]e^{j\omega n} d\omega .$$ (19.8)

19.2.2.2   Feature Selection

The success of any pattern recognition system depends on the choice of features used to represent the continuous time waveforms. In feature extraction stage, eight kinds of parameters of the EMG signal are extracted. The parameters are split into four groups: three integer parameters group, three floating-point parameters group, AR model parameters group, and cepstrum parameters group. They are combined with each other in classification stage to find the highest classification rate in the PC-based discriminative system. A set of features that have the highest classification rate in the PC-based system is then implemented in the DSP-based controller.

19.2.2.3   Classification by Neural Network

The parameters extracted from the feature selection stage can be used to separate different input patterns into different output classes. It needs a powerful classifier to solve the nonlinear discriminative system of the EMG signal. We adopt the backpropagation neural network (BPNN) as the classifier [15]. The classifier has one hidden layer and the entire system is a three-layer perceptron. The gradient steepest descent method is adopted to solve the minimization error of each input pattern.

19.3   DSP-based Prosthetic Controller

The pattern recognition is implemented in the DSP-based prosthetic controller system. To reduce the computation time, a high performance processor, Texas Instruments TMS320C31 DSP, is selected as the kernel of the entire prosthetic controller. For saving the time to develop the prosthetic controller system, a 'C3x DSP starter kit (DSK) is adopted. The DSK is a low-cost, easy-to-use, and expandable development platform. The DSK has a TMS320C31-60 on board to handle signal-processing applications.

19.3.1   Hardware Architecture of the Controller

The development of hardware architecture can be divided into two stages: the off-line stage and the on-line stage. In the off-line stage, the approximate initial weights of the neural network for each amputee are obtained. While in the on-line stage, the weights are finely tuned for the same amputee to gain a high performance classification rate.

19.3.1.1   The Off-line Stage of the Prosthetic Controller

Because every amputee has different characteristics of the muscle, the hand motions must be learned before the controller is used. In this stage, the most important thing is to train the amputee to perform eight prehensile motions. Initial weights can be obtained to represent the characteristics of the muscle for each amputee. The architecture of the off-line stage is given in Figure 2.

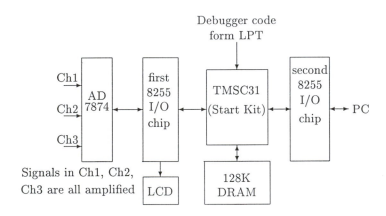

Figure 2   Hardware structure of the off-line stage

19.3.1.2   The On-line Stage of the Prosthetic Controller

This stage has an on-line learning mode and on-line testing mode. A 4X4 keyboard is used to input the learning goal for on-line learning. After on-line learning, the controller changes to the on-line testing mode. The amputee can control the prosthetic hand directly in this mode. If the controller has a bad recognition rate, it can change back to the on-line learning mode for another training. Figure 3 is the hardware structure of the on-line stage.

19.3.2   The Software System of the Controller

The software development tools are TMS320C3x/C4x Assembler/Linker and 'C3x DSK Debugger supported by Texas Instruments. The entire software system is composed of four main parts: EMG signal collection, signal processing, feature extraction, and BPNN classification.

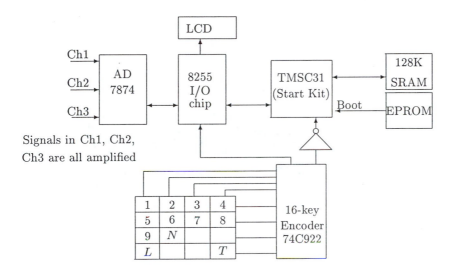

Figure 3   Hardware structure of the on-line stage

19.3.2.1   Signal Collection

After amplifying the EMG signal twenty times using an operational amplifier, the controller is then developed by the DSP code. First, when the ADC chip receives the amplified EMG signals, the EMG signals are converted into a digital version. For getting the stable EMG signals, about 1000 points within 300-400 ms are collected. The sampling rate is about 0.4 ms. However, it is difficult to detect the starting of the muscle contraction. The IEMG was applied well and confidently in many papers [3,11]. Thus, we select the average IEMG as the criterion to verify the muscle contraction for canceling the unstable noise. The average IEMG value of 25 ms raw window length is calculated to judge whether the muscle contracts or not. A bias is added to the threshold of the muscle contraction according to the surrounding environment noise. Because the computation time of DSP is faster than 25 ms raw EMG signal, the raw EMG window can be regarded as a continuous series of 25 ms raw window. When the average IEMG value exceeds the pre-defined threshold, 1000 running EMG data are collected. It is further illustrated by the following equation and Figure 4.

$$\text{if} \quad \overline{IEMG} = \frac{1}{N}\sum_{i=t}^{t+N} |X_i| > T_H$$

$$X_i = \text{collect EMG raw data for analyzing} \quad i = 1 \text{ to } 1000, \qquad (19.9)$$

where N=windows length; T_H = threshold value.

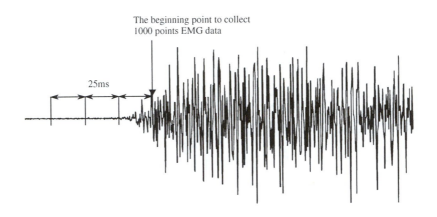

Figure 4   Collecting EMG signal

19.3.2.2   Signal Processing

After collecting 1000 points EMG data, the primary thing is to filter the uncertain signal. The bandwidth of the EMG signal is about 30-400 Hz in previous research and the environment noise is about 60 Hz. Thus, a bandpass filter with bandwidth from 30 Hz to 400 Hz and a 60 Hz notch filter is designed using a Butterworth type filter. Often, the infinite impulse response (IIR) filters are more efficient in getting better magnitude response for a given filter order. In other words, an IIR filter can run faster in lower filter order. The Matlab software is used to design the Butterworth IIR digital filter. The resultant filter is shown below.
• 30 Hz to 400 Hz Butterworth bandpass filter
The system transfer function of a sixth-order badpass filter is designed by

$$H_1(z) = \frac{Y_1(z)}{X_1(z)} \qquad (19.10)$$

$$= \frac{0.047946z^6 - 0.143838z^4 + 0.143838z^2 - 0.047946}{z^6 - 4.015503z^5 + 6.783575z^4 - 6.32227z^3 + 3.503237z^2 - 1.090804z + 0.141969}.$$

The sampling period is 0.4 ms.

• 60 Hz Butterworth notch filter

The system transfer function of a fourth-order notch filter is designed by

$$H_2(z) = \frac{Y_2(z)}{X_2(z)}$$

$$= \frac{z^4 - 3.96z^3 + 5.9256z^2 - 3.970296z + 1.00520676}{z^4 - 3.92z^3 + 5.8045z^2 - 3.847284z + 0.963242}. \qquad (19.11)$$

The sampling period is also 0.4 ms.

Because the bandpass and notch filter functions are both linear systems, they can be combined to form a tenth-order filter that includes both characteristics. The transfer function of the augmented system can be written as

$$H(z) = \frac{Y(z)}{X(z)} = H_1(z) \times H_2(z) \qquad (19.12)$$

The designed filters are coded in the DSP algorithm. For reducing memory requirement and sensitivity of quantization coefficient, both bandpass filter and notch filter are used rather than the tenth-order filter. These two filters are configured in cascade structure. The connected canonical-form structure equation of the filter is given as

$Step\ 1 : k = 1$

$y[k - 1, n] = x[n]$

$w[k, n] = a_{0k}y[k - 1, n] + a_{1k}w[k, n - 1] + a_{2k}w[k, n - 2] + \cdots + a_{6k}y[k, n - 6]$

$y[k, n] = b_{0k}w[k, n] + b_{1k}w[k, n - 1] + b_{2k}w[k, n - 2] + \cdots + b_{6k}y[k, n - 6]$

$Step\ 2 : k = 2$

$w[k, n] = a_{0k}y[k - 1, n] + a_{1k}w[k, n - 1] + a_{2k}w[k, n - 2] + \cdots + a_{4k}y[k, n - 4]$

$y[k, n] = b_{0k}w[k, n] + b_{1k}w[k, n - 1] + b_{2k}w[k, n - 2] + \cdots + b_{4k}y[k, n - 4]$

$y[n] = y[k, n] . \qquad (19.13)$

The frequency response of the desired filter is shown in Figure 5. It possesses the combined characteristics of both the bandpass filter and the notch filter.

19.3.2.3 Feature Extraction

The features of the EMG signal for each channel are extracted from the filtered EMG signals. For the controller design, the features are selected from the PC-based analysis [3]. The combined feature 1 is based on IEMG, Variance, Wave Length and WAMP, while the combined feature 2 is based on IEMG, Variance, Wave Length, WAMP, Zero-crossings and 2nd-order AR. Figure 6 shows the correct rate for different parameter groups. The 4th-order AR parameters group plus three integer parameters group and three floating-point parameters group can represent the meaningful features of the EMG signals for each case.

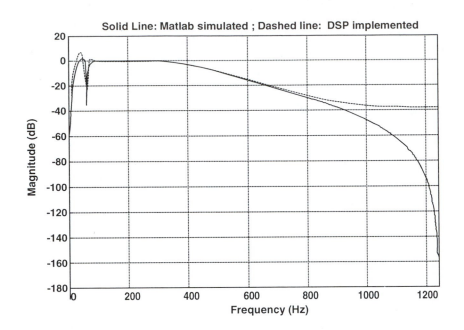

Figure 5   Frequency response comparison of $H(z)$

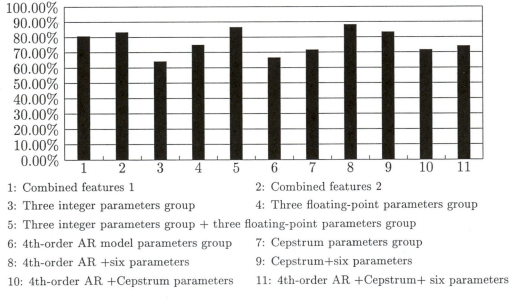

1: Combined features 1   2: Combined features 2
3: Three integer parameters group   4: Three floating-point parameters group
5: Three integer parameters group + three floating-point parameters group
6: 4th-order AR model parameters group   7: Cepstrum parameters group
8: 4th-order AR +six parameters   9: Cepstrum+six parameters
10: 4th-order AR +Cepstrum parameters   11: 4th-order AR +Cepstrum+ six parameters

Figure 6   Comparison of correct rate with different parameter group

In this controller, those best-combined features are selected as the EMG signal features. It turns out that ten parameters for each channel signal and thirty parameters for each motion type are obtained. These thirty parameters are input vectors for BPNN. It is shown in Table 1.

Table 1   Number of Parameters in the Controller

Parameter Group	Each Channel Signal	Type Motion Each
Three integer parameters	3	3×3
Three floating-point parameters	3	3×3
4th-order AR model parameters	4	4×3
Total parameters	10	30

19.3.2.4   BPNN Classification

The BPNN algorithm mentioned before is coded in assembly language. The BPNN is three-layer architecture with input layer, hidden layer, and output layer. The number of the hidden layer node is determined according to PC-based analysis. Here, 15 nodes are used for the hidden layer. The input layer has 30 nodes for representative features of each motion. The output layer has 8 nodes for discriminating 8 motions. The neural network off-line learns the initial weights and maximum/minimum values of each input vector. Then, it is on-line implemented in DSP.

The on-line stage of the system consists of learning mode and testing mode. In the on-line learning mode, the user must perform the motion and go through the desired learning motion. The on-line learning mode supports noise key to cancel the wrong motions occurred in the learning period. After finishing the learning mode, the program is switched to the testing mode. In the testing mode, the user can perform any of 8 motions to control the prosthetic hand directly. If the user has problems in controlling the prosthetic hand, it can be changed back to the learning mode to improve the classification rate of the controller in any time. The system spends most of time in data collection, feature extraction, and BPNN computation. In contrast, it only takes about 0.7 seconds in running testing mode. The time period is calculated from starting execution of motion to identification of the motion type.

19.4   Implementation and Results of the DSP-based Controller

The implementation of the DSP-based system is divided into two stages: off-line learning stage and on-line testing stage. The off-line learning stage is used to get the initial weight, the initial bias values, and extrema of input vectors for neural network classification. The types of input vector are defined in Table 1. They can be regarded as the EMG signal characteristics of the user. The on-line testing stage is used to control the prosthetic hand according to the neural network built in the off-line stage. Each hand motion is trained 20 times (off-line learning stage), and tested 10 times (on-line testing stage). The picture of the controller is shown in Figure 7.

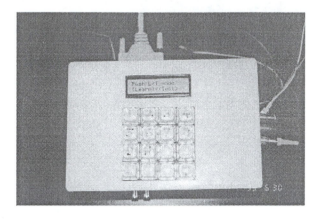

Figure 7   DSP-based myoelectric controller box

19.4.1   Off-line Stage Implementation

The locations of the three-channel electrodes are shown in Figure 8. The off-line stage is to train the BP neural network using eight hand motions, i.e., power grasp, hook grasp, wrist flexion, lateral pinch, flattened hand, index hand, three-jaw chuck, and cylindrical grasp. Each hand-motion is learned twenty times. The features of three-channel EMG signals are calculated by the DSP chip and sent to PC to run BP neural network to get the initial weight, the initial bias values, and extrema of the input vectors. Each feature is normalized so that each input vector has the same range for training in the neural network. The neural network has 30 input nodes, 15 hidden layer nodes, and 8 output nodes. The execution of BP Nets program is shown in Figure 9.

From Figure 9, the learning process is good and converges very fast. Upon finishing off-line learning stage, the initial weight, the bias values, and extrema of the input vectors of the neural network are downloaded into SRAM, and the system is changed to the on-line testing stage.

19.4.2   On-line Stage Implementation

After the off-line learning stage, the system performs the on-line stage and serves as a prosthetic controller. In the on-line stage, the on-line learning feature can improve the controller correct rate. Different situations of the muscles and different noise of environment affect the variation of the EMG signals. The neural network must be adjusted to get better performance. In Figure 3, the keyboard was designed for keying in the learning goal so that an amputee can easily train the motion. In the on-line stage, the controller is used to control the graphic NTU-hand [12], which is a five-finger and seventeen degree-of-freedom dexterous hand, through RS-232. If the classification rate is not satisfactory, the user can push 'L' button on the keyboard to change to the learning stage to re-train the neural network. The amount of learning time for each motion can be defined by the DSP program. After completing the learning process, it returns to the testing stage automatically. The on-line stage analysis of the DSP-based system and its implementation are shown in Figure 10.

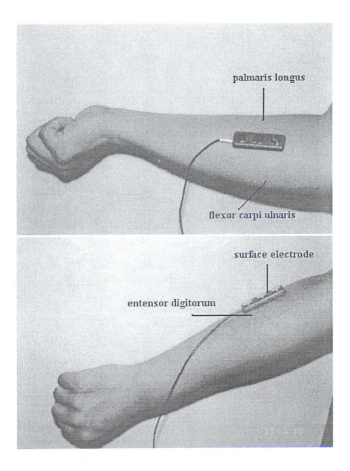

Figure 8   Locations of surface electrodes

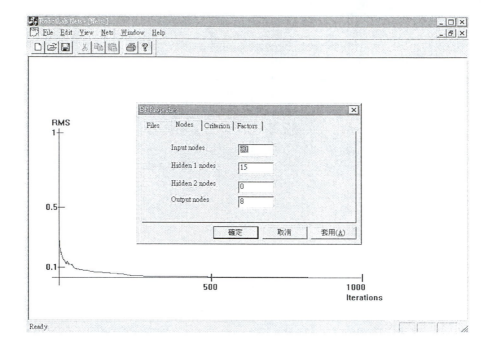

Figure 9   Nets program

Figure 10   On-line stage analysis of the DSP-based system

19.4.3   On-line Analysis Results

Each hand motion is tested 10 times (on-line testing stage). The controller uses the best-combined features as EMG features. The correct rate of the entire system is shown in Table 2. Note that the three-jaw chuck motion is always classified as a lateral pinch. These two hand-motions do look like each other and are imperceptible. Because the training times for the off-line stage are not long enough, we push the key 'L' to run the on-line learning stage. Each motion is learned five more times. After the on-line learning stage, the correct rate of three-jaw chuck is improved to 70%. The result is given in Table 3.

Table 2   Motion Correct Rates of DSP-based System

Movement Name	DSP-based
Power grasp	90.0%
Hook grasp	100.0%
Wrist flexion	80.0%
Lateral pinch	100.0%
Flattened hand	100.0%
Index hand	80.0%
Three-jaw chuck	0.0%
Cylindrical grasp	80.0%
Average rate	78.75%

Table 3   Motion Correct Rates After the On-line Learning

Movement Name	Before Learning	After Learning
Power grasp	90.0%	90.0%
Hook grasp	100.0%	70.0%
Wrist flexion	80.0%	90.0%
Lateral pinch	100.0%	100.0%
Flattened hand	100.0%	90.0%
Index hand	80.0%	100.0%
Three-jaw chuck	0.0%	70.0%
Cylindrical grasp	80.0%	90.0%
Average rate	78.75%	87.5%

From Table 3, the correct rate is improved about 9%. It is clear that the on-line learning stage can update the weight and the bias values to improve the controller performance. The above results show that the DSP-based controller successfully discriminates eight hand motions.

19.5   Conclusions

In this chapter, we discussed a DSP-based system that integrates a signal pre-processing module, a digital filter module, and a pattern recognition module. The DSP-based controller is portable and compact. Furthermore, the on-line learning function is capable of improving the classification rate. After on-line learning, the correct rate of the controller can be up to 87.5% in on-line testing. Note that the best-combined feature is selected as the input vector of the EMG signals and the BPNN as classifier in both the PC-based and DSP-based systems.

References

1. M. Akay, *Biomedical Signal Processing,* Academic Press, San Diego, 1994.

2. D. J. Atkins and R. H. Meier III, *Comprehensive Management of the Upper-Limb Amputee,* Springer-Verlag, New York, 1989.

3. C. Y. Chen, Development of a myoelectric controller for a multi- degree prosthetic hand, *Master's Thesis,* Department of Mechanical Engineering, National Taiwan University, 1998.

4. D. Graupe, J. Magnussen, and A. A. M. Beex, A microprocessor system for multifunctional control of upper limb prostheses via myoelectric signal identification, *IEEE Transactions on Automatic Control,* Vol. 23, No. 4, pp. 538-544, 1978.

5. L. T. Han, Development of a modular prosthetic hand: NTU-Hand III, *Master's Thesis,* Department of Mechanical Engineering, National Taiwan University, 1998.

6. H. P. Huang and C. Y. Chen, Development of a myoelectric discrimination system for a multi-degree prosthetic hand, *1999 IEEE International Conference On Robotics and Automation,* Detroit, USA, May 10-15, 1999.

7. T. Iberall, G. Sukhatme, D. Beattie, and G. A. Bekey, Control philosophy and simulation of a robotic hand as a model for prosthetic hands, *IEEE International Conference on Intelligent Robots and Systems,* Vol. 2, pp. 824 -831, 1993.

8. T. Iberall, G. S. Sukhatme, D. Beattie, and G. A. Bekey, On the development of EMG control for a prosthesis using a robotic hand, *IEEE International Conference on Robotics and Automation,* Vol. 2, pp. 1753-1758, 1994.

9. S. C. Jacobsen, D. F. Knutti, R. T. Johnson, and H. H. Sears, Development of the Utah artificial arm, *IEEE Transactions on Biomedical Engineering,* BME-29, No. 4, pp. 249-269, 1982.

10. F. P. Kendall, E. K. McCreary, and P. G. Provance, *Muscles Testing and Function: with Posture and Pain,* Williams & Wilkins, Baltimore, 1993.

11. C. H. Kuo, The development of a controller for artificial arm via EMG pattern recognition, *Master's Thesis,* Department of Electrical Engineering, National Taiwan University, 1995.

12. L. R. Lin and H. P. Huang, Mechanism design of a new multifingered robot hand, *IEEE International Conference on Robotics and Automation,* Vol.2, pp. 1471-1476, 1996.

13. L. R. Lin and H. P. Huang, Integrating fuzzy control of the dextrous NTU hand, *IEEE/ASME Transactions on Mechatronics,* Vol. 1, No. 3, pp. 219-229, 1996.

14. H. Liu, T. Iberall, and G. A. Bekey, The multi-dimensional quality of task requirements for dextrous robot hand control, *IEEE International Conference on Robotics and Automation,* Vol. 1, pp. 452-457, 1989.

15. D. Rumelhart, G. E. Hinton, and R. J. Williams, Learning internal representations by error propagation, D. E. Rumelhart and J. L. McClelland, eds., *Parallel Distributed Processing,* Cambridge, MA: MIT Press, Vol. 1, pp. 318-362, 1986.

16. H. H. Sears and J. Shaperman, Proportional myoelectric hand control: An evaluation, *American Journal of Physical Medicine and Rehabilitation,* Vol. 70, No. 1, pp. 20-28, 1991.

17. M. I. Vuskovic and A. K. Marjanski, Programmed synergy in dextrous robotic hands, *IEEE International Conference on Robotics and Automation,* Atlanta, GA, pp. 449-455, 1999.

INDEX